DISCARDED

CAMBRIDGE TRACTS IN MATHEMATICS

General Editors

B. BOLLOBAS, W. FULTON, A. KATOK, F. KIRWAN, P. SARNAK

150 Harmonic maps, conservation laws and moving frames

Second edition

Harmonic maps, conservation laws and moving frames

Second edition

Frédéric Hélein
Ecole Normale Supérieure de Cachan

PUBLISHED BY THE PRESS SYNDICATE OF THE UNIVERSITY OF CAMBRIDGE
The Pitt Building, Trumpington Street, Cambridge, United Kingdom

CAMBRIDGE UNIVERSITY PRESS
The Edinburgh Building, Cambridge CB2 2RU, UK
40 West 20th Street, New York, NY 10011-4211, USA
477 Williamstown Road, Port Melbourne, VIC 3207, Australia
Ruiz de Alarcón 13, 28014, Madrid, Spain
Dock House, The Waterfront, Cape Town 8001, South Africa

http://www.cambridge.org

© Frédéric Hélein 2002

The first edition of this work was published in French as *Applications harmoniques, lois de conservation et repères mobiles* by Diderot Editeur in 1996. An English translation, translated by L. Almeida, was published by Diderot Editeur in 1997.

This book is in copyright. Subject to statutory exception
and to the provisions of relevant collective licensing agreements,
no reproduction of any part may take place without
the written permission of Cambridge University Press.

Second edition published 2002

Printed in the United Kingdom at the University Press, Cambridge

A catalogue record of this book is available from the British Library

ISBN 0 521 81160 0 hardback

to Henry Wente

Contents

Foreword		*page* ix
Introduction		xiii
Acknowledgements		xxii
Notation		xxiii
1	Geometric and analytic setting	1
	1.1 The Laplacian on $(\mathcal{M}, g)$	2
	1.2 Harmonic maps between two Riemannian manifolds	5
	1.3 Conservation laws for harmonic maps	11
	1.3.1 Symmetries on $\mathcal{N}$	12
	1.3.2 Symmetries on $\mathcal{M}$: the stress–energy tensor	18
	1.3.3 Consequences of theorem 1.3.6	24
	1.4 Variational approach: Sobolev spaces	31
	1.4.1 Weakly harmonic maps	37
	1.4.2 Weakly Noether harmonic maps	42
	1.4.3 Minimizing maps	42
	1.4.4 Weakly stationary maps	43
	1.4.5 Relation between these different definitions	43
	1.5 Regularity of weak solutions	46
2	Harmonic maps with symmetry	49
	2.1 Bäcklund transformation	50
	2.1.1 S^2-valued maps	50
	2.1.2 Maps taking values in a sphere S^n, $n \geq 2$	54
	2.1.3 Comparison	56
	2.2 Harmonic maps with values into Lie groups	58
	2.2.1 Families of curvature-free connections	65
	2.2.2 The dressing	72
	2.2.3 Uhlenbeck factorization for maps with values in $U(n)$	77

		2.2.4 S^1-action	79
	2.3	Harmonic maps with values into homogeneous spaces	82
	2.4	Synthesis: relation between the different formulations	95
	2.5	Compactness of weak solutions in the weak topology	101
	2.6	Regularity of weak solutions	109
3		Compensations and exotic function spaces	114
	3.1	Wente's inequality	115
		3.1.1 The inequality on a plane domain	115
		3.1.2 The inequality on a Riemann surface	119
	3.2	Hardy spaces	128
	3.3	Lorentz spaces	135
	3.4	Back to Wente's inequality	145
	3.5	Weakly stationary maps with values into a sphere	150
4		Harmonic maps without symmetry	165
	4.1	Regularity of weakly harmonic maps of surfaces	166
	4.2	Generalizations in dimension 2	187
	4.3	Regularity results in arbitrary dimension	193
	4.4	Conservation laws for harmonic maps without symmetry	205
		4.4.1 Conservation laws	206
		4.4.2 Isometric embedding of vector-bundle-valued differential forms	211
		4.4.3 A variational formulation for the case $m = n = 2$ and $p = 1$	215
		4.4.4 Hidden symmetries for harmonic maps on surfaces?	218
5		Surfaces with mean curvature in L^2	221
	5.1	Local results	224
	5.2	Global results	237
	5.3	Willmore surfaces	242
	5.4	Epilogue: Coulomb frames and conformal coordinates	244
References			254
Index			263

Foreword

Harmonic maps between Riemannian manifolds provide a rich display of both differential geometric and analytic phenomena. These aspects are inextricably intertwined — a source of undiminishing fascination.

Analytically, the problems belong to elliptic variational theory: *harmonic maps* are the solutions of the Euler–Lagrange equation (section 1.2)

$$\Delta_g u^i + g^{\alpha\beta}(x)\Gamma^i_{jk}(u(x))\frac{\partial u^j}{\partial x^\alpha}\frac{\partial u^k}{\partial x^\beta} = 0 \tag{1}$$

associated to the Dirichlet integral (section 1.1)

$$E(u) = \int_{\mathcal{M}} \frac{|du(x)|^2}{2} d\mathrm{vol}_g.$$

Surely that is amongst the simplest — and yet general — intrinsic variational problems of Riemannian geometry. The system (1) is second order elliptic of divergence type, with linear principal parts in diagonal form with the same Laplacian in each entry; and whose first derivatives have quadratic growth. That is quite a restrictive situation; indeed, those conditions ensure the regularity of continuous weak solutions of (1).

The entire harmonic mapping scene (as of 1988) is surveyed in the articles [50] and [51].

2-DIMENSIONAL DOMAINS

Harmonic maps $u : \mathcal{M} \longrightarrow \mathcal{N}$ with 2-dimensional domains $\mathcal{M}$ present special features, crucial to their applications to minimal surfaces (i.e. conformal harmonic maps) and to deformation theory of Riemann surfaces. Amongst these, as they appear in this monograph:

(i) The Dirichlet integral is a conformal invariant of $\mathcal{M}$. Consequently, harmonicity of u (characterized via the Euler–Lagrange operator associated to E) depends only on the conformal structure of $\mathcal{M}$ (section 1.1).

(ii) Associated with a harmonic map is a holomorphic quadratic differential on $\mathcal{M}$ (locally represented by the function f of section 1.3).

(iii) The inequality of Wente. Qualitatively, that ensures that the Jacobian determinant of a map u (a special quadratic expression involving first derivatives of u) may have slightly more differentiability than might be expected (section 3.1).

(iv) The $\mathcal{C}^2$ maps are dense in $H^1(\mathcal{M}, \mathcal{N})$†

To gain a perspective on the use of harmonic maps of surfaces, the reader is advised to consult [48] and [116] for minimal surfaces and the problem of Plateau. Applications to the theory of deformations of Riemann surfaces can be found in [68] and [49]. The book [98] provides an introduction to all these questions.

REGULARITY

A key step in Morrey's solution of the Plateau problem is his

Theorem 1 (Morrey) *Let $\mathcal{M}$ be a Riemann surface, and $u : \mathcal{M} \longrightarrow \mathcal{N}$ a map with $E(u) < +\infty$. Suppose that u minimizes the Dirichlet integral E_B on every disk B of $\mathcal{M}$ (with respect to the Dirichlet problem induced by the trace of u on the boundary of B). Then u is Hölder continuous.*

In particular, u is harmonic (and as regular as the data permits).

The proof is based on Morrey's Dirichlet growth estimate — related to the growth estimates in section 3.5.

The main goal of the present monograph is the following result, giving a definitive generalization of Theorem 1:

Theorem 2 (Hélein) *Let $(\mathcal{M}, g)$ be a Riemann surface, and $(\mathcal{N}, h)$ a compact Riemannian manifold without boundary. If $u : \mathcal{M} \longrightarrow \mathcal{N}$ is a weakly harmonic map with $E(u) < +\infty$, then u is harmonic.*

† See the proof of lemma 4.1.6 and [145].

That is indeed a major achievement, made some fifty years after Morrey's special case. Hélein first established his theorem in certain particular cases ($\mathcal{N} = S^n$ and various Riemannian homogeneous spaces); then he announced Theorem 2 in [85]. That Note includes a beautifully clear sketch of the proof, together with a description of the new ideas — an absolute gem of presentation!

The high quality is maintained here:

COMMENTARY ON THE TEXT

First of all, the author's exposition requires only a few formalities from differential geometry and variational theory. Secondly, the pace is leisurely and well motivated throughout.

For instance: chapter 1 develops the required background for harmonic maps. The author is satisfied with maps and Riemannian metrics of differentiability class $\mathcal{C}^2$; higher differentiability then follows from general principles. Various standard conservation laws are derived. All that is direct and efficient.

As a change of scene, chapter 2 is an excursion into the methods of completely integrable systems, as applied to harmonic maps of a Riemann surface into S^n (or a Lie group; or a homogeneous space), via conservation laws. One purpose is to illustrate hidden symmetries of Lax form (e.g. related to dressing action). Another is to provide motivations for the methods and constructions used in chapter 4 — especially the role of symmetry in the range.

Chapter 3 describes various spaces of functions — Hardy and Lorentz spaces, in particular — as an exposition specially designed for applications in chapters 4 and 5. Those include refinements and modifications of Wente's inequality; and come under the heading of compensation phenomena — certainly delicate and lovely mathematics!

Chapter 4 is the heart of the monograph — as already noted. There are two new steps required as preparation for the proof of theorem 2:

(i) Lemma 4.1.2, which reduces the problem to the case in which $(\mathcal{N}, h)$ is a Riemannian manifold diffeomorphic to a torus.

(ii) Careful construction of a special frame field on $(\mathcal{N}, h)$ — called a Coulomb frame. Equations (4.10) are derived, serving as some sort of conservation law. When the spaces of Hardy and Lorentz enter the scene, they produce a gain of regularity (see lemma 4.1.7).

Finally, in chapter 5 the methods of Coulomb frames and compensation techniques are applied to problems of surfaces in Euclidean spaces whose second fundamental form or mean curvatures are square-integrable.

James Eells

Introduction

The contemplation of the atlas of an airline company always offers us something puzzling: the trajectories of the airplanes look curved, which goes against our basic intuition, according to which the shortest path is a straight line. One of the reasons for this paradox is nothing but a simple geometrical fact: on the one hand our earth is round and on the other hand the shortest path on a sphere is an arc of great circle: a curve whose projection on a geographical map rarely coincides with a straight line. Actually, choosing the trajectories of airplanes is a simple illustration of a classical variational problem in differential geometry: finding the geodesic curves on a surface, namely paths on this surface with minimal lengths.

Using water and soap we can experiment an analogous situation, but where the former path is now replaced by a soap film, and for the surface of the earth — which was the ambient space for the above example — we substitute our 3-dimensional space. Indeed we can think of the soap film as an excellent approximation of some ideal elastic matter, infinitely extensible, and whose equilibrium position (the one with lowest energy) would be either to shrink to one point or to cover the least area. Thus such a film adopts a minimizing position: it does not minimize the length but the area of the surface. Here is another classical variational problem, the study of minimal surfaces.

Now let us try to imagine a 3-dimensional matter with analogous properties. We can stretch it inside any geometrical manifold, as for instance a sphere: although our 3-dimensional body will be confined — since generically lines will shrink to points — it may find an equilibrium configuration. Actually the mathematical description of such a situation, which is apparently more abstract than the previous ones, looks like the mathematical description of a nematic liquid crystal in equilibrium.

Such a bulk is made of thin rod shaped molecules (*nema* means thread in Greek) which try to be parallel each to each other. Physicists have proposed different models for these liquid crystals where the mean orientation of molecules around a point in space is represented by a vector of norm 1 (hence some point on the sphere). Thus we can describe the configuration of the material using a map defined in the domain filled by the liquid crystal, with values into the sphere. We get a situation which is mathematically analogous to the abstract experiment described above, by imagining we are trying to imprison a piece of perfectly elastic matter inside the surface of a sphere. The physicists Oseen and Frank proposed a functional on the set of maps from the domain filled with the material into the sphere, which is very close to the elastic energy of the abstract ideal matter.

What makes all these examples similar (an airplane, water with soap and a liquid crystal)? We may first observe that these three situations illustrate variational problems. But the analogy is deeper because each of these examples may be modelled by a map (describing the deformation of some body inside another one) which maps a differential manifold into another one, and which minimizes a quantity which is more or less close to a perfect elastic energy. To define that energy, we need to measure the infinitesimal stretching imposed by the mapping and to define a measure on the source space. Such definitions make sense provided that we use Riemannian metrics on the source and target manifolds.

Let $\mathcal{M}$ denote the source manifold, $\mathcal{N}$ the target manifold and u a differentiable map from $\mathcal{M}$ into $\mathcal{N}$. Given Riemannian metrics on these manifolds we may define the *energy* or *Dirichlet integral*

$$E(u) = \frac{1}{2} \int_{\mathcal{M}} |du|^2 d\text{vol},$$

where $|du|$ is the Hilbert–Schmidt norm of the differential du of u and $d\text{vol}$ is the Riemannian measure on $\mathcal{M}$. If we think of the map u as the way to confine and stretch an elastic $\mathcal{M}$ inside a rigid $\mathcal{N}$, then $E(u)$ represents an elastic deformation energy. Smooth maps (i.e. of class $\mathcal{C}^2$) which are critical points of the Dirichlet functional are called *harmonic maps*. For the sake of simplicity, let us assume that $\mathcal{N}$ is a submanifold of a Euclidean space. Then the equation satisfied by a harmonic map is

$$\Delta u(x) \perp T_{u(x)} \mathcal{N},$$

where Δ is the Laplacian on $\mathcal{M}$ associated to the Riemannian metric, and $T_{u(x)}\mathcal{N}$ is the tangent space to $\mathcal{N}$ at the point $u(x)$. For different choices on $\mathcal{M}$ and $\mathcal{N}$, a harmonic map will be a constant speed parametrization of a geodesic (if the dimension of $\mathcal{M}$ is 1), a harmonic function (if $\mathcal{N}$ is the real line) or something hybrid.

It is possible to extend the notion of harmonic maps to much less regular maps, which belong to the Sobolev space $H^1(\mathcal{M},\mathcal{N})$ of maps from $\mathcal{M}$ into $\mathcal{N}$ with finite energy. The above equation is true but only in the distribution sense and we speak of *weakly harmonic maps*.

Because of the simplicity of this definition, we can meet examples of harmonic maps in various situations in geometry as well as in physics. For example, any submanifold $\mathcal{M}$ of an affine Euclidean space has a constant mean curvature (or more generally a parallel mean curvature) if and only if its Gauss map is a harmonic map. A submanifold $\mathcal{M}$ of a manifold $\mathcal{N}$ is minimal if and only if the immersion of $\mathcal{M}$ in $\mathcal{N}$ is harmonic. In condensed matter physics, harmonic maps between a 3-dimensional domain and a sphere have been used as a simplified model for nematic liquid crystals. In theoretical physics, harmonic maps between surfaces and Lie groups are extensively studied, since they lead to properties which are strongly analogous to (anti)self-dual Yang–Mills connections on 4-dimensional manifolds, but they are simpler to handle. In such a context they correspond to the so-called σ-models. Recently, the interest of physicists in these objects has been reinforced since their quantization leads to examples of conformal quantum field theories — an extremely rich subject. In some sense the quantum theory for harmonic maps between a surface and an Einstein manifold (both endowed with Minkowski metrics) corresponds to string theory (in the absence of supersymmetries). Other models used in physics, such as the Skyrme model, Higgs models or Ginzburg–Landau models [12], show strong connections with the theory of harmonic maps into a sphere or a Lie group.

Despite their relatively universal character, harmonic maps became an active topic for mathematicians only about four decades ago. One of the first questions was motivated by algebraic topology: given two Riemannian manifolds and a homotopy class for maps between these manifolds, does there exist a harmonic map in this homotopy class? In the case where the sectional curvature of the target manifold is negative, James Eells and Joseph Sampson showed in 1964 that this is true, using the heat equation. Then the subject developed in many different directions and aroused many fascinating questions in topology, in differential geometry, in algebraic geometry and in the analysis of partial differential equations. Important generalizations have been proposed, such as the evolution equations for harmonic maps between manifolds (heat equation or wave equation) or the p-harmonic maps (i.e. the critical points of the integral of the p-th power of $|du|$). During the same period, and essentially independently, physicists also developed many interesting ideas on the subject.

The present work does not pretend to be a complete presentation of the theory of harmonic maps. My goal is rather to offer the reader an introduction to this subject, followed by a communication of some recent results. We will be motivated by some fundamental questions in analysis, such as the compactness in the weak topology of the set of weakly harmonic maps, or their regularity. This is an opportunity to explore some ideas and methods (symmetries, compensation phenomena, the use of *moving frames* and of *Coulomb moving frames*), the scope of which is, I believe, more general than the framework of harmonic maps.

The regularity problem is the following: is a weakly harmonic map u smooth? (for instance if $\mathcal{N}$ is of class $\mathcal{C}^{k,\alpha}$, is u of class $\mathcal{C}^{k,\alpha}$, for $k \geq 2$, $0 < \alpha < 1$?). The (already) classical theory of quasilinear elliptic partial differential equation systems ([117], [103]) teaches us that any *continuous* weakly harmonic map is automatically as regular as allowed by the the regularity of the Riemannian manifolds involved. The critical step is thus to know whether or not a weakly harmonic map is continuous. Answers are extremely different according to the dimension of the source manifold, the curvature of the target manifold, its topology or the type of definition chosen for a weak solution.

The question of compactness in the weak topology of weakly harmonic maps is the following problem. Given a sequence $(u_k)_{k \in \mathbb{N}}$ of weakly harmonic maps which converge in the weak topology of $H^1(\mathcal{M}, \mathcal{N})$ towards a map u, can we deduce that the limit u is a weakly harmonic map? Such a question arises when, for instance, one wants to prove the existence of solutions to evolution problems for maps between manifolds. This is a very disturbing problem: we will see that the answer is yes in the case where the target manifold is symmetric, but we do not know the answer in the general situation.

The first idea which this book stresses is the role of symmetries in a variational problem. It is based on the following observation, due to Emmy Noether: if a variational problem is invariant under the action of a continuous group of symmetries, we can associate to each solution of this variational problem a system of conservation laws, i.e. one, or several, divergence-free vector fields defined on the source domain. The number of independent conservation laws is equal to the dimension of the group of symmetries. The importance of this result has been celebrated for years in theoretical physics. For example, in the particular case where the variational problem involves one variable (the time) the conservation law is just the prediction that a scalar quantity is constant in time (the conservation of the energy comes from the invariance under time translations, the conservation of the momentum is a consequence of the invariance under translations in space...). One of the goals of this book is to convince you that Noether's theorem is also fundamental in the study of partial differential equations such as harmonic maps.

In a surprising way, the exploitation of symmetries for analytical purposes is strongly related to compensation phenomena: by handling conservation laws, remarkable non-linear quantities (for an analyst) naturally appear. The archetype of this kind of quantity is the Jacobian determinant

$$\{a,b\} := \frac{\partial a}{\partial x}\frac{\partial b}{\partial y} - \frac{\partial a}{\partial y}\frac{\partial b}{\partial x},$$

where a and b are two functions whose derivatives are square-integrable (i.e. a and b belong to the Sobolev space H^1). Since Charles B. Morrey, it is known that such a quantity enjoys the miraculous property of being simultaneously non-linear and continuous with respect to the weak topology in the space H^1; if a_ϵ and b_ϵ converge *weakly* in H^1 towards a and b, then $\{a_\epsilon, b_\epsilon\}$ converges towards $\{a,b\}$ in the sense of distributions. This is the subject of the theory of compensated compactness of François Murat and Luc Tartar. Moreover the same quantity $\{a,b\}$ possesses regularity or integrability properties slightly better than any other bilinear function of the partial derivatives of a and b. It seems that this result of "compensated regularity" was observed for the first time by Henry Wente in 1969. For twenty years this phenomenon was used only in the context of constant mean curvature surfaces, by H. Wente and by Haïm Brezis and Jean-Michel Coron (further properties were pointed out by these last two authors and also by L. Tartar). But more recently, at the end of the 1980s, works of Stefan Müller, followed by Ronald Coifman, Pierre-Louis Lions, Yves Meyer and Stephen Semmes, shed a new light on the quantity $\{a,b\}$, and in particular it was established that this Jacobian determinant belongs to the *Hardy space*, a slightly improved version of the space of integrable functions L^1. All these results played a vital role in the progress which has been obtained recently in the regularity theory of harmonic maps, and are the companion ingredients to the conservation laws.

The limitation of techniques which use conservation laws is that symmetric variational problems are exceptions. Thus the above methods are not useful, a priori, for the study of harmonic maps with values into a non-symmetric manifold. We need then to develop new techniques. One idea is the use of moving frames. It consists in giving, for each point y in $\mathcal{N}$, an orthonormal basis $(e_1, \ldots, e_n)$ of the tangent space to $\mathcal{N}$ at y, that depends smoothly on the point y. This system of coordinates on the tangent space $T_y\mathcal{N}$ was first developed by Gaston Darboux and mainly by Elie Cartan. These moving frames turn out to be extremely suitable in differential geometry and allow a particularly elegant presentation of the Riemannian geometry (see [37]). But in the problems with which we are concerned, we will use a particular class of moving frames, satisfying an extra differential equation. It consists essentially of a condition which expresses that the moving frame is a harmonic section (a generalization of harmonic maps to the case of fiber bundles) of a fiber bundle over $\mathcal{M}$ whose fiber at x is precisely the set of orthonormal bases of the tangent space to $\mathcal{N}$ at $u(x)$. Since the rotation group $SO(n)$ is a symmetry group for that bundle and for the associated variational problem, our condition gives rise to conservation laws, thanks to Noether's theorem. We call such a moving frame a *Coulomb moving frame*, inspired by the analogy with the use of Coulomb gauges by physicists for gauge theories. The use of such a system of privileged coordinates is crucial for the study of the regularity of weakly harmonic maps, with values into an arbitrary manifold. It leads to the appearance of these magical quantities similar to $\{a, b\}$, that we spoke about before.

The first chapter of this book presents a description of harmonic maps and of various notions of weak solutions. We will emphasize Noether's theorem through two versions which play an important role for harmonic maps. In the (exceptional but important) case where the target manifold $\mathcal{N}$ possesses symmetries, the conservation laws lead to very particular properties which will be presented in the second chapter. But in constrast, there is a symmetry which is observed in general cases and which is related to invariance under change of coordinates on the source manifold $\mathcal{M}$. It is not really a geometrical symmetry in general and it will lead to some covariant version of Noether's theorem: the stress–energy tensor

$$S_{\alpha\beta} = \frac{|du|^2}{2} g_{\alpha\beta} - \left\langle \frac{\partial u}{\partial x^\alpha}, \frac{\partial u}{\partial x^\beta} \right\rangle$$

always has a vanishing covariant divergence. This equation has a consequence which is very important for the theory of the regularity of weak solutions: the monotonicity formula. In the case where there is a geometrical symmetry acting on $\mathcal{M}$, some of the covariant conservation laws specialize and become true conservation laws. One particular case is when the dimension of $\mathcal{M}$ is 2, since then the harmonic map problem is invariant under conformal transformations of $\mathcal{M}$, and hence the stress–energy tensor coincides with the *Hopf differential* and is holomorphic. We end this chapter by a quick survey of the regularity results which are known concerning weak solutions.

The second chapter is a suite of variations on the version of Noether's theorem which concerns harmonic maps with values into a symmetric manifold $\mathcal{N}$. We present various kinds of results but they are all consequences of the same conservation law. If for instance $\mathcal{N}$ is the sphere S^2 in 3-dimensional space, we start from

$$\mathrm{div}(u^i \nabla u^j - u^j \nabla u^i) = 0, \ \forall i, j = 1, 2, 3.$$

Using this conservation law, we will see that it is easy to exhibit the relations between harmonic maps from a surface into S^2, and surfaces of constant mean curvature or positive constant Gauss curvature in 3-dimensional space. We hence recover the construction due to Ossian Bonnet of families of parallel surfaces with constant mean curvature and constant Gauss curvature. Moreover, we can deduce from this conservation law a formulation (which was probably discovered by K. Pohlmeyer and by V.E. Zhakarov and A.B. Shabat) using loop groups, of the harmonic maps problem between a surface and a symmetric manifold. Such a formulation is a feature of completely integrable systems, like the Korteweg–de Vries equation (see [150]). Many authors have used this theory during the last decade in a spectacular way: Karen Uhlenbeck deduced a classification of all harmonic maps from the sphere S^2 into the group $U(n)$ [174]. After Nigel Hitchin, who obtained all harmonic maps from a torus into S^3 by algebraico-geometric methods [94], Fran Burstall, Dirk Ferus, Franz Pedit and Ulrich Pinkall were able to construct all harmonic maps from a torus into a symmetric manifold (the symmetry group of which is compact semi-simple) [24] and more recently an even more general construction has been obtained by Joseph

Dorfmeister, Franz Pedit and HongYu Wu [46]. We will give a brief description of some of these results.

In another direction, the same conservation law allows one to prove in a few lines some analysis results such as the compactness in the weak topology of the set of weakly harmonic maps with values into a symmetric manifold, or their regularity (complete or partial depending on other hypotheses): we present the existence result for solutions to the wave equation for maps with values in a symmetric manifold due to Jalal Shatah, and my regularity result for weakly harmonic maps between a surface and a sphere.

The third chapter, which is essentially devoted to compensation phenomena and to Hardy and Lorentz spaces, brings very different ingredients by constrast with the previous chapter, but complementary. The main object is the Jacobian determinant $\{a, b\}$. We begin by showing the following result due to H. Wente: if a and b belong to the Sobolev space $H^1(\Omega, \mathbb{R})$, where Ω is a domain in the plane $\mathbb{R}^2$, and if ϕ is the solution on Ω of

$$\begin{cases} -\Delta \phi = \{a, b\} & \text{in } \Omega \\ \phi = 0 & \text{on } \partial\Omega, \end{cases}$$

then ϕ is continuous and is in $H^1(\Omega, \mathbb{R})$. Moreover, we can estimate the norm of ϕ in the spaces involved as a function of the norms of da and db in L^2. Then we will discuss some optimal versions of this theorem and its relations with the isoperimetric inequality and constant mean curvature surfaces. Afterwards we will introduce Hardy and Lorentz spaces and see how they can be used to refine Wente's theorem. As an illustration of these ideas, the chapter ends with the proof of a result of Lawrence Craig Evans on the partial regularity of weakly stationary maps with values into the sphere.

The fourth chapter deals with harmonic maps with values into manifolds without symmetry. We thus need to work without the conservation laws which were at the origin of the results of chapter 2. For the regularity problem we substitute for the conservation laws the use of *Coulomb moving frames* on the target manifold $\mathcal{N}$. Given a map u from $\mathcal{M}$ into $\mathcal{N}$, a Coulomb moving frame consists in an orthonormal frame field on $\mathcal{M}$ which is a *harmonic section* of the pull-back by u of the orthonormal tangent frame bundle on $\mathcal{N}$ (i.e. the fiber bundle whose base manifold is $\mathcal{M}$, obtained by attaching to each point x in $\mathcal{M}$ the set of (direct) orthonormal bases of the tangent space to $\mathcal{N}$ at $u(x)$). Using to this construction and the analytical tools introduced in chapter 3, we may extend the regularity results obtained in the two previous chapters, by dropping the symmetry hypothesis on $\mathcal{N}$: we prove my theorem on the regularity of weakly harmonic maps on a surface, then a generalization of it due to Philippe Choné and lastly the generalization of the result of L.C. Evans proved in chapter 3, obtained by Fabrice Bethuel. Strangely, we are not able to present a definite answer to the compactness problem in the weak topology of the set of weakly harmonic maps. Motivated by this question we end chapter 4 by studying the possibility of building conservation laws without symmetries. It leads us to "isometric embedding" problems for covariantly closed differential forms, with coefficients in a vector bundle equipped with a connection. Such problems look interesting by themselves, as this class of questions offers a hybrid generalization of Poincaré's lemma for closed differential forms, and the isometric embedding problem for Riemannian manifolds.

The fifth chapter does not directly concern harmonic maps, but is an excursion into the study of conformal parametrizations of surfaces. The starting point is a result of Tatiana Toro which established the remarkable fact that an embedded surface in Euclidean space which has a square-integrable second fundamental form is Lipschitz. Soon after, Stefan Müller and Vladimir Šverák proved that any conformal parametrization of such a surface is bilipschitz. Their proof relies in a clever way on the compensation results described in chapter 3 about the quantity $\{a, b\}$, and on the use of Hardy space. We give here a slightly different presentation of the result and of the proof of their result: we do not use Hardy space but only Wente's inequality and Coulomb moving frames. More precisely, we study the space of conformal parametrizations of surfaces in Euclidean space with second fundamental form bounded in L^2, and we show a compactness result for this space. This tour will naturally bring us to an amusing interpretation of Coulomb moving frames: a Coulomb moving frame associated to the identity map from a surface to itself corresponds essentially to a system of conformal coordinates.

Acknowledgements

First of all, I thank the *Fondation Peccot* which accorded me the honor of giving a series of lectures at the *Collège de France* in 1994. At this time indeed I planned to present a few notes on these lectures. But rapidly this project was transformed into the writing of this book. Some of the expounded ideas were conceived in 1994, when I was a guest of Mariano Giaquinta and Giuseppe Modica in the University of Florence, and I am grateful to them. I also thank the Carnegie-Mellon University and Irene Fonsecca, David Kinderlehrer and Luc Tartar for their hospitality in 1995, which gave me the opportunity to develop this project extensively.

I also wish to express my gratitude and friendship to all people who contributed to the improvement of the text through their remarks, their advice and their encouragement, and more particularly: François Alouges, Fabrice Bethuel, Fran Burstall, Gilles Carbou, Jean-Michel Coron, Françoise Demengel, Yuxin Ge, Jean-Michel Ghidaglia, Ian Marshall, Frank Pacard, Laure Quivy, Tristan Rivière, Pascal Romon, Peter Topping, Tatiana Toro, Dong Ye ... without forgetting Jim Eells who gave me precious information and offered us the preface of this book.

Lastly I want to thank Luís Almeida who helped me a lot, by translating large parts of this text into english.

Notation

Ω will denote an open subset of $\mathbb{R}^m$.

- $L^p(\Omega)$: Lebesgue space. For $1 \leq p \leq \infty$, $L^p(\Omega)$ is the set (of equivalence classes) of measurable functions f from Ω to $\mathbb{R}$ such that $||f||_{L^p} < +\infty$, where

$$||f||_{L^p} := \left(\int_\Omega |f(x)|^p dx^1 ... dx^m \right)^{\frac{1}{p}}, \text{ if } 1 \leq p < \infty,$$

$$||f||_{L^\infty} := \inf\{M \in [0, +\infty] \mid |f(x)| \leq M \text{ a.e.}\}.$$

- $L^p_{loc}(\Omega)$: space of measurable functions f from Ω to $\mathbb{R}$ such that for every compact subset K of Ω, the restriction of f to K, $f_{|K}$, belongs to $L^p(K)$.
- $W^{k,p}(\Omega)$: Sobolev space. For each multi-index $s = (s_1, ..., s_m) \in \mathbb{N}^m$, we define $|s| = \sum_{\alpha=1}^m s_\alpha$, and $D_s = \frac{\partial^{|s|}}{(\partial x^1)^{s_1}...(\partial x^m)^{s_m}}$. Then, for $k \in \mathbb{R}$ and $1 \leq p \leq \infty$,

$$W^{k,p}(\Omega) := \{f \in L^p(\Omega) \mid \forall s, |s| \leq k, D_s f \in L^p(\Omega)\}.$$

Here, $D_s f$ is a derivative of order $|s|$ of f, in the sense of distributions. On this space we have the norm

$$||f||_{W^{k,p}} := \sum_{|s| \leq k} ||D_s f||_{L^p}.$$

- $W^{-k,p}(\Omega)$: the dual space of $W^{k,p}(\Omega)$.
- $H^k(\Omega) := W^{k,2}(\Omega)$. On this space we have the norm (equivalent to

$\|f\|_{W^{k,2}}$)

$$\|f\|_{H^k} := \left(\sum_{|s|\leq k} \|D_s f\|_{L^2}^2 \right)^{\frac{1}{2}}.$$

- $\mathcal{C}^k(\Omega)$: set of continuous functions on Ω which are k times differentiable and whose derivatives up to order k are continuous (for $k \in \mathbb{N}$ or $k = \infty$).
- $\mathcal{C}^k_c(\Omega)$: set of functions in $\mathcal{C}^k(\Omega)$ with compact support in Ω.
- $\mathcal{D}'(\Omega)$: space of distributions over Ω (it is the dual of $\mathcal{D}(\Omega) := \mathcal{C}^\infty_c(\Omega)$).
- $\mathcal{C}^{0,\alpha}(\Omega)$: Hölder space. Set of functions f, continuous over Ω, such that

$$\sup_{x,y\in\Omega} \frac{|f(x)-f(y)|}{|x-y|^\alpha} < +\infty,$$

(for $0 < \alpha < 1$).
- $\mathcal{C}^{k,\alpha}(\Omega)$: set of k times differentiable continuous functions such that all derivatives up to order k belong to $\mathcal{C}^{0,\alpha}(\Omega)$.
- $\mathcal{H}^1$: Hardy space, see definitions 3.2.4, 3.2.5 and 3.2.8.
- $BMO(\Omega)$: space of functions with bounded mean oscillation, see definition 3.2.7.
- $L^{(p,q)}(\Omega)$: Lorentz space, see definition 3.3.2.
- $\mathcal{L}^{q,\lambda}(\Omega)$: Morrey–Campanato space, see definition 3.5.9.
- $E_{x,r}$: see example 1.3.7, section 4.3 and section 3.5.
- The scalar product between two vectors X and Y is denoted by $\langle X, Y \rangle$ or $X \cdot Y$.
- $\{a,b\} := \frac{\partial a}{\partial x}\frac{\partial b}{\partial y} - \frac{\partial a}{\partial y}\frac{\partial b}{\partial x}$, see section 3.1.
- $\{u \cdot v\}$: if u and v are two maps from a domain in $\mathbb{R}^2$ with values into a Euclidean vector space $(V, \langle .,. \rangle)$, $\{u \cdot v\} := \langle \frac{\partial u}{\partial x}, \frac{\partial v}{\partial y} \rangle - \langle \frac{\partial u}{\partial y}, \frac{\partial v}{\partial x} \rangle$.
- $\{a,b\}_{\alpha\beta}$: see section 4.3.
- $\widehat{ab}, \widetilde{ab}_\Omega$: see section 3.1.
- $\Lambda^p \mathbb{R}^m$: algebra of p-forms with constant coefficients over $\mathbb{R}^m$ (p-linear skew-symmetric forms over $\mathbb{R}^m$). $\Lambda \mathbb{R}^m = \bigoplus_{p=0}^m \Lambda^p \mathbb{R}^m$.
- $\wedge$: wedge product in the algebra $\Lambda \mathbb{R}^m$ (see [47] or [183]).
- d: exterior differential, acting linearly over $\mathcal{D}'(\Omega) \otimes \Lambda \mathbb{R}^m$ and such that $\forall \phi \in \mathcal{C}^\infty(\Omega), \forall \alpha \in \Lambda \mathbb{R}^m$, $d(\phi \otimes \alpha) = \sum_{\alpha=1}^m \frac{\partial \phi}{\partial x^\alpha} dx^\alpha \wedge \alpha$.
- $\times$: vector product in $\mathbb{R}^3$:

$$\begin{pmatrix} x^1 \\ x^2 \\ x^3 \end{pmatrix} \times \begin{pmatrix} y^1 \\ y^2 \\ y^3 \end{pmatrix} = \begin{pmatrix} x^2 y^3 - x^3 y^2 \\ x^3 y^1 - x^1 y^3 \\ x^1 y^2 - x^2 y^1 \end{pmatrix}.$$

- ${}^t u$: for any vector $u = \begin{pmatrix} u^1 \\ \vdots \\ u^n \end{pmatrix}$, ${}^t u = (u^1, \ldots, u^n)$.
- $GL(E)$: if E is a vector space, $GL(E)$ is the group of invertible endomorphisms of E.
- $M(n \times n, \mathbb{R})$ or $M(n \times n, \mathbb{C})$: algebra of (real or complex) square matrices with n rows (and n columns).
- $\mathbb{1}$: identity matrix.
- δ_b^a: Kronecker symbol, its value is 1 if $a = b$ and 0 if $a \neq b$.
- $O(n) := \{R \in M(n \times n, \mathbb{R}) \mid {}^t R R = \mathbb{1}\}$.
- $SO(n) := \{R \in M(n \times n, \mathbb{R}) \mid {}^t R R = \mathbb{1}, \det R = 1\}$.
- $SO(n)^{\mathbb{C}} := \{R \in M(n \times n, \mathbb{C}) \mid {}^t R R = \mathbb{1}, \det R = 1\}$.
- $so(n) := \{A \in M(n \times n, \mathbb{R}) \mid {}^t A + A = 0\}$.
- $SU(n) := \{R \in M(n \times n, \mathbb{C}) \mid {}^t \bar{R} R = \mathbb{1}, \det R = 1\}$.
- $su(n) := \{A \in M(n \times n, \mathbb{C}) \mid {}^t \bar{A} + A = 0, \operatorname{tr} A = 0\}$.
- $Spin(3)$: (2-fold) universal covering of $SO(3)$. It is identified with $SU(2)$.
- $S^{n-1} := \{y \in \mathbb{R}^n \mid |y| = \sqrt{(y^1)^2 + \cdots + (y^n)^2} = 1\}$.

1
Geometric and analytic setting

This chapter essentially describes the objects and properties that will interest us in this work. For a more detailed exposition of the general background in Riemannian geometry and in analysis on manifolds, one may refer for instance to [183] and [98]. After recalling how to associate, to each Riemannian metric on a manifold, a Laplacian operator on the same manifold, we will give a definition of smooth harmonic map between two manifolds. Very soon, we will use the variational framework, which consists in viewing harmonic maps as the critical points of the Dirichlet functional.

Next, we introduce a frequently used ingredient in this book: Noether's theorem. We present two versions of it: one related to the symmetries of the image manifold, and the other which is a consequence of an invariance of the problem under diffeomorphisms of the domain manifold (in this case it is not exactly Noether's theorem, but a "covariant" version).

These concepts may be extended to contexts where the map between the two manifolds is less regular. In fact, a relatively convenient space is that of maps with finite energy (Dirichlet integral), $H^1(\mathcal{M}, \mathcal{N})$. This space appears naturally when we try to use variational methods to construct harmonic maps, for instance the minimization of the Dirichlet integral. The price to pay is that when the domain manifold has dimension larger than or equal to 2, maps in $H^1(\mathcal{M}, \mathcal{N})$ are not smooth, in general. Moreover, $H^1(\mathcal{M}, \mathcal{N})$ does not have a differentiable manifold structure. This yields that several non-equivalent generalizations of the notion of harmonic function coexist in $H^1(\mathcal{M}, \mathcal{N})$ (weakly harmonic, stationary harmonic, minimizing, ...). We will conclude this chapter with a brief survey of the known results on weakly harmonic maps in $H^1(\mathcal{M}, \mathcal{N})$. As we will see, the results are considerably different according to which definition of critical point of the Dirichlet integral we adopt.

NOTATION: $\mathcal{M}$ and $\mathcal{N}$ are differentiable manifolds. Most of the time, $\mathcal{M}$ plays the role of domain manifold, and $\mathcal{N}$ that of image manifold; we will suppose $\mathcal{N}$ to be compact without boundary. In case they are abstract manifolds (and not submanifolds) we may suppose that they are $\mathcal{C}^\infty$ (in fact, thanks to a theorem of Whitney, we may show that every $\mathcal{C}^1$ manifold is $\mathcal{C}^1$-diffeomorphic to a $\mathcal{C}^\infty$ manifold). Unless stated otherwise, $\mathcal{M}$ is equipped with a $\mathcal{C}^{0,\alpha}$ Riemannian metric g, where $0 < \alpha < 1$. For $\mathcal{N}$, we consider two possible cases: either it is an abstract manifold with a $\mathcal{C}^1$ Riemannian metric h, or we will need to suppose it is a $\mathcal{C}^2$ immersed submanifold of $\mathbb{R}^N$. The second situation is a special case of the first one, but nevertheless, Nash's theorem (see [123], [74] and [77]) assures us that if h is $\mathcal{C}^l$ for $l \geq 3$, then there exists a $\mathcal{C}^l$ isometric immersion of $(\mathcal{N}, h)$ in $(\mathbb{R}^N, \langle .,. \rangle)$.

Several regularity results are presented in this book. We will try to present them under minimal regularity hypotheses on $(\mathcal{M}, g)$ and $(\mathcal{N}, h)$, keeping in mind that any improvement of the hypotheses on $(\mathcal{M}, g)$ and $(\mathcal{N}, h)$ automatically implies an improvement of the conclusion, as explained in theorem 1.5.1.

We write $m := \dim \mathcal{M}$ and $n := \dim \mathcal{N}$.

1.1 The Laplacian on $(\mathcal{M}, g)$

For every metric g on $\mathcal{M}$ there exists an associated Laplacian operator Δ_g, acting on all smooth functions on $\mathcal{M}$ taking their values in $\mathbb{R}$ (or any vector space over $\mathbb{R}$ or $\mathbb{C}$). To define it, let us use a local coordinate system $(x^1, \ldots, x^m)$ on $\mathcal{M}$. Denote by

$$g_{\alpha\beta}(x) = g(x)\left(\frac{\partial}{\partial x^\alpha}, \frac{\partial}{\partial x^\beta}\right)$$

the coefficients of the metric, and by $\det g(x)$ the determinant of the matrix whose elements are $g_{\alpha\beta}(x)$. Then, for each real-valued function ϕ defined over an open subset Ω of $\mathcal{M}$, we let

$$\Delta_g \phi = \frac{1}{\sqrt{\det g}} \frac{\partial}{\partial x^\alpha}\left(\sqrt{\det g}\, g^{\alpha\beta}(x) \frac{\partial \phi}{\partial x^\beta}\right) \tag{1.1}$$

where we adopt the convention that repeated indices should be summed over. The metric g induces on the cotangent space $T_x^* \mathcal{M}$ a metric which

1.1 The Laplacian on $(\mathcal{M}, g)$

we denote by $g^{\sharp}$. Its coefficients are given by $g^{\alpha\beta} = g^{\sharp}(dx^{\alpha}, dx^{\beta})$. Recall that $g^{\alpha\beta}(x)$ represents an element of the inverse matrix of $(g_{\alpha\beta})$.

Definition 1.1.1 *Any smooth function ϕ defined over an open subset Ω of $\mathcal{M}$ and satisfying*

$$\Delta_g \phi = 0$$

is called a harmonic function.

We can easily check through a computation that the operator Δ_g does not depend on the choice of the coordinate system, but it will be more pleasant to obtain this as a consequence of a variational definition of Δ_g. Let

$$dvol_g = \sqrt{\det g(x)}\, dx^1 \ldots dx^m, \qquad (1.2)$$

be the Riemannian measure. For each smooth function ϕ from $\Omega \subset \mathcal{M}$ to $\mathbb{R}$, let

$$E_{(\Omega,g)}(\phi) = \int_\Omega e(\phi)\, dvol_g \qquad (1.3)$$

be the energy or Dirichlet integral of ϕ (which may be finite or not). Here, $e(\phi)$ is the energy density of ϕ and is given by

$$e(\phi) = \frac{1}{2} g^{\alpha\beta}(x) \frac{\partial \phi}{\partial x^\alpha} \frac{\partial \phi}{\partial x^\beta}. \qquad (1.4)$$

It is easy to check that the Dirichlet integral does not depend on the choice of the local coordinate system and that, if ψ is a compactly supported smooth function on $\Omega \subset \mathcal{M}$, then for all $t \in \mathbb{R}$,

$$E_{(\Omega,g)}(\phi + t\psi) = E_{(\Omega,g)}(\phi) - t \int_\Omega (\Delta_g \phi)\psi\, dvol_g + O(t^2). \qquad (1.5)$$

Hence, $-\Delta_g$ appears as the variational derivative of E_Ω, which provides us with an equivalent definition of the Laplacian.

Thus, the Laplacian does not depend on the coordinate system used. However, it depends on the metric. For instance, let us consider the effect of a conformal transformation on $(\mathcal{M}, g)$, i.e. compare the Dirichlet integrals and the Laplacians on the manifolds $(\mathcal{M}, g)$ and $(\mathcal{M}, e^{2v}g)$, where v is a smooth real-valued function on $\mathcal{M}$. We have

$$dvol_{e^{2v}g} = e^{mv}\, dvol_g, \qquad (1.6)$$

and for the energy density (1.4)

$$e_{e^{2v}g}(\phi) = e^{-2v} e_g(\phi). \tag{1.7}$$

Thus,

$$E_{(\Omega, e^{2v}g)}(\phi) = \int_\Omega e^{(m-2)v} e_g(\phi) \, d\mathrm{vol}_g. \tag{1.8}$$

However, we notice that in case $m = 2$, the Dirichlet integrals calculated using the metrics g and $e^{2v}g$ coincide, and thus are invariant under a conformal transformation of the metric.

Still in the case $m = 2$, we have

$$\Delta_{e^{2v}g}(\phi) = e^{-2v} \Delta_g \phi. \tag{1.9}$$

Therefore, for $m = 2$, every function which is harmonic over $(\mathcal{M}, g)$ will also be so over $(\mathcal{M}, e^{2v}g)$. More generally, if $(\mathcal{M}, g)$ and $(\mathcal{M}', g')$ are two Riemannian surfaces and Ω and Ω' are two open subsets of $\mathcal{M}$ and $\mathcal{M}'$ respectively, then if $T : (\Omega, g) \longrightarrow (\Omega', g')$ is a conformal diffeomorphism, we have

$$E_{(\Omega, g)}(\phi \circ T) = E_{(\Omega', g')}(\phi), \forall \phi \in \mathcal{C}^1(\Omega', \mathbb{R}) \tag{1.10}$$

and

$$\Delta_g(\phi \circ T) = \lambda (\Delta_{g'} \phi) \circ T, \tag{1.11}$$

where

$$\lambda = \frac{1}{2} g^{\alpha\beta}(x) g'_{ij}(T(x)) \frac{\partial T^i}{\partial x^\alpha} \frac{\partial T^j}{\partial x^\beta}.$$

Thus,

Proposition 1.1.2 *The Dirichlet integral, and the set of harmonic functions over an open subset of a Riemannian surface, depend only on the conformal structure of this surface.*

This phenomenon, characteristic of dimension 2, has many consequences, among them the following, which is very useful: first recall that according to the theorem below, locally all conformal structures are equivalent.

Theorem 1.1.3 *Let $(\mathcal{M}, g)$ be a Riemannian surface. Then, for each point x_0 in $(\mathcal{M}, g)$, there is a neighborhood U of x_0 in $\mathcal{M}$, and a diffeomorphism T from the disk*

$$D = \{(x,y) \in \mathbb{R}^2 | \ x^2 + y^2 < 1\}$$

to U, such that, if c is the canonical Euclidean metric on the disk, $T : (D, c) \longrightarrow (U, g)$ is a conformal map. We say that T^{-1} is a local conformal chart in $(\mathcal{M}, g)$ and that (x, y) are conformal coordinates.

Remark 1.1.4 *There are several proofs of this result, depending on the regularity of g. The oldest supposes g to be analytic. Later methods like that of S.S. Chern (see [36]), where g is supposed to be just Hölder continuous, have given results that are valid under weaker regularity assumptions. At the end of this book (theorem 5.4.3) we can find a proof of theorem 1.1.3 under weaker assumptions.*

Using theorem 1.1.3, we can express the Dirichlet integral over U of a map ϕ from $\mathcal{M}$ to $\mathbb{R}$, simply as

$$\int_U e(\phi) \, d\mathrm{vol}_g = \int_{D^2} \frac{1}{2} \left[\left(\frac{\partial (\phi \circ T)}{\partial x} \right)^2 + \left(\frac{\partial (\phi \circ T)}{\partial y} \right)^2 \right] dxdy,$$

and ϕ will be harmonic if and only if

$$\Delta(\phi \circ T) = \frac{\partial^2 (\phi \circ T)}{\partial x^2} + \frac{\partial^2 (\phi \circ T)}{\partial y^2} = 0.$$

Thus, when studying harmonic functions on a Riemannian surface, we can always suppose, at least locally, that our equations are similar to those corresponding to the case where the domain metric is flat (Euclidean).

1.2 Harmonic maps between two Riemannian manifolds

We now introduce a second Riemannian manifold, $\mathcal{N}$, supposed to be compact and without boundary, which we equip with a metric h. Recall that over any Riemannian manifold $(\mathcal{N}, h)$, there exists a unique *connection* or *covariant derivative*, ∇, having the following properties.

(i) ∇ is a linear operator acting on the set of smooth (at least $\mathcal{C}^1$) tangent vector fields on $\mathcal{N}$. To each $\mathcal{C}^k$ vector field X (where

$k \geq 1$) on $\mathcal{N}$, we associate a field of $\mathcal{C}^{k-1}$ linear maps from $T_y \mathcal{N}$ to $T_y \mathcal{N}$ defined by

$$T_y \mathcal{N} \ni Y \longmapsto \nabla_Y X \in T_y \mathcal{N}.$$

(ii) ∇ is a derivation, i.e. for any smooth function α from $\mathcal{N}$ to $\mathbb{R}$, any vector field X and any vector Y in $T_y \mathcal{N}$,

$$\nabla_Y(\alpha X) = d\alpha(Y)X + \alpha \nabla_Y X.$$

(iii) The metric h is parallel for ∇, i.e. for any vector fields X, Y, and for any vector Z in $T_y \mathcal{N}$,

$$d(h_y(X,Y))(Z) = h_y(\nabla_Z X, Y) + h_y(X, \nabla_Z Y).$$

(iv) ∇ has zero torsion, i.e. for any vector fields X, Y,

$$\nabla_X Y - \nabla_Y X - [X, Y] = 0.$$

∇ is called the Levi-Civita connection.

Let $(y^1, \ldots, y^n)$ be a local coordinate system on $\mathcal{N}$, and $h_{ij}(y)$ the coefficients of the metric h in these coordinates. We can show (see, for instance, [47]) that for any vector field $Y = Y^i \frac{\partial}{\partial y^i}$,

$$\nabla_X \left(Y^i \frac{\partial}{\partial y^i} \right) = \left(X^j \frac{\partial Y^i}{\partial y^j} + \Gamma^i_{jk} X^j Y^k \right) \frac{\partial}{\partial y^i}$$

where

$$\Gamma^i_{jk} = \frac{1}{2} h^{il} \left(\frac{\partial h_{jl}}{\partial x^k} + \frac{\partial h_{kl}}{\partial x^j} - \frac{\partial h_{jk}}{\partial x^l} \right) \tag{1.12}$$

are the Christoffel symbols.

Let $u : \mathcal{M} \longrightarrow \mathcal{N}$ be a smooth map.

Definition 1.2.1 *u is a harmonic map from $(\mathcal{M}, g)$ to $(\mathcal{N}, h)$ if and only if u satisfies at each point x in $\mathcal{M}$ the equation*

$$\Delta_g u^i + g^{\alpha\beta}(x) \Gamma^i_{jk}(u(x)) \frac{\partial u^j}{\partial x^\alpha} \frac{\partial u^k}{\partial x^\beta} = 0. \tag{1.13}$$

1.2 Harmonic maps between two Riemannian manifolds

Once more, the reader may check that this definition is independent of the coordinates chosen on $\mathcal{M}$ and $\mathcal{N}$. However, it is easier to see this once we notice that harmonic maps are critical points of the Dirichlet functional

$$E_{(\mathcal{M},g)}(u) = \int_{\mathcal{M}} e(u)(x)\, d\mathrm{vol}_g, \tag{1.14}$$

where

$$e(u)(x) = \frac{1}{2} g^{\alpha\beta}(x) h_{ij}(u(x)) \frac{\partial u^i}{\partial x^\alpha} \frac{\partial u^j}{\partial x^\beta},$$

and where u is forced to take its values in the manifold $\mathcal{N}$. The proof of this result, in a more general setting, will be given later on, in lemma 1.4.10. When we say that $u : \mathcal{M} \longrightarrow \mathcal{N}$ is a critical point of $E_{(\mathcal{M},g)}$, it is implicit that for each one-parameter family of deformations

$$u_t : \mathcal{M} \longrightarrow \mathcal{N},\ t \in I \subset \mathbb{R},$$

which has a $\mathcal{C}^1$ dependence on t, and is such that $u_0 \equiv u$ on $\mathcal{M}$ and, for every t, $u_t = u$ outside a compact subset K of $\mathcal{M}$, we have

$$\lim_{t \to 0} \frac{E_{(\mathcal{M},g)}(u_t) - E_{(\mathcal{M},g)}(u)}{t} = 0.$$

Different types of deformations will be specified in section 1.4. Notice that, by checking that $E_{(\mathcal{M},g)}(u)$ is invariant under a change of coordinates on $(\mathcal{M}, g)$, we show that definition 1.2.1 does not depend on the coordinates chosen on $\mathcal{M}$ (the same is true for the coordinates on $\mathcal{N}$).

EFFECT OF A CONFORMAL TRANSFORMATION ON $(\mathcal{M}, g)$, IF $m = 2$

As we noticed in the previous section, in dimension 2 (i.e. when $\mathcal{M}$ is a surface), the Dirichlet functional for real-valued functions on $\mathcal{M}$ is invariant under conformal transformations of $(\mathcal{M}, g)$. This property remains true when we replace real-valued functions by maps into a manifold $(\mathcal{N}, h)$. An immediate consequence of this is the following generalization of proposition 1.1.2.

Proposition 1.2.2 *The Dirichlet integral, and the set of harmonic maps on an open subset of a Riemannian surface, depend only on the conformal structure.*

By theorem 1.1.3, we can always suppose that we have locally conformal coordinates $(x, y) \in \mathbb{R}^2$ on $(\mathcal{M}, g)$. In these coordinates equation (1.13) becomes

$$\frac{\partial^2 u^i}{\partial x^2} + \frac{\partial^2 u^i}{\partial y^2} + \Gamma^i_{jk}(u)\left(\frac{\partial u^j}{\partial x}\frac{\partial u^k}{\partial x} + \frac{\partial u^j}{\partial y}\frac{\partial u^k}{\partial y}\right) = 0.$$

ANOTHER DEFINITION

Henceforth, we will not use formulation (1.13), but an alternative one where we think of $\mathcal{N}$ as a submanifold of a Euclidean space. In fact, thanks to the Nash–Moser theorem ([123], [102], [77]), we know that, provided h is $\mathcal{C}^3$, it is always possible to isometrically embed $(\mathcal{N}, h)$ into a vector space $\mathbb{R}^N$, with the Euclidean scalar product $\langle .,. \rangle$. Then, we will obtain a new expression for the Dirichlet integral

$$E_{(\mathcal{M},g)}(u) = \int_{\mathcal{M}} \frac{1}{2} g^{\alpha\beta}(x) \left\langle \frac{\partial u}{\partial x^\alpha}, \frac{\partial u}{\partial x^\beta} \right\rangle dvol_g \qquad (1.15)$$

where now we think of u as a map from $\mathcal{M}$ to $\mathbb{R}^N$ satisfying the constraint

$$u(x) \in \mathcal{N}, \forall x \in \mathcal{M}. \qquad (1.16)$$

Therefore, we have another definition.

Definition 1.2.3 *u is a harmonic map from $(\mathcal{M}, g)$ to $\mathcal{N} \subset \mathbb{R}^N$, if and only if u is a critical point of the functional defined by (1.15), among the maps satisfying the constraint (1.16). We can then see that u satisfies*

$$\Delta_g u \perp T_{u(x)}\mathcal{N}, \ \forall x \in \mathcal{M}. \qquad (1.17)$$

The proof of (1.17) will be given, in a more general setting, in lemma 1.4.10. This equation means that for every point x of $\mathcal{M}$, $\Delta_g u(x)$ is a vector of $\mathbb{R}^N$ belonging to the normal subspace to $\mathcal{N}$ at $u(x)$. At first glance, condition (1.17) seems weaker than equation (1.13), since we just require that the vector $\Delta_g u$ belongs to a subspace of $\mathbb{R}^N$. This imprecision is illusory: by this we mean that it is possible to calculate the normal component of $\Delta_g u$, a priori unknown, using the first derivatives of u.

1.2 Harmonic maps between two Riemannian manifolds

Lemma 1.2.4 *Let u be a $\mathcal{C}^2$ map from $\mathcal{M}$ to $\mathcal{N}$, not necessarily harmonic. For each $x \in \mathcal{M}$, let $P_u^\perp$ be the orthogonal projection from $\mathbb{R}^N$ onto the normal subspace to $T_{u(x)}\mathcal{N}$ in $\mathbb{R}^N$. Then, for every x in $\mathcal{M}$,*

$$P_u^\perp(\Delta_g u) = -g^{\alpha\beta} A(u)\left(\frac{\partial u}{\partial x^\alpha}, \frac{\partial u}{\partial x^\beta}\right), \tag{1.18}$$

where $A(y)$ is an $\mathbb{R}^N$-valued symmetric bilinear form on $T_y\mathcal{N}$ whose coefficients are smooth functions of y. A is the second fundamental form of the embedding of $\mathcal{N}$ into $\mathbb{R}^N$.

A first way of writing A explicitly is to choose over sufficiently small open sets ω of $\mathcal{N}$ an $(N-n)$-tuple of smooth vector fields $(e_{n+1}, \ldots, e_N)$: $\omega \longrightarrow (\mathbb{R}^N)^{N-n}$, such that at each point $y \in \omega$, $(e_{n+1}(y), \ldots, e_n(y))$ is an orthonormal basis of $(T_y\mathcal{N})^\perp$. Then, for each pair of vectors (X, Y) in $(T_y\mathcal{N})^2$,

$$A(y)(X,Y) = \sum_{j=n+1}^{N} \langle X, D_Y e_j \rangle e_j,$$

where $D_Y e_j = \sum_{i=1}^{N} Y^i \frac{\partial e_j}{\partial y^i}$ is the derivative of e_j along Y in $\mathbb{R}^N$. Another possible definition for A is

$$A(y)(X,Y) = D_X P_y^\perp(Y). \tag{1.19}$$

Proof of lemma 1.2.4 We have

$$P_u^\perp \left(g^{\alpha\beta} \sqrt{\det g} \, \frac{\partial u}{\partial x^\beta}\right) = 0,$$

which implies that

$$P_u^\perp \left(\frac{\partial}{\partial x^\alpha}\left(g^{\alpha\beta} \sqrt{\det g} \, \frac{\partial u}{\partial x^\beta}\right)\right) + \frac{\partial P_u^\perp}{\partial x^\alpha}\left(g^{\alpha\beta} \sqrt{\det g} \, \frac{\partial u}{\partial x^\beta}\right) = 0.$$

Thus,

$$\begin{aligned}P_u^\perp(\Delta_g u) &= \frac{1}{\sqrt{\det g}} P_u^\perp \left(\frac{\partial}{\partial x^\alpha}\left(g^{\alpha\beta} \sqrt{\det g} \, \frac{\partial u}{\partial x^\beta}\right)\right) \\ &= -g^{\alpha\beta} D_{\frac{\partial u}{\partial x^\alpha}} P_u^\perp \left(\frac{\partial u}{\partial x^\beta}\right).\end{aligned}$$

And we conclude that

$$P_u^\perp(\Delta_g u) = -g^{\alpha\beta} A(u)\left(\frac{\partial u}{\partial x^\alpha}, \frac{\partial u}{\partial x^\beta}\right), \qquad (1.20)$$

where A is given by (1.19). $\qquad\square$

We come back to harmonic maps according to definition 1.2.3 and denote, for each $y \in \mathcal{N}$, by P_y the orthogonal projection of $\mathbb{R}^N$ onto $T_y\mathcal{N}$. Since $P_y + P_y^\perp = \mathbb{1}$, from lemma 1.2.4 we deduce that for every harmonic map u from $(\mathcal{M}, g)$ to $\mathcal{N}$,

$$\Delta_g u + g^{\alpha\beta} A(u)\left(\frac{\partial u}{\partial x^\alpha}, \frac{\partial u}{\partial x^\beta}\right) = 0. \qquad (1.21)$$

Example 1.2.5 $\mathbb{R}^n$-VALUED MAPS
If the image manifold is a Euclidean vector space, such as $(\mathbb{R}^n, \langle ., .\rangle)$, then a map $u : (\mathcal{M}, g) \longrightarrow \mathbb{R}^n$ is harmonic if and only if each of its components u^i is a real-valued harmonic function on $(\mathcal{M}, g)$.

Example 1.2.6 GEODESICS
If the domain manifold $\mathcal{M}$ has dimension 1 (i.e. is either an interval in $\mathbb{R}$, or a circle), equation (1.21) becomes, denoting by t the variable on $\mathcal{M}$,

$$\frac{d^2 u}{dt^2} + A(u)\left(\frac{du}{dt}, \frac{du}{dt}\right) = 0,$$

which is the equation satisfied by a constant speed parametrization of a geodesic in $(\mathcal{N}, h)$.

Example 1.2.7 MAPS TAKING THEIR VALUES IN THE UNIT SPHERE OF $\mathbb{R}^3$
In this case we have

$$\mathcal{N} = S^2 = \{y \in \mathbb{R}^3 | \, |y| = 1\},$$

where $|y| = \left(\sum_{i=1}^{3}(y^i)^2\right)^{\frac{1}{2}}$ is the norm of y. Notice that for each map $u : (\mathcal{M}, g) \longrightarrow S^2$, we have

$$0 = \Delta_g |u|^2 = 2 \langle u, \Delta_g u \rangle + 4e(u),$$

where

$$e(u) = \frac{1}{2}g^{\alpha\beta}(x)\left\langle \frac{\partial u}{\partial u^\alpha}, \frac{\partial u}{\partial x^\beta}\right\rangle,$$

and thus, since the normal space to S^2 at u is $\mathbb{R}u$,

$$\begin{aligned}P_u^\perp(\Delta_g u) &= \langle u, \Delta_g u\rangle u \\ &= -2e(u)u\,.\end{aligned} \qquad (1.22)$$

But if u is an S^2-valued harmonic map,

$$P_u(\Delta_g u) = 0,$$

which, together with (1.22), yields

$$\Delta_g u + 2e(u)u = 0. \qquad (1.23)$$

Exercises

1.1 Let $(\mathcal{M}, g)$ be a Riemannian manifold. Show that a map $u : (\mathcal{M}, g) \longrightarrow S^1$ is harmonic if and only if for any simply connected subset Ω of $\mathcal{M}$, there exists a harmonic function $f : (\Omega, g) \longrightarrow \mathbb{R}$ such that

$$u(x) = e^{if(x)}, \ \forall x \in \Omega.$$

1.3 Conservation laws for harmonic maps

Noether's theorem is a very general result in the calculus of variations. It enables us to construct a divergence-free vector field on the domain space, from a solution of a variational problem, provided we are in the presence of a continuous symmetry. For 1–dimensional variational problems, the divergence-free vector field is just a quantity which is conserved in time (= the variable): for instance in mechanics, the energy or the momentum. We can find in [124] a presentation of Noether's theorem, and of its extensions (see also [140]). We will present here two versions which are frequently used for harmonic maps. The first is obtained in the case where the symmetry group acts on the image manifold, and the second is connected to the symmetries of the domain.

1.3.1 Symmetries on $\mathcal{N}$

We start with a simple example, that of harmonic maps from an open set Ω of $\mathbb{R}^m$, taking values in the sphere $S^2 \subset \mathbb{R}^3$. In this case equation (1.23) of the previous section can be written as

$$\Delta u + u|du|^2 = 0, \qquad (1.24)$$

where

$$|du|^2 = \sum_{i=1}^{3}\sum_{\alpha=1}^{m}\left(\frac{\partial u^i}{\partial x^\alpha}\right)^2 = 2e(u),$$

and

$$\Delta u = \sum_{\alpha=1}^{m}\frac{\partial^2 u}{(\partial x^\alpha)^2}.$$

Taking the vector product of equation (1.24) by u, we obtain

$$u \times \Delta u = 0, \qquad (1.25)$$

or equivalently,

$$\sum_{\alpha=1}^{m}\frac{\partial}{\partial x^\alpha}\left(u \times \frac{\partial u}{\partial x^\alpha}\right) = \sum_{\alpha=1}^{m}\frac{\partial u}{\partial x^\alpha} \times \frac{\partial u}{\partial x^\alpha} + u \times \Delta u = 0. \qquad (1.26)$$

Several authors (see [99], [35], [153]) have noticed and used independently equation (1.26). Its interest is that it singles out a vector field $j = (j^1, \ldots, j^m)$ on Ω, given by

$$j^\alpha = u \times \frac{\partial u}{\partial x^\alpha}$$

which is divergence-free. It allows us to rewrite (1.24) in the form (1.26) where the derivatives of u appear in a linear and not quadratic way, which is extremely useful when we are working with weak regularity hypotheses on the solution (see the works of Luís Almeida [2] and Yuxin Ge [65] for an example of how to take advantage of this equation under very weak regularity hypotheses).

We will now see that the existence of this conservation law (div $j = 0$) is a general phenomenon, which is due to the symmetries of the sphere S^2. Let Ω be an open subset of $\mathcal{M}$ and L be a Lagrangian defined for maps from Ω to $\mathcal{N}$: we suppose that L is a $\mathcal{C}^1$ function defined on

$T\mathcal{N} \otimes_{\mathcal{M} \times \mathcal{N}} T^*\mathcal{M} := \{(x, y, A) \mid (x, y) \in \mathcal{M} \times \mathcal{N}, A \in T_y\mathcal{N} \otimes T_x^*\mathcal{M}\}$ with values in $\mathbb{R}$ (here A can be seen as a linear map from $T_x\mathcal{M}$ to $T_y\mathcal{N}$). We take a ($\mathcal{C}^1$ density) measure $d\mu(x)$ on Ω. It is then possible to define a functional $\mathcal{L}$ on the set of maps $\mathcal{C}^1(\Omega, \mathcal{N})$ by letting

$$\mathcal{L}(u) = \int_\Omega L(x, u(x), du(x)) d\mu(x).$$

For instance, if we are given metrics g and h over $\mathcal{M}$ and $\mathcal{N}$, we may choose

$$L(x, u(x), du(x)) = e(u)(x) = \frac{1}{2} g^{\alpha\beta}(x) h_{ij}(u(x)) \frac{\partial u^i}{\partial x^\alpha} \frac{\partial u^j}{\partial x^\beta}$$

and

$$d\mu(x) = d\mathrm{vol}_g(x).$$

Let X be a tangent vector field on $\mathcal{N}$. We say that X is an infinitesimal symmetry for L if and only if

$$\frac{\partial L}{\partial y^i}(x, y, A) X^i(y) + \frac{\partial L}{\partial A_\alpha^i}(x, y, A) \frac{\partial X^i}{\partial y^j}(y) A_\alpha^j = 0. \qquad (1.27)$$

This implies, in particular, the following relation. Suppose that the vector field X is Lipschitz. It is then possible to integrate the flow of X for all time ($\mathcal{N}$ is compact!). For any $y \in \mathcal{N}$, $t \in \mathbb{R}$, write

$$\exp_y tX = \gamma(t) \in \mathcal{N},$$

where γ is the solution of

$$\begin{cases} \gamma(0) &= y \\ \frac{d\gamma}{dt} &= X(\gamma). \end{cases}$$

A consequence of (1.27) is that for every map u from Ω to $\mathcal{N}$,

$$L(x, \exp_u tX, d(\exp_u tX)) = L(x, u, du). \qquad (1.28)$$

To check it, it suffices to differentiate equation (1.28) w.r.t. t and use (1.27).

Theorem 1.3.1 *Let X be a Lipschitz tangent vector field on $\mathcal{N}$, which is an infinitesimal symmetry for L. If $u : \Omega \longrightarrow \mathcal{N}$ is a critical point of $\mathcal{L}$, then*

$$\operatorname{div}\left(\frac{\partial L}{\partial A} \cdot X(u)\right) = 0 \qquad (1.29)$$

or equivalently, using the coordinates $(x^1, \ldots, x^m)$ on Ω such that $d\mu = \rho(x)dx^1 \ldots dx^m$,

$$\sum_{\alpha=1}^{m} \frac{\partial}{\partial x^\alpha}\left(\rho(x)X^i(u)\frac{\partial L}{\partial A^i_\alpha}(x, u, du)\right) = 0. \qquad (1.30)$$

Remark 1.3.2

  (i) *In the case where L is the Lagrangian of the harmonic map, this result was first obtained in [134].*
  (ii) *The vector field J defined over Ω by*

$$J^\alpha = \rho(x)\frac{\partial L}{\partial A^i_\alpha}(x, u, du)X^i(u)$$

is often called the Noether current by physicists. Equations (1.29) and (1.30) are called conservation laws.

Proof of theorem 1.3.1 To lighten the notation, we can always suppose that the coordinates $(x^1, \ldots, x^m)$ on Ω are such that $d\mu = dx^1 \ldots dx^m$. This corresponds to fixing an arbitrary coordinate system, in which $d\mu$ has the density $\rho(x)$, and then changing $L(x, u, du)\rho(x)$ into $L(x, u, du)$. The proof will be the same. With an analogous simplification purpose, we will replace $\exp_u tX$ by $u + tX(u) + o(t)$ in the calculations (for small t). If we choose to use local charts on $\mathcal{N}$, and hence view u as a map from Ω to an open subset of $\mathbb{R}^n$, this will not be a big problem. In the case where we choose to represent $\mathcal{N}$ as a submanifold of $\mathbb{R}^N$, we should extend L, a priori defined over $T\mathcal{N} \otimes_{\mathcal{M}\times\mathcal{N}} T^*\mathcal{M}$, to a C^1 Lagrangian function on $\Omega \times \mathbb{R}^N \times (\mathbb{R}^N \otimes T^*_x\mathcal{M})$. Such a construction does not pose any particular problem. The fact of u being a critical point of $\mathcal{L}$ implies, in particular, that for every test function $\phi \in C^\infty_c(\Omega, \mathbb{R})$,

$$\begin{aligned}\mathcal{L}(\exp_u(t\phi X)) &= \mathcal{L}(u + t\phi X + o(t)) \\ &= \mathcal{L}(u) + o(t).\end{aligned} \qquad (1.31)$$

But

$$\mathcal{L}(u + t\phi X + o(t))$$

$$= \int_\Omega L(x, u + t\phi X(u), du + t\phi d(X(u)) + td\phi \cdot X(u)) dx^1 \ldots dx^m$$
$$+ o(t)$$
$$= \int_\Omega L(x, u + t\phi X(u), du + t\phi d(X(u))) dx^1 \ldots dx^m$$
$$+ t \int_\Omega X^i \frac{\partial \phi}{\partial x^\alpha} \frac{\partial L}{\partial A^i_\alpha}(x, u, du) dx^1 \ldots dx^m + o(t).$$

By relation (1.28),

$$L(x, u + t\phi X(u), du + t\phi d(X(u))) = L(x, u, du) + o(t),$$

and so

$$\mathcal{L}(u + t\phi X + o(t)) = \mathcal{L}(u) + \int_\Omega (X^i \frac{\partial L}{\partial A^i_\alpha}(x, u, du)) \frac{\partial \phi}{\partial x^\alpha} dx^1 \ldots dx^m + o(t). \tag{1.32}$$

Comparing (1.31) and (1.32), we obtain

$$\int_\Omega \frac{\partial \phi}{\partial x^\alpha}(X^i \frac{\partial L}{\partial A^i_\alpha}(x, u, du)) dx^1 \ldots dx^m = 0, \quad \forall \phi \in \mathcal{C}^\infty_c(\Omega, \mathbb{R}),$$

which is the variational formulation of (1.30). □

As an example of applying this result, let us consider the case of harmonic maps. We have

$$\begin{aligned} L(x, y, A) &= \tfrac{1}{2} g^{\alpha\beta}(x) h_{ij}(y) A^i_\alpha A^j_\beta \\ &= \tfrac{1}{2} g^{\alpha\beta}(x) \langle A_\alpha, A_\beta \rangle, \end{aligned}$$

and

$$\rho(x) = \sqrt{\det g}\, dx^1 \ldots dx^m.$$

Then X is an infinitesimal symmetry if the flow generated by X is a family of isometries of $(\mathcal{N}, h)$, i.e. if X is a Killing vector field. Such fields are characterized by the fact that

$$L_X h = 0,$$

where L is the Lie derivative. In local coordinates, this is written as

$$\begin{aligned}(L_X h)_{ij} &= X^k \frac{\partial h_{ij}}{\partial y^k} + \frac{\partial X^k}{\partial y^i} h_{kj} + \frac{\partial X^k}{\partial y^j} h_{ik} \\ &= h_{jk} \nabla_{\frac{\partial}{\partial y^i}} X^k + h_{ik} \nabla_{\frac{\partial}{\partial y^j}} X^k = 0\,.\end{aligned}$$

The Noether current is then the vector field

$$J = J^\alpha \frac{\partial}{\partial x^\alpha} = g^{\alpha\beta}(x) \left\langle X(u), \frac{\partial u}{\partial x^\beta} \right\rangle \sqrt{\det g}\, \frac{\partial}{\partial x^\alpha}\,.$$

Example 1.3.3 *Let us come back to the sphere – this time of arbitrary dimension n:*

$$\mathcal{N} = S^n = \{y \in \mathbb{R}^{n+1} \mid |y| = 1\}\,.$$

The group of rotations of $\mathbb{R}^{n+1}$, $SO(n+1)$, leaves S^n invariant, and acts isometrically on S^n. The set of Killing vector fields on S^n can be identified with the Lie algebra $so(n+1)$ of $SO(n+1)$. We will systematically identify $so(n+1)$ with the set of $(n+1) \times (n+1)$ skew-symmetric real matrices, and

$$SO(n+1) = \{R \in M((n+1) \times (n+1), \mathbb{R}) \mid {}^tRR = \mathbb{1}, \det R = 1\}\,.$$

In fact, to each element $a \in so(n+1)$, we may associate the tangent vector field on S^n given by

$$X_a : y \longmapsto a.y\,,$$

and we can obtain all Killing vector fields in this way. The previous theorem states that the Noether current

$$J = J^\alpha \frac{\partial}{\partial x^\alpha} = \left\langle a.u, \frac{\partial u}{\partial x^\beta} \right\rangle g^{\alpha\beta}(x) \sqrt{\det g}\, \frac{\partial}{\partial x^\alpha} \qquad (1.33)$$

is divergence-free, i.e.

$$\sum_{\alpha=1}^m \frac{\partial J^\alpha}{\partial x^\alpha} = \sum_{\alpha=1}^m \frac{\partial}{\partial x^\alpha} \left(g^{\alpha\beta}(x) \left\langle a.u, \frac{\partial u}{\partial x^\beta} \right\rangle \sqrt{\det g} \right) = 0\,.$$

1.3 Conservation laws for harmonic maps

Example 1.3.4 *(continuation of the previous example).* We restrict ourselves to the case where Ω is an open subset of $\mathbb{R}^m$ equipped with the Euclidean metric, and where $n = 2$. By successively choosing for $a \in so(3)$

$$a_1 = \begin{pmatrix} 0 & 0 & 0 \\ 0 & 0 & -1 \\ 0 & 1 & 0 \end{pmatrix},$$

$$a_2 = \begin{pmatrix} 0 & 0 & 1 \\ 0 & 0 & 0 \\ -1 & 0 & 0 \end{pmatrix},$$

$$a_3 = \begin{pmatrix} 0 & -1 & 0 \\ 1 & 0 & 0 \\ 0 & 0 & 0 \end{pmatrix},$$

we obtain the currents (using (1.30))

$$j_{1,\alpha} = u^2 \frac{\partial u^3}{\partial x^\alpha} - u^3 \frac{\partial u^2}{\partial x^\alpha},$$

$$j_{2,\alpha} = u^3 \frac{\partial u^1}{\partial x^\alpha} - u^1 \frac{\partial u^3}{\partial x^\alpha},$$

$$j_{3,\alpha} = u^1 \frac{\partial u^2}{\partial x^\alpha} - u^2 \frac{\partial u^1}{\partial x^\alpha}.$$

We recognize that $(j_{1,\alpha}, j_{2,\alpha}, j_{3,\alpha})$ are the three components of the vector $u \times \frac{\partial u}{\partial x^\alpha}$, and Noether's theorem yields

$$\sum_{\alpha=1}^{3} \frac{\partial}{\partial x^\alpha} \left(u \times \frac{\partial u}{\partial x^\alpha} \right) = 0.$$

Thus, we recover (1.26).

This equation is particularly interesting in the case of a sphere (also for other homogeneous manifolds), since the isometry group $SO(n+1)$ acts transitively on S^n and, at an infinitesimal scale, this yields that at each point y of S^n, the set of all $a.y$, for $a \in so(n+1)$, is equal to $T_y S^n$ (equivalently if a_i is a basis of $so(n+1)$, then the $a_i.y$'s span $T_y S^n$). This implies that equation (1.26) is equivalent to the original S^2-valued harmonic map equation, (1.24). We will take advantage of this in sections 2.5 and 2.6.

1.3.2 Symmetries on $\mathcal{M}$: the stress–energy tensor

Let us start with an example: when the manifold $\mathcal{M}$ is 2-dimensional, surprising conservation laws occur. Let us place ourselves in local conformal coordinates (x, y) on $\mathcal{M}$, and consider the $\mathbb{C}$-valued function defined by

$$f = \left|\frac{\partial u}{\partial x}\right|^2 - \left|\frac{\partial u}{\partial y}\right|^2 - 2i\left\langle \frac{\partial u}{\partial x}, \frac{\partial u}{\partial y}\right\rangle.$$

One can check that when u is a harmonic map taking values in a Riemannian manifold $(\mathcal{N}, h)$, f is a holomorphic function of the variable $z = x + iy$. This was first noticed, in some special cases, by H.E. Rauch and K. Shibata, then by J. Sampson ([143]), in the 1960s, before people realized around 1980 that this result is general and related (through Noether's theorem) to the problem's invariance under conformal transformations ([5], [127]).

Remark 1.3.5 *A geometric characterization of f is the following: the inverse image (or pull-back) by u of the metric on $\mathcal{N}$,*

$$u^*h = |u_x|^2 (dx)^2 + \langle u_x, u_y\rangle (dx \otimes dy + dy \otimes dx) + |u_y|^2 (dy)^2,$$

may be decomposed over the complexified tangent space $T_x\mathcal{M} \otimes \mathbb{C}$, as

$$u^*h = \frac{1}{4}[f(dz)^2 + |du|^2(dz \otimes d\bar{z} + d\bar{z} \otimes dz) + \overline{f}(d\bar{z})^2].$$

This decomposition does not depend on the local (conformal) coordinate system chosen. Thus, we can define f by the relation $4(u^*h)^{(2,0)} = f(dz)^2$.

We will come back to this in definition 1.3.10. First we will study the general case where $(\mathcal{M}, g)$ is an arbitrary manifold of any dimension.

Let Φ be a diffeomorphism of the open subset $\Omega \subset (\mathcal{M}, g)$ onto itself, and u be a map from Ω to $\mathcal{N}$. The pull-back of g by Φ is the tensor Φ^*g defined by

$$(\Phi^*g)_{\alpha\beta} = g_{\gamma\delta}(\Phi(x))\frac{\partial \Phi^\gamma}{\partial x^\alpha}\frac{\partial \Phi^\delta}{\partial x^\beta}. \tag{1.34}$$

Φ^*g defines a new metric over Ω, such that (Ω, Φ^*g) and (Ω, g) are "geometrically identical" manifolds, in the sense that

$$\Phi : (\Omega, \Phi^*g) \longrightarrow (\Omega, g)$$

1.3 Conservation laws for harmonic maps

is an isometry. This is why we consider that the two maps

$$u : (\Omega, g) \longrightarrow (\mathcal{N}, h)$$

and

$$u \circ \Phi : (\Omega, \Phi^* g) \longrightarrow (\mathcal{N}, h)$$

are two different representations of a same geometric object. Moreover, the transformation of u into $u \circ \Phi$, which is just a change of coordinates, preserves the Dirichlet integral, provided that we also transform the metric. In fact, a change of variable,

$$\phi = \Phi(x),$$

yields

$$\begin{aligned} E_{(\Omega, \Phi^* g)}(u \circ \Phi) &= \int_\Omega g^{\gamma\delta}(\Phi(x)) \frac{\partial x^\alpha}{\partial \Phi^\gamma} \frac{\partial x^\beta}{\partial \Phi^\delta} \left\langle \frac{\partial(u \circ \Phi)}{\partial x^\alpha}, \frac{\partial(u \circ \Phi)}{\partial x^\beta} \right\rangle \\ &\quad \sqrt{\det \Phi^* g}\, dx^1 \ldots dx^m \\ &= \int_\Omega g^{\gamma\delta}(\phi) \left\langle \frac{\partial u}{\partial \phi^\alpha}, \frac{\partial u}{\partial \phi^\beta} \right\rangle \sqrt{\det g(\phi)}\, d\phi^1 \ldots d\phi^m \\ &= E_{(\Omega, g)}(u). \end{aligned}$$

Hence, we notice the invariance of the Dirichlet integral under the action of the diffeomorphism group

$$\text{Diff}(\Omega) = \{\Phi \in \mathcal{C}^1(\Omega, \Omega)|\ \Phi \text{ is invertible and }\ \Phi^{-1} \in \mathcal{C}^1(\Omega, \Omega)\}$$

on the pairs (g, u). We expect to find conservation laws. Nevertheless, two remarkable differences should be pointed out, when compared to the classical setting of Noether's theorem (like the one we saw before). First, the group Diff(Ω) is infinite dimensional. Next, the Lagrangian of the harmonic map is not exactly invariant under this group action, to the extent that we need to change the metric g, and thus the Lagrangian L, at the same time as we change the map u, in order to obtain the invariance. The result we will obtain will not concern the existence of divergence-free vector fields, but that of an order 2 symmetric tensor, whose covariant derivative (w.r.t. the metric g) will vanish. Such a tensor S will be defined by

$$S = e(u)g - u^* h, \tag{1.35}$$

where $e(u)$ is the energy density. In local coordinates

$$S_{\alpha\beta}(x) = \frac{1}{2}h_{ij}(u(x))g^{\gamma\delta}(x)\frac{\partial u^i}{\partial x^\gamma}\frac{\partial u^j}{\partial x^\delta}g_{\alpha\beta} - h_{ij}(u(x))\frac{\partial u^i}{\partial x^\alpha}\frac{\partial u^j}{\partial x^\beta}$$

$$= \frac{1}{2}g^{\gamma\delta}(x)\left\langle\frac{\partial u}{\partial x^\gamma},\frac{\partial u}{\partial x^\delta}\right\rangle g_{\alpha\beta}(x) - \left\langle\frac{\partial u}{\partial x^\alpha},\frac{\partial u}{\partial x^\beta}\right\rangle.$$

The tensor S is called the stress–energy tensor (this name comes from general relativity where its variational origin was first pointed out by David Hilbert [91]). The following result is due to Paul Baird, James Eells and A.I. Pluzhnikov.

Theorem 1.3.6 *[5], [127] Let $u : (\mathcal{M}, g) \longrightarrow (\mathcal{N}, h)$ be a smooth harmonic map. Then, the stress–energy tensor of u, S, has vanishing covariant divergence. This can be stated in two equivalent ways:*

(i) *For every compactly supported tangent vector field X on $\mathcal{M}$,*

$$\int_{\mathcal{M}} (L_X g^\sharp)^{\alpha\beta} S_{\alpha\beta} \, d\mathrm{vol}_g = 0, \qquad (1.36)$$

where

$$(L_X g^\sharp)^{\alpha\beta} = X^\gamma \frac{\partial g^{\alpha\beta}}{\partial x^\gamma} - \frac{\partial X^\alpha}{\partial x^\gamma} g^{\gamma\beta} - \frac{\partial X^\beta}{\partial x^\gamma} g^{\alpha\gamma}$$

is the Lie derivative of $g^\sharp$ with respect to X.

(ii) *The tensor S satisfies the following equation for $\beta = 1, \ldots, m$,*

$$\mathrm{div}_g(g^{\alpha\beta} S_{\gamma\beta}) = \nabla_{\frac{\partial}{\partial x^\alpha}}(g^{\alpha\gamma} S_{\gamma\beta}) = 0. \qquad (1.37)$$

Proof We first prove (i). Let X be a compactly supported Lipschitz vector field on $\mathcal{M}$, and Φ_t the flow generated by X,

$$\begin{cases} \Phi_0(x) = x, \ \forall x \in \mathcal{M} \\ \dfrac{\partial(\Phi_t(x))}{\partial t} = X(\Phi_t(x)). \end{cases}$$

For all t, Φ_t exists and is an element of $\mathrm{Diff}(\mathcal{M})$.

1.3 Conservation laws for harmonic maps

Recall that for any tensor field $T = T^{\alpha_1...\alpha_a}_{\beta_1...\beta_b} \frac{\partial}{\partial x^{\alpha_1}} \otimes \cdots \otimes \frac{\partial}{\partial x^{\alpha_a}} \otimes dx^{\beta_1} \otimes \cdots \otimes dx^{\beta_b}$ on $\mathcal{M}$, the Lie derivative of T w.r.t. X is defined by

$$(L_X T)_x = \lim_{t \to 0} \frac{\Phi_t^*(T_{\Phi_t(x)}) - T_x}{t}. \tag{1.38}$$

Here, $\Phi_t^* T$ denotes the tensor field such that

$$(\Phi_t^* T)^{\alpha_1...\alpha_a}_{x,\beta_1...\beta_b} \frac{\partial \Phi_t^{\alpha'_1}}{\partial \xi^{\alpha_1}} \cdots \frac{\partial \Phi_t^{\alpha'_a}}{\partial \xi^{\alpha_a}} = T^{\alpha'_1...\alpha'_a}_{\Phi_t(x),\beta'_1...\beta'_b} \frac{\partial \Phi_t^{\beta'_1}}{\partial \xi^{\beta_1}} \cdots \frac{\partial \Phi_t^{\beta'_b}}{\partial \xi^{\beta_b}}. \tag{1.39}$$

Replacing in (1.39) $\Phi_t(x)$ by $x + tX(x) + o(t)$, and $(\Phi_t^* T)_x$ by $T_x + t(L_X T)_x + o(t)$ (a consequence of (1.38)), we obtain

$$
\begin{aligned}
(L_X T)^{\alpha_1...\alpha_a}_{x\beta_1...\beta_b} &= X^\gamma \frac{\partial T^{\alpha_1...\alpha_a}_{x\beta_1...\beta_b}}{\partial x^\gamma} + T^{\alpha_1...\alpha_a}_{x\gamma...\beta_b} \frac{\partial X^\gamma}{\partial x^{\beta_1}} + \cdots + T^{\alpha_1...\alpha_a}_{x\beta_1...\gamma} \frac{\partial X^\gamma}{\partial x^{\beta_b}} \\
&\quad - T^{\gamma...\alpha_a}_{x\beta_1...\beta_b} \frac{\partial X^{\alpha_1}}{\partial x^\gamma} - \cdots - T^{\alpha_1...\gamma}_{x\beta_1...\beta_b} \frac{\partial X^{\alpha_b}}{\partial x^\gamma}.
\end{aligned}
$$

Since u is harmonic, i.e. a critical point of $E_{(\mathcal{M},g)}$, we have

$$\frac{d}{dt}\left(E_{(\mathcal{M},g)}(u \circ \Phi_{-t})\right)_{|t=0} = 0. \tag{1.40}$$

We will take advantage of this relation. We start by noticing that performing the change of variable

$$\phi = (\Phi_{-t})^{-1}(x) = \Phi_t(x),$$

we have, writing $E_g = E_{(\mathcal{M},g)}$,

$$
\begin{aligned}
E_g(u \circ \Phi_{-t}) &= E_{\Phi_t^* g}(u \circ \Phi_{-t} \circ \Phi_t) \\
&= E_{\Phi_t^* g}(u).
\end{aligned}
$$

We need to estimate $E_{\Phi_t^* g}(u)$ for small t, and in order to do so, we first calculate $\Phi_t^* g$, up to first order in t:

$$
\begin{aligned}
(\Phi_t^* g)_{\alpha\beta} &= g_{ab}(x + tX(x))(\delta^a_\alpha + t\frac{\partial X^a}{\partial x^\alpha})(\delta^b_\beta + t\frac{\partial X^b}{\partial x^\beta}) + o(t) \\
&= g_{\alpha\beta}(x) + t(L_X g)_{\alpha\beta}(x) + o(t),
\end{aligned}
$$

where

$$(L_X g)_{\alpha\beta} = X^\gamma \frac{\partial g_{\alpha\beta}}{\partial x^\gamma} + \frac{\partial X^\gamma}{\partial x^\alpha} g_{\gamma\beta} + \frac{\partial X^\gamma}{\partial x^\beta} g_{\alpha\gamma}.$$

Thus, we deduce that

$$
\begin{aligned}
\det\left(\Phi_t^* g\right) &= \det\left(g_{\alpha\beta} + t(L_X g)_{\alpha\beta}\right) + o(t) \\
&= \det\left(g_{\alpha\beta}\right) \det\left(\delta_\beta^\alpha + t\, g^{\alpha\gamma}(L_X g)_{\gamma\beta}\right) + o(t) \\
&= (\det g)\left(1 + t\, tr_g(L_X g)\right) + o(t). \qquad (1.41)
\end{aligned}
$$

We also need to estimate $\Phi_t^* g^\sharp$:

$$
\begin{aligned}
(\Phi_t^* g^\sharp) &= g^{ab}(x + tX(x))\left(\delta_a^\alpha - t\frac{\partial X^\alpha}{\partial x^a}\right)\left(\delta_b^\beta - t\frac{\partial X^\beta}{\partial x^b}\right) + o(t) \\
&= g^{\alpha\beta}(x) + t(L_X g^\sharp)^{\alpha\beta}(x) + o(t). \qquad (1.42)
\end{aligned}
$$

Now we obtain, thanks to (1.41) and (1.42),

$$
\begin{aligned}
E_{\Phi_t^* g}(u) &= \int_{\mathcal{M}} (g^{\alpha\beta}(x) + t(L_X g^\sharp)^{\alpha\beta}) \left\langle \frac{\partial u}{\partial x^\alpha}, \frac{\partial u}{\partial x^\beta} \right\rangle \\
&\quad \sqrt{\det g}\,\sqrt{1 + t\, g^{\alpha\beta}(L_X g)_{\alpha\beta}}\, dx^1 \ldots dx^m + o(t) \\
&= \int_{\mathcal{M}} g^{\alpha\beta}(x) \left\langle \frac{\partial u}{\partial x^\alpha}, \frac{\partial u}{\partial x^\beta} \right\rangle \sqrt{\det g}\, dx^1 \ldots dx^m \\
&\quad + t\int_{\mathcal{M}} \left[(L_X g^\sharp)^{\alpha\beta} \left\langle \frac{\partial u}{\partial x^\alpha}, \frac{\partial u}{\partial x^\beta} \right\rangle \right. \\
&\quad \left. + \frac{1}{2} g^{\alpha\beta} \left\langle \frac{\partial u}{\partial x^\alpha}, \frac{\partial u}{\partial x^\beta} \right\rangle g^{\gamma\delta}(L_X g)_{\gamma\delta} \right] \sqrt{\det g}\, dx^1 \ldots dx^m \\
&\quad + o(t).
\end{aligned}
$$

But since

$$
0 = L_X(g^{\gamma\beta} g_{\gamma\beta}) = (L_X g^\sharp)^{\alpha\beta} g_{\alpha\beta} + g^{\alpha\beta}(L_X g)_{\alpha\beta},
$$

we finally have

$$E_{\Phi_t^*g}(u) = E_g(u) + t\int_{\mathcal{M}} (L_X g^{\sharp})^{\alpha\beta}\left[\left\langle \frac{\partial u}{\partial x^{\alpha}}, \frac{\partial u}{\partial x^{\beta}}\right\rangle\right.$$

$$\left. - \frac{1}{2}g^{\gamma\delta}\left\langle \frac{\partial u}{\partial x^{\gamma}}, \frac{\partial u}{\partial x^{\delta}}\right\rangle g_{\alpha\beta}\right] \sqrt{\det g}\, dx^1 \ldots dx^m + o(t)$$

$$= E_g(u) - t\int_{\mathcal{M}} (L_X g^{\sharp})^{\alpha\beta} S_{\alpha\beta}\, d\mathrm{vol}_g + o(t).$$

Thus, for (1.40) to be true it is necessary (and sufficient) that

$$\int_{\mathcal{M}} (L_X g^{\sharp})^{\alpha\beta} S_{\alpha\beta}\, d\mathrm{vol}_g = 0.$$

This proves (1.36). □

Proof of (ii) First, we notice that for any tangent vector field X,

$$g^{\alpha\gamma}\nabla_{\frac{\partial}{\partial x^{\gamma}}} X^{\beta} + g^{\beta\gamma}\nabla_{\frac{\partial}{\partial x^{\gamma}}} X^{\alpha} = g^{\alpha\gamma}\frac{\partial X^{\beta}}{\partial x^{\gamma}} + g^{\beta\gamma}\frac{\partial X^{\alpha}}{\partial x^{\gamma}} - \frac{\partial g^{\alpha\beta}}{\partial x^{\gamma}} X^{\gamma}$$

$$= -(L_X g^{\sharp})^{\alpha\beta}. \quad (1.43)$$

To check this, it suffices to develop the l.h.s. of (1.43). It follows that, since $S_{\alpha\beta} = S_{\beta\alpha}$,

$$2\, g^{\alpha\gamma}\nabla_{\frac{\partial}{\partial x^{\gamma}}} X^{\beta} S_{\alpha\beta} = -(L_X g^{\sharp})^{\alpha\beta} S_{\alpha\beta}. \quad (1.44)$$

Suppose that X is compactly supported. Then, using (1.44) we have (by Stokes' formula)

$$0 = \int_{\mathcal{M}} \frac{\partial}{\partial x^{\gamma}}\left(g^{\gamma\alpha} X^{\beta} S_{\alpha\beta}\sqrt{\det g}\right) dx^1 \ldots dx^m$$

$$= \int_{\mathcal{M}} \nabla_{\frac{\partial}{\partial x^{\gamma}}}\left(g^{\gamma\alpha} X^{\beta} S_{\alpha\beta}\right)\sqrt{\det g}\, dx^1 \ldots dx^m$$

$$= \int_{\mathcal{M}} \left(g^{\gamma\alpha}\left(\nabla_{\frac{\partial}{\partial x^{\gamma}}} X\right)^{\beta} S_{\alpha\beta} + X^{\beta}\nabla_{\frac{\partial}{\partial x^{\gamma}}}\left(g^{\gamma\alpha} S_{\alpha\beta}\right)\right) d\mathrm{vol}_g$$

$$= \int_{\mathcal{M}} -\frac{1}{2}(L_X g^{\sharp})^{\alpha\beta} S_{\alpha\beta}\, d\mathrm{vol}_g + \int_{\mathcal{M}} X^{\beta}\nabla_{\frac{\partial}{\partial x^{\gamma}}}\left(g^{\gamma\alpha} S_{\alpha\beta}\right) d\mathrm{vol}_g.$$

This identity clearly shows that (1.36) is equivalent to

$$\int_{\mathcal{M}} X^\beta \mathrm{div}_g(g^{\gamma\alpha} S_{\alpha\beta})\, d\mathrm{vol}_g = 0, \ \forall X \text{ compactly supported},$$

which is the same as (1.37). □

1.3.3 Consequences of theorem 1.3.6

Case A EXISTENCE OF A KILLING VECTOR FIELD

Suppose that there is a tangent vector field X on $(\mathcal{M}, g)$ whose flow is a family of isometries, i.e. that X is Killing. Such a field is characterized by the equation $L_X g = 0$. Hence, X is an infinitesimal symmetry for $E_{(\mathcal{M},g)}$, in the sense that

$$E_g(u \circ \Phi_{-t}) = E_{\Phi_t^* g}(u) = E_g(u).$$

We are precisely in a case where Noether's theorem applies (see [124]): the Noether current

$$J^\alpha = g^{\alpha\beta} S_{\beta\gamma} X^\gamma \sqrt{\det g} \tag{1.45}$$

satisfies

$$\mathrm{div}\, J = \sum_{\alpha=1}^m \frac{\partial J^\alpha}{\partial x^\alpha} = 0. \tag{1.46}$$

This can be deduced from the previous theorem since from (1.44) it follows that

$$\begin{aligned}
\nabla_{\frac{\partial}{\partial x^\alpha}} (g^{\alpha\beta} S_{\beta\gamma} X^\gamma) &= g^{\alpha\beta} (\nabla_{\frac{\partial}{\partial x^\alpha}} X)^\gamma S_{\alpha\beta} + X^\gamma \nabla_{\frac{\partial}{\partial x^\alpha}} (g^{\alpha\beta} S_{\beta\gamma}) \\
&= -\frac{1}{2} (L_X g^\sharp)^{\alpha\beta} S_{\alpha\beta} + X^\gamma \nabla_{\frac{\partial}{\partial x^\alpha}} (g^{\alpha\beta} S_{\beta\gamma}) \\
&= 0,
\end{aligned}$$

using (1.37) and the fact that $L_X g^\sharp = 0$.

Case B THE METRIC g IS EUCLIDEAN

1.3 Conservation laws for harmonic maps

Suppose that $\mathcal{M}$ is an open subset Ω of $\mathbb{R}^m$ and that g is the Euclidean metric. Then, all translations in $\mathbb{R}^m$ are symmetries (generated by constant vector fields). Thus, using the preceding results we obtain

$$\sum_{\alpha=1}^{m} \frac{\partial S_{\alpha\beta}}{\partial x^\alpha} = 0, \forall \beta. \tag{1.47}$$

This equation gives rise to the following formula: let X be a $\mathcal{C}^1$ tangent vector field on Ω, and consider the tangent vector field $Y = Y^\alpha \frac{\partial}{\partial x^\alpha}$, defined on Ω by

$$Y^\alpha(x) = S_{\alpha\beta}(x) X^\beta(x).$$

We integrate div Y over an open subset ω of Ω, with smooth boundary, and use Stokes' formula

$$\int_\omega \operatorname{div} Y \, dx^1 \ldots dx^m = \int_{\partial \omega} Y.n \, d\sigma(x).$$

Using (1.47), we obtain

$$\int_\omega S_{\alpha\beta} \frac{\partial X^\beta}{\partial x^\alpha} dx^1 \ldots dx^m = \int_{\partial \omega} S_{\alpha\beta} X^\beta . n^\alpha d\sigma(x). \tag{1.48}$$

Recall that in $\mathbb{R}^m$, S can be written as

$$S_{\alpha\beta} = \frac{1}{2} |du|^2 \delta_{\alpha\beta} - \left\langle \frac{\partial u}{\partial x^\alpha}, \frac{\partial u}{\partial x^\beta} \right\rangle,$$

where $|du|^2 = \sum_{\alpha=1}^{m} \sum_{i=1}^{N} \left(\frac{\partial u^i}{\partial x^\alpha} \right)^2$. We give two examples of applying (1.48).

Example 1.3.7 As ω we choose the ball

$$B(x_0, r) = \{ x \in \mathbb{R}^m | \, |x - x_0| < r \},$$

and as X, $X(x) = x - x_0$. Then (1.48) yields

$$(2 - m) \int_{B(x_0,r)} |du|^2 dx^1 \ldots dx^m = r \int_{\partial B(x_0,r)} \left(2 \left| \frac{\partial u}{\partial n} \right|^2 - |du|^2 \right) d\sigma(x). \tag{1.49}$$

For $x_0 \in \Omega$, and r such that $B(x_0, r) \subset \Omega$, let

$$E_{x_0,r}(u) = r^{2-m} \int_{B(x_0,r)} |du|^2 dx^1 \ldots dx^m.$$

Differentiating w.r.t. r,

$$\begin{aligned}\frac{d}{dr} E_{x_0,r}(u) &= (2-m) r^{1-m} \int_{B(x_0,r)} |du|^2 dx^1 \ldots dx^m \\ &\quad + r^{2-m} \int_{\partial B(x_0,r)} |du|^2 d\sigma(x),\end{aligned}$$

and using (1.49), we obtain

$$\frac{d}{dr} E_{x_0,r}(u) = r^{2-m} \int_{\partial B(x_0,r)} 2 \left| \frac{\partial u}{\partial r} \right|^2 d\sigma(x). \quad (1.50)$$

This identity was first noticed by [131] for the Yang–Mills problem (see [146]). In the context of harmonic maps it is due to [149]. In particular, it implies the monotonicity formula $\frac{d}{dr} E_{x_0,r}(u) \geq 0$, which is an essential ingredient of the regularity theory for minimizing harmonic maps [146], and of theorems 3.5.1 and 4.3.1 of this book (see [54], [10]).

Example 1.3.8 We suppose that ω and X are such that

(i) the tensor

$$A_{\alpha\beta}(x) = \frac{1}{2} (\operatorname{div} X) \delta_{\alpha\beta} - \frac{1}{2} \left(\frac{\partial X^\alpha}{\partial x^\beta} + \frac{\partial X^\beta}{\partial x^\alpha} \right)$$

is positive definite everywhere (which excludes the case $m = 2$, where tr $A = 0$).

(ii) on $\partial \omega$, $X \cdot n \geq 0$. Then (1.48) yields

$$\int_\omega \left\langle \frac{\partial u}{\partial x^\alpha}, \frac{\partial u}{\partial x^\beta} \right\rangle A_{\alpha\beta} dx^1 \ldots dx^m$$

$$= \int_{\partial \omega} \left(\frac{1}{2} |du|^2 X \cdot n - \left\langle \frac{\partial u}{\partial x^\alpha}, \frac{\partial u}{\partial x^\beta} \right\rangle X^\alpha n^\beta \right) d\sigma(x). \quad (1.51)$$

Now suppose that u is constant on the boundary of ω. This implies that $\frac{\partial u}{\partial x^\alpha} = \frac{\partial u}{\partial n} n^\alpha$, and that the r.h.s. of (1.51) is equal to

$$-\int_{\partial\omega}\frac{1}{2}|du|^2 X\cdot n\, d\sigma(x),$$

which is negative. Thus, we obtain the inequality

$$\int_\omega \left\langle \frac{\partial u}{\partial x^\alpha}, \frac{\partial u}{\partial x^\beta}\right\rangle A_{\alpha\beta}\, dx^1\ldots dx^m \leq 0.$$

But since the tensor $\left\langle \frac{\partial u}{\partial x^\alpha}, \frac{\partial u}{\partial x^\beta}\right\rangle$ is non-negative and $A_{\alpha\beta}$ is positive, this implies that

$$\left\langle \frac{\partial u}{\partial x^\alpha}, \frac{\partial u}{\partial x^\beta}\right\rangle = 0$$

and thus u is constant over ω.

This type of result was first obtained in [129], in the case of the Yamabe equation. In the harmonic map context it was settled by John C. Wood, who showed that every C^2 harmonic map over an open star-shaped set of dimension greater than 2, which is constant on the boundary of this set, is also constant in its interior [184]. Recall that ω is star-shaped if and only if there is a point x_0 in ω, such that for all $x \in \partial\omega$, the segment $[x_0, x)$ belongs to ω. Wood's result follows from the preceding analysis by choosing as the vector field $X(x) = x - x_0$. In the case where $m = 2$, this method no longer works, but Luc Lemaire proved a similar result which is valid for every simply connected open set [106].

Case C $\mathcal{M}$ is a surface

Let us come back to identity (1.44) which we used in Case A to deduce a conservation law in the case where there exists a Killing vector field. Writing $S^\alpha_\beta = g^{\alpha\gamma}S_{\gamma\beta}$, this identity states that, for any vector field X,

$$2S^\beta_\alpha \nabla_{\frac{\partial}{\partial x^\beta}} X^\alpha = -(L_X g^{\alpha\beta})S_{\alpha\beta}.$$

If u is harmonic, theorem 1.3.6 (1.37) implies that $\nabla_{\frac{\partial}{\partial x^\alpha}} S^\alpha_\beta = 0$ for all β, and thus

$$\begin{aligned}\nabla_{\frac{\partial}{\partial x^\beta}}(S^\beta_\alpha X^\alpha) &= S^\beta_\alpha \nabla_{\frac{\partial}{\partial x^\beta}} X^\alpha + \left(\nabla_{\frac{\partial}{\partial x^\beta}} S^\beta_\alpha\right) X^\alpha \\ &= -\frac{1}{2}(L_X g^{\alpha\beta})S_{\alpha\beta}.\end{aligned} \qquad (1.52)$$

We see that for the vector field $Y = S_\alpha^\beta X^\alpha \frac{\partial}{\partial x^\beta}$ to be covariant divergence-free, it is necessary and sufficient that

$$(L_X g^{\alpha\beta}) S_{\alpha\beta} = 0. \tag{1.53}$$

For instance we know that this condition is satisfied when X is an isometry flow. Let us see what happens when X generates a flow of conformal diffeomorphisms of $(\mathcal{M}, g)$. We call such a field a conformal Killing field. It is characterized by the fact that there exists a real function λ on $\mathcal{M}$ such that

$$L_X g = \lambda g$$

and

$$L_X g^\sharp = -\lambda g^\sharp.$$

For such a flow,

$$\begin{aligned}(L_X g^{\alpha\beta}) S_{\alpha\beta} &= -\lambda g^{\alpha\beta} S_{\alpha\beta} \\ &= -\lambda(m-2) e(u).\end{aligned}$$

It follows that for $m = 2$ we have

$$\nabla_{\frac{\partial}{\partial x^\beta}}(S_\alpha^\beta X^\alpha) = 0.$$

Consequently we obtain

Theorem 1.3.9 *Suppose $m = 2$. Then for every conformal Killing field X and for any harmonic map $u : (\mathcal{M}, g) \longrightarrow (\mathcal{N}, h)$ we have*

$$\nabla_{\frac{\partial}{\partial x^\beta}}(S_\alpha^\beta X^\alpha) = \frac{1}{\sqrt{\det g}} \frac{\partial}{\partial x^\beta}(\sqrt{\det g}\, S_\alpha^\beta X^\alpha) = 0. \tag{1.54}$$

The interest of this theorem is that the group of conformal transformations of a Riemannian surface is very big. In fact, thanks to proposition 1.1.2 and theorem 1.1.3, it is always possible to use local conformal coordinates, $z = x + iy \in \omega \subset \mathbb{C}$, on $\mathcal{M}$, in which the metric can be written as

$$g = e^{2f}((dx)^2 + (dy)^2),$$

and every holomorphic (or anti-holomorphic) diffeomorphism over ω induces a conformal transformation. Thus, for every holomorphic function $F : \omega \longrightarrow \mathbb{C}$, the vector field

$$X = \Re(F)\frac{\partial}{\partial x} + \Im(F)\frac{\partial}{\partial y}$$

is a conformal Killing field.

In these coordinates

$$\begin{aligned}(S_{\alpha\beta}) &= \frac{|du|^2}{2}\begin{pmatrix} 1 & 0 \\ 0 & 1 \end{pmatrix} - \begin{pmatrix} |u_x|^2 & \langle u_x, u_y\rangle \\ \langle u_x, u_y\rangle & |u_y|^2 \end{pmatrix} \\ &= \frac{1}{2}\begin{pmatrix} -\Re\phi & \Im\phi \\ \Im\phi & \Re\phi \end{pmatrix},\end{aligned}$$

where

$$|du|^2 = \left|\frac{\partial u}{\partial x}\right|^2 + \left|\frac{\partial u}{\partial y}\right|^2,$$

and

$$\phi = \left|\frac{\partial u}{\partial x}\right|^2 - \left|\frac{\partial u}{\partial y}\right|^2 - 2i\left\langle \frac{\partial u}{\partial x}, \frac{\partial u}{\partial y}\right\rangle.$$

It follows that $S_\beta^\alpha = e^{-2f} S_{\alpha\beta}$ and $\sqrt{\det g}\, S_\beta^\alpha = S_{\alpha\beta}$. By applying theorem 1.3.9 successively with $X = \frac{\partial}{\partial x}$ and $Y = \frac{\partial}{\partial y}$, we obtain

$$-\frac{\partial}{\partial x}\Re\phi + \frac{\partial}{\partial y}\Im\phi = 0,$$

and

$$\frac{\partial}{\partial x}\Im\phi + \frac{\partial}{\partial y}\Re\phi = 0.$$

Thus, it follows that

$$\frac{\partial \phi}{\partial \bar{z}} = 0. \tag{1.55}$$

We see a holomorphic function appear, which can "generate" all conservation laws due to conformal transformations. If we make a conformal change of coordinates, i.e. if we make a change of variables

$$z = x + iy = T(X + iZ) = T(Z),$$

where T is an orientation–preserving conformal transformation, then the new holomorphic function obtained from the stress–energy tensor in the Z coordinates is given by

$$\phi \circ T(Z) \left(\frac{\partial T}{\partial Z}\right)^2.$$

This is the reason why we usually consider the quadratic differential form

$$\omega = \phi (dz)^2 \qquad (1.56)$$

which is invariant under a holomorphic coordinate change.

Definition 1.3.10 *Let u be a C^1 map between a Riemannian surface $(\mathcal{M}, g)$ and a Riemannian manifold $(\mathcal{N}, h)$. By the Hopf differential of u we mean the $\mathbb{C}$-valued bilinear form ω, defined on $\mathcal{M}$ by the following property. In every local conformal coordinate system $z = x + iy$, on $(\mathcal{M}, g)$, ω is written as*

$$\omega = \left(\left|\frac{\partial u}{\partial x}\right|^2 - \left|\frac{\partial u}{\partial y}\right|^2 - 2i \left\langle \frac{\partial u}{\partial x}, \frac{\partial u}{\partial y} \right\rangle\right) dz \otimes dz.$$

A more intrinsic way of defining ω is $\omega = \frac{1}{4}(u^*h)^{(2,0)}$ (see remark 1.3.5).

Exercises

Let Ω be an open subset of $\mathbb{R}^m$ and $\mathcal{N}$ a manifold without boundary. $x = (x^1, \ldots, x^m)$ are the coordinates on $\mathbb{R}^m$, and $y = (y^1, \ldots, y^m)$ the local coordinates on $\mathcal{N}$. For each map u from Ω to $\mathcal{N}$, we define

$$\mathcal{L}(u) = \int_\Omega L(u, du) dx = \int_\Omega L\left(u^i, \frac{\partial u^i}{\partial x^\alpha}\right) dx,$$

where the Lagrangian L is a C^2 function on the set

$$\{(y, A) \mid y \in \mathcal{N},\ A \text{ is a linear map from } \mathbb{R}^m \text{ to } T_y \mathcal{N}\}.$$

Given that L does not depend on x, the variational problem associated to $\mathcal{L}$ is invariant under the action of the translations of $\mathbb{R}^m$. We define the Hamiltonian tensor as

$$H^\alpha_\beta(y, A) = \frac{\partial L}{\partial A^i_\alpha}(y, A) A^i_\beta - \delta^\alpha_\beta L(y, A)$$

(see [140]).

1.2 Show that if h is a Riemannian metric on $\mathcal{N}$ and $L(y, A) = \frac{1}{2} h_{ij}(y) A^i_\alpha A^j_\alpha$, then for each map u from Ω to $\mathcal{N}$

$$S^\alpha_\beta = -H^\alpha_\beta(u, du).$$

1.3 Show that if $u : \Omega \longrightarrow \mathcal{N}$ is a critical point of $\mathcal{L}$, then

$$\frac{\partial}{\partial x^\alpha}(H^\alpha_\beta(u, du)) = 0, \ \forall \beta.$$

1.4 Study the extensions of the results stated above in example 1.3.7 and example 1.3.8 to maps u which are critical points of $\mathcal{L}$.

1.5 Assume further that $m = 2$. Give a characterization of Lagrangians which are invariant under conformal transformations of $\mathbb{R}^2$ (answer: see the end of section 4.2). For critical points, prove the existence of a holomorphic generalized Hopf differential.

1.6 Find all Lagrangians which are invariant under volume–preserving transformations of Ω and characterize the corresponding Hamiltonian tensor. Interpret in this situation the conservation law proved in exercise 1.3.

1.4 Variational approach: Sobolev spaces

Up to now, we have only considered smooth maps, i.e. at least $\mathcal{C}^2$. If we want to use variational methods, it is necessary to extend the definitions given, to the case where the maps belong to a "Hilbert space" derived from the Dirichlet integral by taking the closure of $\mathcal{C}^\infty(\mathcal{M}, \mathcal{N})$ relative to "weak" norms derived from the energy. It then becomes very convenient (although it is not essential, see [80]) to suppose that $(\mathcal{N}, h)$ is isometrically embedded in an Euclidean space $(\mathbb{R}^N, \langle.,.\rangle)$, in order to be able to give the following definition.

Definition 1.4.1

$$H^1(\mathcal{M},\mathcal{N}) = \{u \in L^1_{loc}(\mathcal{M},\mathbb{R}^N) | \int_\mathcal{M} |du|^2 dvol_g < +\infty, u(x) \in \mathcal{N} \text{ a.e.}\}.$$

Remark 1.4.2 *Since $\mathcal{N}$ is compact and $u(x) \in \mathcal{N}$ a.e., the condition $u \in L^1_{loc}(\mathcal{M},\mathbb{R}^N)$ could be replaced by $u \in L^\infty(\mathcal{M},\mathbb{R}^N)$, without changing the definition of $H^1(\mathcal{M},\mathcal{N})$.*

It is natural to ask if this definition depends on the isometric embedding of $(\mathcal{N},h)$ into $\mathbb{R}^N$ used. We have:

Lemma 1.4.3 *Let $J_1 : (\mathcal{N},h) \longrightarrow \mathcal{N}_1 \subset \mathbb{R}^{N_1}$ and $J_2 : (\mathcal{N},h) \longrightarrow \mathcal{N}_2 \subset \mathbb{R}^{N_2}$ be two $\mathcal{C}^1$ isometric embeddings. Then, these two isometric embeddings lead to equivalent definitions of $H^1(\mathcal{M},\mathcal{N})$ in the following sense: there exists a $\mathcal{C}^1$ map, Φ, from $\mathbb{R}^{N_1}$ to $\mathbb{R}^{N_2}$ (bounded in $\mathcal{C}^1(\mathbb{R}^{N_1})$), coinciding with $J_2 \circ J_1^{-1}$ on $\mathcal{N}_1$, and such that*

$$u \in H^1(\mathcal{M},\mathcal{N}_1) \iff \Phi \circ u \in H^1(\mathcal{M},\mathcal{N}_2) \tag{1.57}$$

and

$$\int_\mathcal{M} |du|^2 \, dvol_g = \int_\mathcal{M} |d(\Phi \circ u)|^2 \, dvol_g. \tag{1.58}$$

Proof By considering $\mathbb{R}^{N_1}$ and $\mathbb{R}^{N_2}$ as subspaces of $\mathbb{R}^N$, where $N = \sup(N_1, N_2)$, we can always reduce to the case where $N_1 = N_2 = N$. Let us construct Φ. Let $V_{2\delta}(\mathcal{N}_1)$ be a tubular neighborhood of $\mathcal{N}_1$ in $\mathbb{R}^N$, and let P be a projection from $V_{2\delta}(\mathcal{N}_1)$ to $\mathcal{N}_1$. By this we mean that P maps $V_{2\delta}(\mathcal{N}_1)$ onto $\mathcal{N}_1$, and that its restriction to $\mathcal{N}_1$ coincides with the identity map. Moreover, we will assume P to be $\mathcal{C}^1$, and that all over $V_{2\delta}(\mathcal{N}_1)$, the derivative of P, dP, has rank $n = \dim \mathcal{N}$ (P will then be a submersion). We leave to the reader the task of showing that for δ sufficiently small, it is possible to construct such a projection by using the inverse function theorem and a partition of unity argument (see exercises 1.7 to 1.11).

We choose a function $\eta \in \mathcal{C}^\infty([0,+\infty),[0,1])$ such that

$$\eta(x) = 1 \quad \text{on} \quad [1,\delta]$$
$$\eta(x) = 0 \quad \text{on} \quad [2\delta,+\infty),$$

and define, for each $y \in \mathbb{R}^N$,

$$\begin{cases} \Phi(y) = \eta(\text{dist}(y, \mathcal{N}_1)) J_2 \circ J_1^{-1} \circ P(y) & \text{if } y \in V_{2\delta}\mathcal{N}_1 \\ \Phi(y) = 0 & \text{if } y \notin V_{2\delta}\mathcal{N}_1 \end{cases}$$

It is clear that

$$\Phi = J_2 \circ J_1^{-1} \text{ over } \mathcal{N}_1 \tag{1.59}$$

and

$$\Phi(V_\delta(\mathcal{N}_1)) = \mathcal{N}_2. \tag{1.60}$$

Let $u \in H^1(\Omega, \mathcal{N}_1)$; then it is obvious that $\Phi \circ u(x) \in \mathcal{N}_2$ a.e. Moreover, for any $y \in \mathcal{N}_1$, the restriction $d\Phi|_{T_y\mathcal{N}_1}$ of $d\Phi$ to $T_y\mathcal{N}_1$ is an isometry from $T_y\mathcal{N}_1$ to $T_{\Phi(y)}\mathcal{N}_2$. This yields

$$|du(x)|^2 = |d(\Phi \circ u)(x)|^2 \text{ a.e.}$$

which proves (1.58) and, consequently, also (1.57). $\square$

Remark 1.4.4 *In the proof of this lemma we used, not the orthogonal projection from $V_{2\delta}(\mathcal{N}_1)$ onto $\mathcal{N}_1$ (though it is easy to define), but rather a non-orthogonal projection. This stems from the fact that if $\mathcal{N}_1$ is $\mathcal{C}^1$, then the orthogonal projection is generally only $\mathcal{C}^0$.*

Remark 1.4.5 *Another possibility would have been to use, instead of $H^1(\mathcal{M}, \mathcal{N})$, the space*

$$\overline{\mathcal{C}^2(\mathcal{M}, \mathcal{N})}^{H^1} = \{u \in L^\infty(\mathcal{M}, \mathcal{N}) |\ \exists (\phi_k)_{k \in \mathbb{N}},\ \text{a sequence in } \mathcal{C}^2(\mathcal{M}, \mathcal{N})$$
$$\text{such that } \lim_{k \to +\infty} \|u - \phi_k\|_{H^1} = 0\}.$$

Nevertheless, it happens that, in general, this space is different from $H^1(\mathcal{M}, \mathcal{N})$, unless $m = 2$ (see [146], [15], [8]). Moreover, $\overline{\mathcal{C}^2(\mathcal{M}, \mathcal{N})}^{H^1}$ seems less appropriate for the calculus of variations than $H^1(\mathcal{M}, \mathcal{N})$ since, unlike $H^1(\mathcal{M}, \mathcal{N})$, in general it is not closed for the H^1 weak topology (see [8]).

Thus, we have a function space $H^1(\mathcal{M}, \mathcal{N})$, over which we can define a functional, the Dirichlet integral $E_{(\mathcal{M},g)}$. In order to prove the existence of harmonic maps from $\mathcal{M}$ to $\mathcal{N}$, it is natural to attempt to study the critical points of $E_{(\mathcal{M},g)}$ in $H^1(\mathcal{M},\mathcal{N})$. This involves the following three steps:

- to define what we mean by weak solutions;
- to prove the existence of weak solutions, critical points of $E_{(\mathcal{M},g)}$ (existence);
- to prove that a weak solution is a smooth harmonic map from $\mathcal{M}$ to $\mathcal{N}$ (regularity).

Problems will show up at each of these steps. A specialist in non-linear variational problems will expect to find difficulties, and sometimes barriers, to proving existence and regularity, and this is, in fact, the case. But it happens that even the definition of what should be a weak solution is not totally obvious. The reason is that, unless $m = 1$, $H^1(\mathcal{M},\mathcal{N})$ is not a manifold, and thus a good notion of tangent space to $H^1(\mathcal{M},\mathcal{N})$ does not exist. Consequently, we have some problems in defining the variational derivative of E. This leads to the possibility of giving different definitions of weak solutions.

It would be difficult to present the main points of this theory in a book. Below, we will briefly present the simplest approach to proving the existence of weak solutions: the minimization of $E_{(\mathcal{M},g)}$. We will then give a list of the different notions of critical points. In the following section we will present some regularity results, some of which will be proved in later chapters.

An example of a method to prove the existence of weak solutions: the minimization of $E_{(\mathcal{M},g)}$

It consists of choosing a subset $\mathcal{E}$ of $H^1(\mathcal{M},\mathcal{N})$, defined by imposing a constraint such as the boundary condition in a domain, or the homotopy class, and showing that the infimum

$$\inf_{u \in \mathcal{E}} E_{(\mathcal{M},g)}(u)$$

is attained. The map that achieves the minimum is called minimizing (see below). We can easily show (see exercise 1.12) that $H^1(\mathcal{M},\mathcal{N})$ is closed for the weak topology of $H^1(\mathcal{M},\mathbb{R}^N)$, and that $E_{(\mathcal{M},g)}$ is coercive and lower semi-continuous. Nevertheless, as we will see below, we will come across several problems that are connected to the choice we made of the class $\mathcal{E}$.

The Dirichlet problem

It consists, in the case where $\partial \mathcal{M} \neq \emptyset$, of fixing a boundary condition, i.e. a map

$$g : \partial \mathcal{M} \longrightarrow \mathcal{N},$$

and considering the space

$$H^1_g(\mathcal{M}, \mathcal{N}) = \{u \in H^1(\mathcal{M}\mathcal{N}) |\ u = g \text{ on } \partial \mathcal{M}\}.$$

We then minimize the Dirichlet integral over $H^1_g(\mathcal{M}, \mathcal{N})$ in order to find a harmonic map from $\mathcal{M}$ to $\mathcal{N}$ that coincides with g on $\partial \mathcal{M}$. The first thing to check is whether $H^1_g(\mathcal{M}, \mathcal{N}) \neq \emptyset$. Some obstructions may show up.

Example 1.4.6

$$\mathcal{M} = B^2 = \{(x,y) \in \mathbb{R}^2 |\ x^2 + y^2 < 1\},$$
$$\mathcal{N} = S^1 = \{(x,y) \in \mathbb{R}^2 |\ x^2 + y^2 = 1\} = \partial B^2, \text{ and}$$
$$g(x,y) = (x,y).$$

Then, there is no finite energy extension of g in $H^1_g(B^2, S^1)$. The reason is a topological obstruction. For each C^1 map $g : S^1 \longrightarrow S^1$ (in fact it would suffice to suppose that $g \in H^{\frac{1}{2}}(S^1)$), the following quantity is called the topological degree (or winding number):

$$\deg(g) = \frac{1}{2\pi} \int_{\partial B^2} g^1 dg^2 - g^2 dg^1.$$

It is an integer which represents the number of times the point $g(x,y)$ goes around S^1, when (x,y) goes once around S^1. We will see that if $u \in H^1(B^2, S^1)$, then the degree of $u_{|\partial B^2}$ is necessarily equal to 0. Since for $g(x,y) = (x,y)$, we have $\deg(g) = 1$, this implies that $H^1_g(B^2, S^1)$ is empty.

(i) On the one hand, for any map $u \in H^1(B^2, \mathbb{R}^2)$, we have, by Stokes' formula,

$$\int_{B^2} d\alpha = \int_{\partial B^2} \alpha, \qquad (1.61)$$

where $\alpha = u^1 du^2 - u^2 du^1$. This is a well-known formula in the case where u is of class C^2. It extends to $H^1(B^2, \mathbb{R}^2)$, by using the density of $C^2(B^2, \mathbb{R}^2)$ in $H^1(B^2, \mathbb{R}^2)$, and showing that the two integrals in (1.61) are continuous functionals over $H^1(B^2, \mathbb{R}^2)$ and $H^{\frac{1}{2}}(\partial B^2, \mathbb{R}^2)$, respectively.

(ii) On the other hand, if $u \in H^1(B^2, S^1)$, we deduce from $|u(x,y)|^2 = 1$ a.e. that
$$\langle u, du \rangle = 0 \quad a.e. \text{ on } B^2.$$
This equation implies that
$$du^1 = \sum_{i=1}^{2} u^i (u^i du^1 - u^1 du^i) = -u^2 \alpha \quad a.e.$$
and
$$du^2 = \sum_{i=1}^{2} u^i (u^i du^2 - u^2 du^i) = u^1 \alpha \quad a.e.$$
and thus
$$du^1 \wedge du^2 = -u^1 u^2 \alpha \wedge \alpha = 0 \quad a.e. \tag{1.62}$$

(iii) Therefore, if $u \in H^1(B^2, S^1)$, we deduce from (1.61) and (1.62) that the degree of $u_{|\partial B^2}$ is zero.

Other difficulties appear when we only have $g \in H^{\frac{1}{2}}(\partial \mathcal{M}, \mathcal{N})$ (the natural trace space). In fact, in this case, we still do not know exactly if $H^1_g(\mathcal{M}, \mathcal{N})$ is empty or not (see [82], [13], [138]).

Example 1.4.7 (*Case where it works*). *If g is $\mathcal{C}^1$, and is homotopic to a constant, we can show that $H^1_g(\mathcal{M}, \mathcal{N}) \neq \emptyset$. Likewise, in case $\mathcal{M} = B^3$, the unit ball in $\mathbb{R}^3$, we may extend any map $g \in H^1(\partial B^3, \mathcal{N})$ to all of B^3 by letting*
$$u(x) = g\left(\frac{x}{|x|}\right).$$
We can thus show that $H^1_g(B^3, S^2) \neq \emptyset$.

Let us recall that if E and F are two topological spaces (in fact, we are interested in the case where E and F are manifolds), and if u_0 and u_1 are two continuous maps from E to F, we say that u_0 and u_1 are homotopic if there exists a continuous map $H : E \times [0,1] \longrightarrow F$, called a homotopy, such that $H(x,0) = u_0(x)$ and $H(x,1) = u_1(x)$.

We can show that homotopy is an equivalence relation, and the equivalence classes are called homotopy classes. (In the case $E = F = S^1$, the homotopy classes are precisely the subsets of maps in $\mathcal{C}(S^1, S^1)$ which

1.4 Variational approach: Sobolev spaces

have the same degree. Likewise, it is possible to characterize the homotopy classes of maps from S^2 to S^2 by defining a topological degree for maps from S^2 to S^2).

Once we know that $H^1_g(\mathcal{M}, \mathcal{N}) \neq \emptyset$, it is easy to show that $E_{(\mathcal{M},g)}$ attains its minimum in $H^1_g(\mathcal{M}, \mathcal{N})$ (see exercises 1.12 to 1.14). The map thus obtained is said to be minimizing (see below).

HARMONIC REPRESENTATIVES OF A HOMOTOPY CLASS

The problem here is, given a map ϕ from $\mathcal{M}$ to $\mathcal{N}$, to find a harmonic map homotopic to ϕ. To simplify, suppose that $\partial \mathcal{M} = \emptyset$, and denote the homotopy class of ϕ by $[\phi]$. It is then natural to try to minimize $E_{(\mathcal{M},g)}$ over $\overline{[\phi]}^{H^1}$, the closure of $[\phi]$ in $H^1(\mathcal{M}, \mathcal{N})$. The difficulty here is that $\overline{[\phi]}^{H^1}$ may include smooth maps which are not homotopic to ϕ.

Example 1.4.8 $\mathcal{M} = \mathcal{N} = S^3$, the unit sphere in $\mathbb{R}^4$. Then, if $\phi = 1\!\!1$, $\overline{[\phi]}^{H^1}$ contains all the constant functions.

In fact, the set $\overline{[\phi]}^{H^1}$ was thoroughly characterized by [180]. In general, the only case where homotopy is preserved under H^1 strong convergence is when $\mathcal{M}$ is a surface. But even then homotopy is not preserved under the weak topology of H^1 (see [141]).

We will now stop drawing the pessimistic picture of all the difficulties that show up. Depressed readers may refer to [51] to convince themselves of the efficacy of the variational techniques, and have an overview of the set of results obtained. As we mentioned, in general the solutions obtained using variational methods in $H^1(\mathcal{M}, \mathcal{N})$ are not smooth and hence are "weak solutions". This is why we need to give a more precise definition of weak solutions.

WEAK SOLUTIONS

1.4.1 Weakly harmonic maps

Suppose that $\mathcal{N}$ is of class $\mathcal{C}^2$. Let $V_\delta \mathcal{N}$ be the tubular neighborhood of radius δ of $\mathcal{N}$ in $\mathbb{R}^N$, and P the projection from $V_\delta \mathcal{N}$ onto $\mathcal{N}$, defined as in lemma 1.4.3. We suppose δ to be sufficiently small for P to be defined and $\mathcal{C}^1$.

Definition 1.4.9 *We say that $u \in H^1(\mathcal{M}, \mathcal{N})$ is weakly harmonic if and only if, for any map v in $H_0^1(\mathcal{M}, \mathbb{R}^N) \cap L^\infty(\mathcal{M}, \mathbb{R}^N)$,*

$$\lim_{t \to 0} \frac{E_{(\mathcal{M},g)}(P(u+tv)) - E_{(\mathcal{M},g)}(u)}{t} = 0. \quad (1.63)$$

Here $H_0^1(\mathcal{M}, \mathbb{R}^N)$ is the closure of $\mathcal{C}_c^\infty(\mathcal{M}, \mathbb{R}^N)$ in $H^1(\mathcal{M}, \mathbb{R}^N)$.

We remark that (1.63) has a meaning for sufficiently small t since $v \in L^\infty(\mathcal{M}, \mathbb{R}^N)$ and thus $u + tv$ belongs to $V_\delta \mathcal{N}$ if $|t| < \delta |v|_{L^\infty}^{-1}$. In case $\mathcal{N}$ is $\mathcal{C}^2$, we can write an Euler equation for u.

Lemma 1.4.10 *Suppose that $\mathcal{N}$ is $\mathcal{C}^2$. Then, every weakly harmonic map $u \in H^1(\mathcal{M}, \mathcal{N})$ satisfies*

(i) *the equation*

$$-\Delta_g u \perp T_u \mathcal{N}$$

weakly, i.e. $\forall v \in H_0^1(\mathcal{M}, \mathbb{R}^N) \cap L^\infty(\mathcal{M}, \mathbb{R}^N)$, if $v(x) \perp T_{u(x)} \mathcal{N}$ a.e., then

$$\int_\mathcal{M} g^{\alpha\beta}(x) \left\langle \frac{\partial u}{\partial x^\alpha}, \frac{\partial v}{\partial x^\beta} \right\rangle \, d\mathrm{vol}_g = 0, \quad (1.64)$$

or equivalently,

(ii) *the equation*

$$\Delta_g u + g^{\alpha\beta} A(u) \left(\frac{\partial u}{\partial x^\alpha}, \frac{\partial u}{\partial x^\beta} \right) = 0$$

weakly, i.e. $\forall v \in H_0^1(\mathcal{M}, \mathbb{R}^N) \cap L^\infty(\mathcal{M}, \mathbb{R}^N)$,

$$\int_\mathcal{M} \left[-g^{\alpha\beta}(x) \left\langle \frac{\partial u}{\partial x^\alpha}, \frac{\partial v}{\partial x^\beta} \right\rangle \right.$$
$$\left. + g^{\alpha\beta}(x) \left\langle A(u)\left(\frac{\partial u}{\partial x^\alpha}, \frac{\partial u}{\partial x^\beta} \right), v \right\rangle \right] d\mathrm{vol}_g = 0, \quad (1.65)$$

where A is the second fundamental form of the immersion of $\mathcal{N}$ in $\mathbb{R}^N$ (see lemma 1.2.4).

1.4 Variational approach: Sobolev spaces

Proof Suppose $v \in H^1_0(\mathcal{M}, \mathbb{R}^N) \cap L^\infty(\mathcal{M}, \mathbb{R}^N)$. We have

$$P(u + tv) = u + tw_t,$$

where $w_t = \int_0^1 \frac{\partial P}{\partial y^i}(u + stv)v^i ds$. Hence

$$E_{(\mathcal{M},g)}(P(u+tv)) = E_{(\mathcal{M},g)}(u) + t\int_\mathcal{M} g^{\alpha\beta}(x) \left\langle \frac{\partial u}{\partial x^\alpha}, \frac{\partial w_t}{\partial x^\beta} \right\rangle dvol_g + o(t)$$

and thus (1.63) is equivalent to

$$\lim_{t \to 0} \int_\mathcal{M} g^{\alpha\beta}(x) \left\langle \frac{\partial u}{\partial x^\alpha}, \frac{\partial w_t}{\partial x^\beta} \right\rangle dvol_g = 0.$$

Since P is C^2, it follows that

$$w_t \to w_0 = dP(u).(v) \text{ in } H^1 \cap L^\infty,$$

and therefore,

$$\int_\mathcal{M} g^{\alpha\beta}(x) \left\langle \frac{\partial u}{\partial x^\alpha}, \frac{\partial}{\partial x^\beta}(dP(u)(v)) \right\rangle dvol_g = 0. \qquad (1.66)$$

Since $v(x) \perp T_{u(x)}\mathcal{N}$ a.e. implies that $dP(u)(v) = v$ a.e., equation (1.64) follows immediately. We leave to the reader the task of checking that (1.64) implies (1.65) (see the proof of lemma 1.2.4). $\square$

As in lemma 1.4.3, we will check that definition 1.4.9 depends on the isometric immersion used for $(\mathcal{N}, h)$.

Lemma 1.4.11 *Let $J_1 : (\mathcal{N}, h) \to \mathcal{N}_1 \subset \mathbb{R}^{N_1}$, and $J_2 : (\mathcal{N}, h) \to \mathcal{N}_2 \subset \mathbb{R}^{N_2}$, be two C^2 isometric immersions of $(\mathcal{N}, h)$. Let $\Phi : \mathbb{R}^{N_1} \to \mathbb{R}^{N_2}$ be the extension of $J_2 \circ J_1^{-1} : \mathcal{N}_1 \to \mathcal{N}_2$ defined in lemma 1.4.3. Then, Φ is C^2, and $u \in H^1(\mathcal{M}, \mathcal{N}_1)$ is weakly harmonic if and only if $\Phi \circ u \in H^1(\mathcal{M}, \mathcal{N}_2)$ is weakly harmonic.*

Proof As in lemma 1.4.3, we suppose that $\mathcal{N}_1 = \mathcal{N}_2 = \mathcal{N}$. It is easy to check that Φ is C^2. Since the desired equivalence is symmetric w.r.t. $\mathcal{N}_1$ and $\mathcal{N}_2$, it suffices to show that $\Phi \circ u$ weakly harmonic $\Rightarrow u$ weakly harmonic. Let $u \in H^1(\mathcal{M}, \mathcal{N}_1)$. Suppose that $\Phi \circ u$ is weakly harmonic in $H^1(\mathcal{M}, \mathcal{N}_2)$. Let $v \in H^1_0(\mathcal{M}, \mathbb{R}^N)$, and t be a sufficiently small real number to ensure that $u + tv$ belongs to $V_\delta \mathcal{N}_1$ a.e. (see lemma 1.4.3). Let

$$u_t = P(u + tv).$$

Notice that $d(P(u+tv))$ takes its values in $T_{u_t}\mathcal{N}_1$, and that $d\Phi_{u_t} \mid T_{u_t}\mathcal{N}_1$ is an isometry from $T_{u_t}\mathcal{N}_1$ to $T_{\Phi(u_t)}\mathcal{N}_2$. It follows that

$$|d\Phi_{u_t}[d(P(u+tv))]|^2 = |d(P(u+tv))|^2 \text{ a.e.}$$

But since $\Phi \circ P = \Phi$ on $V_\delta \mathcal{N}_1$, $d\Phi_{u_t} \circ P_{u+tv} = d\Phi_{u+tv}$, and hence

$$|d(P(u+tv))|^2 = |d(\Phi(u+tv))|^2 \text{ a.e.} \tag{1.67}$$

Moreover,

$$\Phi(u+tv) = \Phi(u) + tw_t, \tag{1.68}$$

where $w_t = \int_0^1 \frac{\partial \Phi}{\partial y^i}(u + stv)v^i ds \in H_0^1(\mathcal{M}, \mathbb{R}^N) \cap L^\infty(\mathcal{M}, \mathbb{R}^N)$.

From (1.67) and (1.68), it follows that

$$\begin{aligned}
E_{(\mathcal{M},g)}(P(u+tv)) &= \int_\mathcal{M} \frac{1}{2}|d(P(u+tv))|^2 \, d\mathrm{vol}_g \\
&= \int_\mathcal{M} \frac{1}{2}|d(\Phi(u+tv))|^2 \, d\mathrm{vol}_g \\
&= \int_\mathcal{M} \frac{1}{2}|d(\Phi(u))|^2 \, d\mathrm{vol}_g \\
&\quad + t\int_\mathcal{M} g^{\alpha\beta}(x) \left\langle \frac{\partial \Phi(u)}{\partial x^\alpha}, \frac{\partial w_t}{\partial x^\beta} \right\rangle d\mathrm{vol}_g + O(t^2).
\end{aligned}$$

Thus,

$$\lim_{t \to 0} \frac{E_{(\mathcal{M},g)}(P(u+tv)) - E_{(\mathcal{M},g)}(u)}{t}$$

$$= \lim_{t \to 0} \int_{\mathcal{M}} g^{\alpha\beta}(x) \left\langle \frac{\partial \Phi(u)}{\partial x^\alpha}, \frac{\partial w_t}{\partial x^\beta} \right\rangle d\text{vol}_g$$

$$= \int_{\mathcal{M}} g^{\alpha\beta}(x) \left\langle \frac{\partial \Phi(u)}{\partial x^\alpha}, \frac{\partial d\Phi(u)(v)}{\partial x^\beta} \right\rangle d\text{vol}_g,$$

since $w_t \to w_0 = d\Phi(u)(v)$ in $H^1 \cap L^\infty$. We deduce from this identity that if $\Phi \circ u$ is weakly harmonic, then u is also weakly harmonic. □

Remark 1.4.12 *Actually, it is possible to assume that $\mathcal{N}$ is only of class $W^{2,\infty}$ (i.e. the sectional curvature is bounded) in lemmas 1.4.10 and 1.4.11. But the definition of weakly harmonic maps with values in manifolds which are not $W^{2,\infty}$ is a delicate question to which a satisfactory answer is yet to be given. For instance, if the image manifold is a cone*

$$C_n = \{(y, z) \in \mathbb{R}^n \times \mathbb{R} \mid z = \tau |y|\},$$

where $\tau \in (0, +\infty)$, or if the image manifold is compact but has a "cone-like" point, there is no satisfactory definition of a projection from a tubular neighborhood of C_n onto C_n, and definition 1.4.9 is meaningless. Likewise, we do not know a generalization of lemma 1.4.11 to the case where the image manifold is not $W^{2,\infty}$.

In case the image manifold has symmetries, it is possible to extend Noether's theorem to weakly harmonic maps.

Theorem 1.4.13 *Suppose that $\mathcal{N}$ is C^2. Let X be a Lipschitz vector field on $\mathcal{N}$, which generates an isometry flow (a so-called Killing vector field). Then, for each weakly harmonic map $u \in H^1(\mathcal{M}, \mathcal{N})$, the vector field on $\mathcal{M}$*

$$J = J^\alpha \frac{\partial}{\partial x^\alpha} = g^{\alpha\beta}(x) \sqrt{\det g(x)} \left\langle X(u), \frac{\partial u}{\partial x^\beta} \right\rangle \frac{\partial}{\partial x^\alpha}$$

is weakly divergence-free on $\mathcal{M}$, i.e. for any function $\phi \in H_0^1(\mathcal{M}, \mathbb{R}) \cap L^\infty(\mathcal{M}, \mathbb{R})$,

$$\int_{\mathcal{M}} g^{\alpha\beta}(x) \left\langle X(u), \frac{\partial u}{\partial x^\beta} \right\rangle \frac{\partial \phi}{\partial x^\alpha} \sqrt{\det g}\, dx^1 \ldots dx^m = 0.$$

Proof The proof is identical to that of theorem 1.3.1. It suffices to check that all calculations make sense in $H^1(\mathcal{M},\mathcal{N})$. □

On the contrary, in general theorem 1.3.6 does not extend to weakly harmonic maps (see the counterexample 1.4.19 below).

1.4.2 Weakly Noether harmonic maps

Let $(\Phi_t)_{t\in\mathbb{R}}$ be a family of diffeomorphisms from $\mathcal{M}$ to $\mathcal{M}$, depending on t in a $\mathcal{C}^1$ way, such that $\Phi_t(x) = x$ outside a compact subset of $\mathcal{M}$, and $\Phi_0(x) = x$ on $\mathcal{M}$.

Definition 1.4.14 *We say that the map $u \in H^1(\mathcal{M},\mathcal{N})$ is weakly Noether harmonic if and only if for each family of diffeomorphisms of $\mathcal{M}$, of the type described above,*

$$\lim_{t\to 0} E_{(\mathcal{M},g)}(u \circ \Phi_t) = 0. \tag{1.69}$$

Copying the proof of theorem 1.3.6 for Noether harmonic maps, we prove the following.

Theorem 1.4.15 *The map $u \in H^1(\mathcal{M},\mathcal{N})$ is weakly Noether harmonic, if and only if the distributional covariant derivative of the stress–energy tensor*

$$S_{\alpha\beta} = e(u)g_{\alpha\beta} - \left\langle \frac{\partial u}{\partial x^\alpha}, \frac{\partial u}{\partial x^\beta} \right\rangle$$

vanishes. This means that for each $\mathcal{C}^1$ vector field X with compact support (in $\mathcal{M}$),

$$\int_{\mathcal{M}} (L_X g^{\alpha\beta}) S_{\alpha\beta} \, d\mathrm{vol}_g = 0.$$

1.4.3 Minimizing maps

Definition 1.4.16 *A map $u \in H^1(\mathcal{M},\mathcal{N})$ is minimizing if and only if for any point $x \in \mathcal{M}$, there exists a compact neighborhood K_x of x, such that $\forall v \in H^1(\mathcal{M},\mathcal{N})$, such that $u = v$ a.e. on $\mathcal{M} \setminus K_x$,*

$$E_{(\mathcal{M},g)}(v) \geq E_{(\mathcal{M},g)}(u). \tag{1.70}$$

1.4.4 Weakly stationary maps

Definition 1.4.17 *A map $u \in H^1(\mathcal{M},\mathcal{N})$ is weakly stationary if and only if it is both weakly harmonic and weakly Noether harmonic.*

1.4.5 Relation between these different definitions

What is the relation between these different definitions? It is easy to show that every minimizing map is weakly stationary. We thus have the inclusions

$$\{Minimizing\ Maps\}$$
$$\subset \{Weakly\ Stationary\ Maps\}$$
$$\subset \{Weakly\ Harmonic\ Maps\}.$$

We can also ask ourselves which of the two notions, weakly Noether harmonic or weakly harmonic, is the "strongest". The answer depends on the regularity of u.

IF u IS $\mathcal{C}^2$

(i) Every $\mathcal{C}^2$ weakly harmonic map is Noether harmonic. The reason is that a $\mathcal{C}^2$ weakly harmonic map is in fact a classical harmonic map, and thus we can apply theorem 1.3.6. In fact, for any tangent field X on $\mathcal{M}$, $u_*X = X^\alpha \frac{\partial u}{\partial x^\alpha}$ is a $\mathcal{C}^1$ map from $\mathcal{M}$ to $\mathbb{R}^N$, which can be used as test function in (1.63) (definition 1.4.9).
(ii) The converse is false unless u is a diffeomorphism from $\mathcal{M}$ to $\mathcal{N}$.

Example 1.4.18 *Let $u \in \mathcal{C}^2(\mathcal{M},\mathcal{N})$ be a conformal map, and suppose that $\mathcal{M}$ is a surface. Then, we see that the conformality of u is equivalent to $S_{\alpha\beta} = 0$, and thus u is Noether harmonic. However, u is not harmonic in general, unless its image $u(\mathcal{M})$ is a minimal surface of $\mathcal{N}$.*

In fact, condition (1.69) just means that a Noether harmonic map "*parametrizes its image in a harmonic way* ".

IF u IS NOT $\mathcal{C}^2$

There is no connection between being weakly harmonic and being weakly Noether harmonic.

Example 1.4.19 Choose $\mathcal{M} = B^3$, the unit ball in $\mathbb{R}^3$. Let

$$u: S^2 \setminus \left\{ \begin{pmatrix} 0 \\ 0 \\ -1 \end{pmatrix} \right\} \longrightarrow \mathbb{C}$$

$$\begin{pmatrix} y^1 \\ y^2 \\ y^3 \end{pmatrix} \longmapsto \frac{y^1 + iy^2}{1 + y^3}$$

be the stereographic projection. u is a conformal diffeomorphism which extends to a diffeomorphism from S^2 to $\mathbb{C} \cup \{\infty\}$. For $\lambda \in (0, +\infty)$, let

$$v_\lambda: B^3 \longrightarrow S^2$$
$$x \longmapsto u^{-1}\left(\lambda u\left(\frac{x}{|x|}\right)\right).$$

In spite of having a singularity at 0, v_λ belongs to $H^1(B^3, S^2)$ and is weakly harmonic. However, v_λ is not weakly Noether harmonic, unless $\lambda = 1$. In fact, the stress–energy tensor of v_λ satisfies the following equation in the sense of distributions:

$$\sum_{\alpha=1}^{3} \frac{\partial S_{\alpha\beta}}{\partial x^\alpha} = \delta_0 V^\beta,$$

where δ_0 is the Dirac mass at 0, and $V = \begin{pmatrix} 0 \\ 0 \\ f(\lambda) \end{pmatrix}$ with

$$\begin{cases} f(\lambda) &= -\dfrac{8\pi}{(\lambda^2 - 1)^2} \cdot (\lambda^4 - 4\lambda^2 \log\lambda - 1) \text{ if } \lambda \neq 1 \\ f(1) &= 0. \end{cases}$$

Exercises

The purpose of these exercises is to show that if $k \geq 1$ and $\mathcal{N}$ is a compact $\mathcal{C}^k$ submanifold of $\mathbb{R}^N$, then there is a neighborhood $V\mathcal{N}$ of $\mathcal{N}$ in $\mathbb{R}^N$ and a $\mathcal{C}^k$ submersion $P: V\mathcal{N} \longrightarrow \mathcal{N}$ coinciding with the identity map over $\mathcal{N}$. Let $p = N - n$ be the codimension of $\mathcal{N}$. We denote by $Gr_p(N)$ the Grassmanian of p-dimensional vector subspaces $\mathbb{R}^N$, and

1.4 Variational approach: Sobolev spaces

by $M(N \times N, \mathbb{R})$ the set of $N \times N$ real matrices (equipped with the Euclidean distance, d, given by the Hilbert–Schmidt norm).

1.7 Show that we can identify $Gr_p(N)$ with
$$\mathcal{P}_p(N) = \{A \in M(N \times N, \mathbb{R}) \mid A = {}^t A = A^2,\ \mathrm{tr}\,A = p\}.$$
Show that there is a neighborhood $\mathcal{V}$ of $\mathcal{P}_p(N)$ in $M(N \times N, \mathbb{R})$ over which the projection $\Pi : \mathcal{V} \to \mathcal{P}_p(N)$, defined by $d(A, \Pi(A)) = d(A, \mathcal{P}_p(N))$, exists and is of class $\mathcal{C}^\infty$.

1.8 For any point $y \in \mathcal{N}$, we define $A_y = (T_y \mathcal{N})^\perp \in \mathcal{P}_p(N)$. Prove that for every real number $\epsilon > 0$, we can cover $\mathcal{N}$ by a finite union of balls $B(y_\alpha, r_\alpha)$ in $\mathbb{R}^N$, centered at points in $\mathcal{N}$, such that $\forall y \in B(y_\alpha, r_\alpha) \cap \mathcal{N},\ d(A_{y_\alpha}, A_y) < \frac{\epsilon}{2}$.

1.9 We consider a partition of unity $(\chi_\alpha)_{1 \le \alpha \le s}$ defined over $\mathcal{VN}$, associated to the balls $B(y_\alpha, r_\alpha)$ (i.e. $\mathrm{Supp}(\chi_\alpha) \subset B(y_\alpha, r_\alpha)$ and $\sum_{\alpha=1}^{s} \chi_\alpha = 1$ in $\mathcal{N}$).

Check that if ϵ is sufficiently small, then we can define a map γ from $\mathcal{N}$ to $\mathcal{P}_p(N)$ by
$$\gamma(y) = \Pi\left(\sum_{\alpha=1}^{s} A_{y_\alpha} \chi_\alpha(y)\right),$$
that γ is $\mathcal{C}^k$ and that $d(\gamma(y), A_y) < \epsilon$.

1.10 Deduce, using the implicit function theorem, that if ϵ is sufficiently small, the family $\{(y, \gamma(y)) \mid y \in \mathcal{N}\}$ defines a $\mathcal{C}^k$ foliation of $\mathcal{VN}$ (up to redefining $\mathcal{VN}$), where $(y, \gamma(y))$ is the affine subspace of $\mathbb{R}^N$ passing through y and parallel to $\gamma(y)$.

1.11 Construct the projection P from $\mathcal{VN}$ to $\mathcal{N}$ such that $P^{-1}(y) = (y, \gamma(y)) \cap \mathcal{VN}$, and check that P satisfies all the required conditions.

Exercises

1.12 Using the Rellich–Kondrakov theorem (see [19]) and a diagonal argument, show that for every sequence $(u_n)_{n \in \mathbb{N}}$ of $H^1(\mathcal{M}, \mathcal{N})$ which converges to a map $u \in H^1(\mathcal{M}, \mathbb{R}^N)$, in the weak topology of $H^1(\mathcal{M}, \mathbb{R}^N)$, we have that $u \in H^1(\mathcal{M}, \mathcal{N})$.

1.13 Let $\mathcal{E}$ be a subset of $H^1(\mathcal{M}, \mathcal{N})$, closed in the weak topology of $H^1(\mathcal{M}, \mathbb{R}^N)$, and let $(u_n)_{n \in \mathbb{N}}$ be a minimizing sequence for $E_{(\mathcal{M}, g)}$ in $\mathcal{E}$.

Suppose that u_n converges to u in the weak topology of $H^1(\mathcal{M}, \mathcal{N})$. Show that u is minimizing.

1.14 We suppose that $\partial \mathcal{M} \neq \emptyset$ and that there exists a map $g : \partial \mathcal{M} \to \mathcal{N}$ such that $H^1_g(\mathcal{M}, \mathcal{N}) \neq \emptyset$. Prove the existence of a minimizing map in $H^1_g(\mathcal{M}, \mathcal{N})$.

1.5 Regularity of weak solutions

What can we say about the regularity of the weak solutions presented above? First of all we have the following result:

Theorem 1.5.1 *Every continuous weakly harmonic map $u \in H^1(\mathcal{M}, \mathcal{N})$ is smooth. More precisely, if we suppose that g is $\mathcal{C}^{k,\alpha}$ (where $k \geq 0$ and $0 < \alpha < 1$), and that h (the metric on $\mathcal{N}$) is $\mathcal{C}^{l,\alpha}$ (where $l \geq 1$) or, which is equivalent, that $\mathcal{N}$ is a $\mathcal{C}^{l+1,\alpha}$ submanifold, then u will be $\mathcal{C}^{\inf(k+1,l+1),\alpha}$.*

The proof of this theorem is the result of contributions made over several years. The first step is to show that every weakly harmonic map u in $H^1(\mathcal{M}, \mathcal{N}) \cap \mathcal{C}^0(\mathcal{M}, \mathcal{N})$ belongs to $\mathcal{C}^{0,\alpha}$. This is obtained by adapting a theorem of Olga Ladyzhenskaya and Nina Ural'tseva ([103], chapter 8, see also [93] or [98]). The final result then follows using classical techniques (see [103] or [69]).

This does not completely answer the previous question since it remains to be seen whether a weakly harmonic map in $H^1(\mathcal{M}, \mathcal{N})$ is continuous.

The answer differs a lot according as we are considering minimizing, weakly stationary, or weakly harmonic maps. Here is a brief survey of some known results. For more information see [50], [51].

CASE WHERE $m = \dim \mathcal{M} = 1$

In this case, every map $u \in H^1(\mathcal{M}, \mathcal{N})$ is a priori continuous (and even $\mathcal{C}^{0,\frac{1}{2}}$), and it is easy to show that every weakly harmonic map is smooth (as smooth as the smoothness of $\mathcal{N}$ allows it to be). A fortiori, weakly stationary and minimizing harmonic maps are also smooth.

CASE WHERE $m = 2$

This is the limit case, since in dimension 2 the space H^1 is continuously embedded in all L^p spaces for $1 \leq p < +\infty$, and even in BMO (see chapter 3), but not in L^∞ or $\mathcal{C}^0$. Nevertheless, weakly harmonic maps are smooth. This result was obtained for minimizing maps by Charles B. Morrey [116], then by Michael Grüter for the case of conformal weakly harmonic maps [78], and by Richard Schoen for weakly stationary maps [144]. The general proof, due to myself [83], [85], will be presented in chapters 2 and 4 of this book.

CASE WHERE $m \geq 3$

More complex phenomena show up. The first case to be completely solved was that of minimizing maps. Richard Schoen and Karen Uhlenbeck showed that if $u \in H^1(\mathcal{M}, \mathcal{N})$ is minimizing, then there exists a closed subset $\mathcal{S}$ of $\mathcal{M}$ (called a singular set of u), such that

(i) The Hausdorff $(m - 3)$–dimensional measure of $\mathcal{S}$, $\mathcal{H}^{m-3}(\mathcal{S})$, is finite. This means, for instance if $m = 3$, that $\mathcal{S}$ consists of a finite union of points (see section 3.5 for a definition of Hausdorff measure).
(ii) u is smooth over $\mathcal{M} \setminus \mathcal{S}$ (see [145], [146]).

This result had previously been obtained by Mariano Giaquinta and Enrico Giusti for the case where the image of u is contained in an open subset diffeomorphic to an open subset of $\mathbb{R}^n$ [70].

This theorem is optimal since we can give examples of minimizing maps having such a singular set. For instance, if $\mathcal{M} = B^m = \{x \in \mathbb{R}^m \mid |x| < 1\}$ and $\mathcal{N} = S^{m-1} = \partial B^n$, then the map

$$u : B^m \longrightarrow S^{m-1}$$
$$x \longmapsto \frac{x}{|x|}$$

is a minimizing map for all $m \geq 3$ (this was shown for $m = 3$ in [22] and for $m \geq 3$ in [108]). For stationary maps, results were obtained by Lawrence Craig Evans for the case where $\mathcal{N} = S^n$ [54] (see theorem 3.5.1), and then by Fabrice Bethuel for the general case [10] (see theorem 4.3.4): every stationary map $u \in H^1(\mathcal{M}, \mathcal{N})$ is smooth outside a singular set $\mathcal{S}$, where $\mathcal{S}$ is a closed set whose $(m - 2)$–dimensional Hausdorff measure, $\mathcal{H}^{m-2}(\mathcal{S})$, is zero. Subsequently Fang Hua Lin has shown that if we also suppose that there are no non-constant harmonic maps from S^2 to $\mathcal{N}$, then $\mathcal{H}^{m-4}(\mathcal{S})$ is finite [109]. Improvements of these results with more precise estimates were recently obtained by F.H. Lin and T. Rivière [110]. (See also [169] and [139] for similar results on Yang–Mills fields.)

Finally, any hope of establishing a regularity result for weakly harmonic maps in general was dissipated by Tristan Rivière, who constructed everywhere–discontinuous finite energy weakly harmonic maps, taking values in the sphere S^2 [136]. We notice, to conclude, that many more precise results may be obtained if we make stronger hypotheses on the image manifold $\mathcal{N}$, as for instance that of [92] for weakly harmonic maps into a "geodesically convex ball". For more references on all these aspects, see the reports [50], [51], [81] or the book [71].

2
Harmonic maps with symmetry

This chapter will be mainly devoted to maps between a surface and a sphere. This special case is particularly interesting. First, it is the simplest setting, apart from real harmonic functions, which is non-linear. Second, both differential geometry and physics provide numerous examples where sphere-valued harmonic maps play a major role (Gauss maps for constant mean curvature surfaces, σ-models, liquid crystals, etc.). In particular, in the first section we will consider the link with constant mean curvature surfaces. Finally, it is a simple model for understanding the special properties of harmonic maps with values in symmetric manifolds. Recall that over such a manifold $\mathcal{N}$, the isometry group is sufficiently large to act transitively on $\mathcal{N}$, i.e. we can go from any point in $\mathcal{N}$ to any other point in $\mathcal{N}$ through the action of this group. This hypothesis is very strong and, in particular, implies that $\mathcal{N}$ is locally isometric to a quotient $\mathfrak{G}/\mathfrak{K}$ of Lie groups, where $\mathfrak{G}$ is the isometry group of $\mathcal{N}$ and $\mathfrak{K}$ is the subgroup of $\mathfrak{G}$ that fixes a point y_0 arbitrarily chosen in $\mathcal{N}$. If, moreover, we suppose that the domain manifold is 2-dimensional, then the simultaneous action of the conformal transformations of the domain and the isometries of the image creates a very rich setting where numerous geometric, algebraic and analytical miracles occur. These miracles stem from the fact that we will be in the presence of a completely integrable (Hamiltonian) system, as in the famous example of the Korteweg–de Vries equation. We will present some aspects of this complete integrability in sections 2.2, 2.3 and 2.4. Finally, sections 2.5 and 2.6 concern analysis results, more specifically, the weak compactness and regularity of weakly harmonic maps in dimension 2.

NOTATION: We write

$$S^n = \{y \in \mathbb{R}^{n+1} \mid |y| = 1\}$$

the unit sphere, and

$$SO(n+1) = \{R \in M((n+1) \times (n+1), \mathbb{R}) \mid {}^tRR = \mathbb{1}, \det R = 1\}$$

the group of rotations in $\mathbb{R}^{n+1}$, acting transitively and isometrically over S^n.

2.1 Bäcklund transformation

A Bäcklund transformation is a non-local transformation which, to a solution of a system of partial differential equations, associates a function which is itself a solution of a system of partial differential equations. A simple example is that of $\mathbb{R}$-valued harmonic functions on a (simply connected) open subset Ω of $\mathbb{C}$. Let u be such a function. Knowing that

$$\Delta u = \frac{\partial}{\partial y}\left(\frac{\partial u}{\partial y}\right) - \frac{\partial}{\partial x}\left(-\frac{\partial u}{\partial x}\right) = 0, \tag{2.1}$$

we can show that there exists v from Ω to $\mathbb{R}$, unique up to a constant, such that

$$\begin{cases} \dfrac{\partial v}{\partial x} = -\dfrac{\partial u}{\partial y} \\[2mm] \dfrac{\partial v}{\partial y} = \dfrac{\partial u}{\partial x}. \end{cases} \tag{2.2}$$

Furthermore, v will then be harmonic and $F = u + iv$ will be a holomorphic function of $z = x + iy$.

2.1.1 S^2-valued maps

Suppose that u is a harmonic map from a simply connected open subset Ω of $\mathbb{C}$ to S^2. By Noether's theorem 1.3.1, we have

$$\frac{\partial}{\partial x}\left(u \times \frac{\partial u}{\partial x}\right) + \frac{\partial}{\partial y}\left(u \times \frac{\partial u}{\partial y}\right) = 0 \tag{2.3}$$

(see example 1.3.3). We deduce from (2.3) that there exists a map B from Ω to $\mathbb{R}^3$, unique up to a constant vector of $\mathbb{R}^3$, such that

$$\begin{cases} \dfrac{\partial B}{\partial x} = u \times \dfrac{\partial u}{\partial y} \\ \dfrac{\partial B}{\partial y} = -u \times \dfrac{\partial u}{\partial x}. \end{cases} \quad (2.4)$$

Recall that if $v = \begin{pmatrix} v^1 \\ v^2 \\ v^3 \end{pmatrix}$, $w = \begin{pmatrix} w^1 \\ w^2 \\ w^3 \end{pmatrix}$,

$$v \times w = \begin{pmatrix} v^2 w^3 - v^3 w^2 \\ v^3 w^1 - v^1 w^3 \\ v^1 w^2 - v^2 w^1 \end{pmatrix}.$$

What can we say about B? First of all, a direct calculation yields

$$\Delta B = 2 u_x \times u_y = 2 B_x \times B_y, \quad (2.5)$$

where $u_x = \frac{\partial u}{\partial x}$, $u_y = \frac{\partial u}{\partial y}$, etc. We come across this type of equation when studying constant mean curvature surfaces. In fact, for any immersion X of an open subset of $\mathbb{R}^2$ in $\mathbb{R}^3$, if X is conformal, i.e. if

$$\left| \frac{\partial X}{\partial x} \right|^2 - \left| \frac{\partial X}{\partial y} \right|^2 = \left\langle \frac{\partial X}{\partial x}, \frac{\partial X}{\partial y} \right\rangle = 0,$$

then X satisfies the equation

$$\Delta X = 2H \frac{\partial X}{\partial x} \times \frac{\partial X}{\partial y},$$

where H is equal to the mean curvature of the image surface of X at the point $X(x, y)$. Thus the conformal parametrizations of surfaces of constant mean curvature equal to 1 are characterized by the system

$$|X_x|^2 - |X_y|^2 - 2i \langle X_x, X_y \rangle = 0 \quad (2.6)$$

$$\Delta X = 2\, X_x \times X_y. \quad (2.7)$$

The resemblance between (2.5) and (2.7) is evident, but we do not know whether B is conformal, i.e. whether B is a solution of (2.6). In fact, B in general is not conformal except in two special cases:

(i) u is an orientation-preserving conformal map, i.e.

$$\det(u, u_x, u_y) \geq 0.$$

Then $B = -u + A$, $A \in \mathbb{R}^3$ (exercise: check)

(ii) u is an orientation-reversing conformal map, i.e.

$$\det(u, u_x, u_y) \leq 0,$$

and then $B = u + A$, $A \in \mathbb{R}^3$ (exercise: check).

However, we remark that B is an immersion if and only if u is an immersion, and if so the image of B is a surface with constant Gauss curvature equal to 1. In fact, in this case it is easy to check that u is the Gauss map of the immersion B, and that the ratio of area swept by u and by B, given by

$$\frac{\det(u, u_x, u_y)}{\det(u, B_x, B_y)},$$

is equal to 1, which proves that the Gauss curvature of the image of B is equal to 1.

As a matter of fact, constant mean curvature surfaces are not far away. We check that

$$\Delta u = -u|du|^2 = u_x \times B_y + B_x \times u_y. \tag{2.8}$$

Thus, if we write the equations obtained after summing and subtracting (2.5) and (2.8), we have

$$\Delta(B + u) = \frac{\partial}{\partial x}(B + u) \times \frac{\partial}{\partial y}(B + u), \tag{2.9}$$

$$\Delta(B - u) = \frac{\partial}{\partial x}(B - u) \times \frac{\partial}{\partial y}(B - u). \tag{2.10}$$

Therefore, we are led to consider

$$X_+ = B + u$$

and

$$X_- = B - u.$$

2.1 Bäcklund transformation

Moreover, we may check that X_+ and X_- are always solutions of

$$\left|\frac{\partial X_\pm}{\partial x}\right|^2 - \left|\frac{\partial X_\pm}{\partial y}\right|^2 - 2i\left\langle\frac{\partial X_\pm}{\partial x}, \frac{\partial X_\pm}{\partial y}\right\rangle = 0. \tag{2.11}$$

This means that at any point x either $dX_\pm(x) = 0$ or $X_\pm$ is a conformal immersion of a neighborhood of x. We call every solution of (2.11) a weakly conformal map. From (2.9), (2.10) and (2.11) we see that if either X_+ or X_- is an immersion, then the image surface has constant mean curvature equal to $\frac{1}{2}$.

Thus, under non-degeneracy hypotheses (u, X_+ are X_- are immersions), we may associate to a harmonic map u a "sandwich" of three immersions of surfaces, at a constant distance 1 from each other, with $B(\Omega)$ in the middle (having Gauss curvature 1), and $X_+(\Omega)$ and $X_-(\Omega)$ (having mean curvature $\frac{1}{2}$) at either side. Conversely, if we start from one of these three surfaces, it is easy to reconstruct the other two, by adding or subtracting the Gauss map. It seems that this result has been known since Ossian Bonnet (see [18] and [42]).

We conclude this subsection with two remarks.

Remark 2.1.1 *Suppose that B is an immersion of an open subset Ω of $\mathbb{R}^2$. Then, since the Gauss curvature of $B(\Omega)$ is constant, we may regard $B(\Omega)$ as the isometric immersion in $\mathbb{R}^3$ of an open subset of the sphere S^2 (at least locally). Furthermore, equation (2.5) implies that*

$$\Delta B \perp B(\Omega)$$

and hence that $B : \Omega \longrightarrow B(\Omega)$ is a harmonic map with values in (an open subset of) S^2. But this open set is not embedded in $\mathbb{R}^3$ according to the canonical embedding. We encourage the reader to think about this remark until we get to section 2.4 of this chapter, where we will come back to this phenomenon.

Remark 2.1.2 *Imagine now that we replace the Laplacian by an elliptic (or hyperbolic) operator of the form*

$$L(f) = \sum_{\alpha,\beta=1}^{2} \frac{\partial}{\partial x^\alpha}\left(a^{\alpha\beta}(x)\frac{\partial f}{\partial x^\beta}\right), \tag{2.12}$$

where $(a^{\alpha\beta})$ is a symmetric matrix whose elements are $L^\infty(\Omega)$ functions such that $\det(a^{\alpha\beta})$ has a constant sign and never gets too close to zero. We may decide to look for maps u from Ω to S^2 which are solutions of

$$L(u)//u. \tag{2.13}$$

We remark that, thanks to the action of $SO(3)$ on S^2 and Noether's theorem, the solutions of (2.13) satisfy

$$\sum_{\alpha,\beta=1}^{2} \frac{\partial}{\partial x^\alpha}\left(a^{\alpha\beta}(x)u \times \frac{\partial u}{\partial x^\beta}\right) = 0,$$

and hence, if Ω is simply connected, there exists a map B from Ω to $\mathbb{R}^3$ such that

$$\begin{cases} \dfrac{\partial B}{\partial x^1} = \displaystyle\sum_{\beta=1}^{2} a^{1\beta}(x)\ u \times \dfrac{\partial u}{\partial x^\beta} \\[2ex] \dfrac{\partial B}{\partial x^2} = -\displaystyle\sum_{\beta=1}^{2} a^{2\beta}(x)\ u \times \dfrac{\partial u}{\partial x^\beta}. \end{cases} \tag{2.14}$$

By a calculation analogous to the one we saw, we may then check that if B is an immersion, then $B(\Omega)$ has Gauss curvature equal to

$$\det(a^{\alpha\beta}(x))^{-1}.$$

This may give us ways of constructing surfaces of prescribed Gauss curvature, as in the Minkowski problem. Such an approach was used by Yuxin Ge [63].

2.1.2 Maps taking values in a sphere S^n, $n \geq 2$

For each integer n, let

$$S^n = \{y \in \mathbb{R}^{n+1} \mid |y| = 1\}.$$

Let Ω be an open subset of $\mathbb{R}^2$, and $u : \Omega \longrightarrow S^n$ a harmonic map. Then by Noether's theorem 1.3.1, u satisfies

$$\frac{\partial}{\partial x}\left(u^i \frac{\partial u^j}{\partial x} - u^j \frac{\partial u^i}{\partial x}\right) + \frac{\partial}{\partial y}\left(u^i \frac{\partial u^j}{\partial y} - u^j \frac{\partial u^i}{\partial y}\right) = 0, \tag{2.15}$$

2.1 Bäcklund transformation

for each $1 \leq i, j \leq n+1$. In order to have at our disposal an operation analogous to the $\mathbb{R}^3$ vector product, it is convenient to introduce the following notation. Every vector $a \in \mathbb{R}^{n+1}$ is identified with the column vector

$$a = \begin{pmatrix} a^1 \\ \vdots \\ a^{n+1} \end{pmatrix}.$$

If $a, b \in \mathbb{R}^{n+1}$, we write

$$a \times b := a\,{}^tb - b\,{}^ta.$$

Notice that $a \times b$ is an anti-symmetric $(n+1) \times (n+1)$ matrix. Below, we identify the set of $(n+1) \times (n+1)$ matrices with the Lie algebra $so(n+1)$ of $SO(n+1)$. We may rewrite (2.15) as

$$\frac{\partial}{\partial x}\left(u \times \frac{\partial u}{\partial x}\right) + \frac{\partial}{\partial y}\left(u \times \frac{\partial u}{\partial y}\right) = 0. \tag{2.16}$$

This implies that there exists a map $b : \Omega \longrightarrow so(n+1)$ such that

$$\begin{cases} \dfrac{\partial b}{\partial x} = u \times \dfrac{\partial u}{\partial y} \\[2mm] \dfrac{\partial b}{\partial y} = -u \times \dfrac{\partial u}{\partial x}. \end{cases} \tag{2.17}$$

Remark 2.1.3 *If $n = 3$, b can be expressed in terms of B, given by (2.4), as*

$$b = \begin{pmatrix} 0 & B^3 & -B^2 \\ -B^3 & 0 & B^1 \\ B^2 & -B^1 & 0 \end{pmatrix}.$$

A direct calculation using (2.17) yields

$$\Delta b = 2\frac{\partial u}{\partial x} \times \frac{\partial u}{\partial y},$$

and also

$$\Delta b + 2\left[\frac{\partial b}{\partial x}, \frac{\partial b}{\partial y}\right] = 0, \tag{2.18}$$

which generalizes (2.5).

Exercises

2.1 Check that the critical points of the functional

$$F_2(b) = \int_\Omega \left(\frac{|db|^2}{2} + \frac{2}{3}\mathrm{tr}\left(b\left[\frac{\partial b}{\partial x}, \frac{\partial b}{\partial y}\right]\right)\right) dxdy$$

over the set of maps from Ω to $so(n+1)$, are precisely the solutions of (2.18).

Just like the map B, constructed in subsection 2.1.1 for the case of S^2, the map b given by (2.17) is not conformal in general. Nevertheless, there exists a way of associating to u and b a map which possesses interesting properties and which, in particular, is conformal. This construction generalizes in a certain way that of the pair of maps (X_+, X_-) seen above. Let M be the map from Ω to $so(n+2)$ given by

$$M = \begin{pmatrix} 0 & {}^t u \\ -u & b \end{pmatrix}.$$

A direct calculation shows that M satisfies

$$\Delta M + \left[\frac{\partial M}{\partial x}, \frac{\partial M}{\partial y}\right] = 0, \tag{2.19}$$

and hence that M is a critical point of the functional

$$F_1(M) = \int_\Omega \left(\frac{|dM|^2}{2} + \frac{1}{3}\mathrm{tr}\left(M\left[\frac{\partial M}{\partial x}, \frac{\partial M}{\partial y}\right]\right)\right) dxdy.$$

Furthermore, M is conformal.

2.1.3 Comparison

We will see that in case $n=2$, equation (2.19) is equivalent to the system of equations (2.9) and (2.10) for X_+ and X_-. The reason is that the Lie algebra $so(4)$ is isomorphic to the product $so(3) \times so(3) \simeq \mathbb{R}^3 \times \mathbb{R}^3$.

2.1 Bäcklund transformation

A convenient way of representing $so(4)$ is to identify $\mathbb{R}^4$ with the set of quaternions

$$\mathbb{H} := \{t + xi + yj + zk \mid (t, x, y, z) \in \mathbb{R}^4\},$$

where $i^2 = j^2 = k^2 = -1$ and $ij = -ji = k$, $jk = -kj = i$ and $ki = -ik = j$. Then we notice that $a \in so(4)$ has a unique decomposition as

$$a = a_L + a_R$$

where

$$a_L \in so(4)_L = \left\{ \begin{pmatrix} 0 & -p^1 & -p^2 & -p^3 \\ p^1 & 0 & -p^3 & p^2 \\ p^2 & p^3 & 0 & -p^1 \\ p^3 & -p^2 & p^1 & 0 \end{pmatrix} \Big/ (p^1, p^2, p^3) \in \mathbb{R}^3 \right\}$$

and

$$a_R \in so(4)_R = \left\{ \begin{pmatrix} 0 & q^1 & q^2 & q^3 \\ -q^1 & 0 & -q^3 & q^2 \\ -q^2 & q^3 & 0 & -q^1 \\ -q^3 & -q^2 & q^1 & 0 \end{pmatrix} \Big/ (q^1, q^2, q^3) \in \mathbb{R}^3 \right\}.$$

And one may easily see that $so(4)_L$ and $so(4)_R$ are both isomorphic to $so(3)$. Because of the identification of $\mathbb{H}$ with $\mathbb{R}^4$, the map $(t, x, y, z) \longmapsto a_L(t, x, y, z)$ corresponds to the left product

$$t + ix + jy + kz \longmapsto p(t + ix + jy + kz),$$

where p is a pure imaginary quaternion (i.e. of the form $ip^1 + jp^2 + kp^3$). Likewise we can represent a_R by the right product

$$t + ix + jy + kz \longmapsto -(t + ix + jy + kz)q,$$

with a pure imaginary quaternion $-q$.

Therefore, we have: $\forall \xi \in \mathbb{H}$,

$$a(\xi) = p\xi - \xi q. \tag{2.20}$$

It is then easy to see that every element of $so(4)_L$ commutes with any element of $so(4)_R$, because of the associativity of the product in $\mathbb{H}$, since

$$a_L(a_R(\xi)) = -p(\xi q) = -(p\xi)q = a_R(a_L(\xi)).$$

Thus, we have the following isomorphisms:

$$\begin{aligned} \Phi: \quad so(4) &\longrightarrow \Im(\mathbb{H}) \oplus \Im(\mathbb{H}) \\ a &\longmapsto (p,q)\,, \end{aligned}$$

where $\Im(\mathbb{H})$ is the set of pure imaginary quaternions, and p and q are defined by (2.20). Φ is a Lie algebra isomorphism, if we equip $\Im(\mathbb{H}) \oplus \Im(\mathbb{H})$ with the Lie bracket

$$[(p,q),(p',q')] = (pp' - p'p, qq' - q'q).$$

We may also deduce from Φ another Lie algebra isomorphism

$$\begin{aligned} \Psi: \quad so(4) &\longrightarrow \mathbb{R}^3 \times \mathbb{R}^3 \\ a &\longmapsto \left(\begin{pmatrix} p^1 \\ p^2 \\ p^3 \end{pmatrix}, \begin{pmatrix} q^1 \\ q^2 \\ q^3 \end{pmatrix} \right) = (\vec{p}, \vec{q}), \end{aligned}$$

where $\mathbb{R}^3 \times \mathbb{R}^3$ is equipped with the Lie bracket

$$[(\vec{p}, \vec{q}), (\vec{p'}, \vec{q'})] = 2(\vec{p} \times \vec{p'}, \vec{q} \times \vec{q'}),$$

where $\times$ is the vector product in $\mathbb{R}^3$. Let us come back to the map M from Ω to $so(4)$, and calculate $\Psi \circ M$:

$$\Psi \circ M = -\frac{1}{2}(X_+, X_-).$$

We then see that equation (2.19) for M is equivalent to equations (2.9) and (2.10) for X_+ and X_-.

2.2 Harmonic maps with values into Lie groups

To go deeper into the structure of weakly harmonic maps from a surface to a symmetric manifold, it is necessary to introduce one more concept. The idea is to reformulate the harmonic map equation as an integrability condition (vanishing curvature) for a differentiable 1-form

2.2 Harmonic maps with values into Lie groups

with coefficients in the Lie algebra of the symmetry group (a connection form), such a 1-form depending on an auxiliary complex parameter. This may seem very strange at first sight, yet it is this reformulation which opens the gates of the arsenal of techniques developed for completely integrable systems, like the Korteweg–de Vries or the non-linear Schrödinger equations (see [56]).

These ideas were introduced in our context by several authors, both physicists and mathematicians ([128], [185], [174]). In [174], Karen Uhlenbeck uses this formulation to study harmonic maps from a surface to the Lie group $U(n)$, and in particular classifies all harmonic maps from S^2 to $U(n)$.

Another important result in [174] is the existence of a very big group (the group of maps from the circle to $U(n)$), acting on the set of $U(n)$-valued harmonic maps. These symmetries had already been found in their infinitesimal version by [45].

We present here this formulation for the case for harmonic maps with values into a compact Lie group $\mathfrak{G}$ (we shall keep in mind the two cases $\mathfrak{G} = SO(n)$ and $\mathfrak{G} = U(n)$). As a warm-up, we start by considering the elementary example of real-valued harmonic functions. Let $u : \Omega \subset \mathbb{C} \longrightarrow \mathbb{R}$ be such a function. Then the function from Ω to $\mathbb{C}$ defined by

$$f = \frac{\partial u}{\partial z} = \frac{1}{2}\left(\frac{\partial u}{\partial x} - i\frac{\partial u}{\partial y}\right)$$

is holomorphic, i.e. it satisfies

$$\frac{\partial f}{\partial \bar{z}} = 0. \qquad (2.21)$$

Another way to write (2.21) is the system

$$d(du) = 0 \qquad (2.22)$$

$$d(\star du) = 0, \qquad (2.23)$$

where $\star du = -\frac{\partial u}{\partial y}dx + \frac{\partial u}{\partial x}dy$. Likewise, any constant coefficient linear combination of du and $\star du$ will be closed thanks to (2.22) and (2.23). Hence, we can also write (2.21) as

$$dg_\lambda = 0, \quad \forall \lambda \in \mathbb{C}^* \qquad (2.24)$$

where

$$\begin{aligned}g_\lambda &= \frac{1}{2}\left(1 - \frac{\lambda + \lambda^{-1}}{2}\right)du - \frac{1}{2}\frac{\lambda - \lambda^{-1}}{2i}(\star du)\\ &= \frac{1}{2}\left[(1-\lambda^{-1})fdz + (1-\lambda)\overline{f}d\overline{z}\right]\,.\end{aligned} \qquad (2.25)$$

Why do we choose such a g_λ? In fact, in this simple example, considering (2.25) is a useless complication. On the other hand, it is possible to generalize (2.24) and (2.25) to the case of harmonic maps taking values in a non-Abelian group, and g_λ is then the right concept.

Let $\mathfrak{G}$ be some compact Lie group (say $SO(n)$ or $U(n)$) and let $u : \Omega \subset \mathbb{C} \longrightarrow \mathfrak{G}$ be a map, not necessarily harmonic. Let

$$\begin{cases} a_x &= u^{-1}\dfrac{\partial u}{\partial x}\\[6pt] a_y &= u^{-1}\dfrac{\partial u}{\partial y}\,. \end{cases} \qquad (2.26)$$

A direct calculation shows that (2.26) implies

$$\frac{\partial a_y}{\partial x} - \frac{\partial a_x}{\partial y} + [a_x, a_y] = 0\,. \qquad (2.27)$$

Conversely, we can show using the Frobenius theorem that if Ω is simply connected, any pair of $\mathcal{C}^1$ maps (a_x, a_y) from Ω to $M(n,\mathbb{C})$ satisfying (2.27) may be obtained from a map u from Ω to $GL(n,\mathbb{C})$ using (2.26). (To prove it, for instance consider the vector fields $X_1 : (x,y,g) \longmapsto (1,0,ga_x)$ and $X_2 : (x,y,g) \longmapsto (0,1,ga_y)$ tangent to the manifold $\Omega \times GL(n,\mathbb{C})$ and observe that condition (2.27) is equivalent to $[X_1, X_2] = 0$.) In our case, the constraint

$$u \in \mathfrak{G}$$

will be equivalent to (up to left multiplication by a $GL(n,\mathbb{C})$ constant)

$$a_x, a_y \in \mathfrak{g}: \quad \text{the Lie algebra of } \mathfrak{G},$$

i.e. $\mathfrak{g} = so(n)$ for $\mathfrak{G} = SO(n)$ and $\mathfrak{g} = u(n)$ for $\mathfrak{G} = U(n)$. If we now assume u to be harmonic, Noether's theorem implies that

$$\frac{\partial a_x}{\partial x} + \frac{\partial a_y}{\partial y} = 0. \tag{2.28}$$

Hence, the study of harmonic maps from Ω to $\mathfrak{G}$ reduces, if Ω is simply connected, to the study of the pairs $a_x, a_y : \Omega \longrightarrow \mathfrak{g}$ satisfying (2.27) and (2.28). Notice that (2.27) is the non-linear analogue of (2.22), and (2.28) the analogue of (2.23).

We can rewrite this system using the complex variable $z = x + iy$. Writing

$$f = \frac{1}{2}(a_x - ia_y) = u^{-1}\frac{\partial u}{\partial z} \in \mathfrak{g} \otimes \mathbb{C},$$

the system (2.27), (2.28) is equivalent to

$$\frac{\partial f}{\partial \bar{z}} + \frac{1}{2}[\bar{f}, f] = 0 \tag{2.29}$$

which looks like a Lax pair formulation.

Finally, we can interpret equation (2.27) in a more geometric way, in terms of connections on bundles.

CONNECTION AND CURVATURE FORMS ON VECTOR BUNDLES

We briefly recall the notions on bundles that we will need (see [101] or [47] for more details).

If $\mathcal{M}$ is a manifold and $\mathfrak{G}$ is a Lie group, a *fiber bundle* $\mathcal{F}$ over $\mathcal{M}$, with *fiber* F and structure group $\mathfrak{G}$, is a differentiable manifold equipped with a differentiable map P, called the *projection* from $\mathcal{F}$ onto $\mathcal{M}$, such that

(i) There exists a covering of $\mathcal{M}$ by open sets U_i such that for all i, there exists a diffeomorphism $\Phi_i : U_i \times F \longrightarrow P^{-1}(U_i)$ such that $P \circ \Phi_i(M, f) = M$.

(ii) If $U_i \cap U_j \neq \emptyset$, we have for $M \in U_i \cap U_j$ and $f \in F$, $(\Phi_i)^{-1} \circ \Phi_j(M, f) = (M, T_{ij}(M)(f))$, where $T_{ij}(M)$ represents the action of an element of the group $\mathfrak{G}$ on F.

The set $\mathcal{F}_M := P^{-1}(\{M\})$ is the *fiber* over M; it is diffeomorphic to F.

A *section* of the bundle $\mathcal{F}$ over $\mathcal{M}$ is a map σ from $\mathcal{M}$ to $\mathcal{F}$, such that

$P \circ \sigma(x) = x, \forall x \in \mathcal{M}$.

One of the main difficulties of differentiable calculus on bundles, making the difference between a fiber bundle and a Cartesian product of manifolds, is that there is no canonical way of comparing two points "living" in two different fibers, even if they are infinitesimally close. The extra information that we need is the choice of a connection or covariant derivative. Since the situation of interest to us is that when F is a vector space (for instance $\mathbb{R}^n$), and $\mathfrak{G}$ is a subgroup of the endomorphism group of F, we present the definition of a connection only for the special case of vector bundles.

A *connection* or *covariant derivative* ∇ is an operator acting on the differentiable sections of $\mathcal{F}$ and satisfying the following conditions:

(i) ∇ associates in a differentiable fashion to every $M \in \mathcal{M}$, $V \in T_M \mathcal{M}$ and every $\mathcal{C}^1$ section σ of $\mathcal{F}$, a vector in $\mathcal{F}_M$, denoted by $\nabla_V \sigma$, which depends linearly on V and σ.

(ii) ∇ is a derivation: if $\phi \in \mathcal{C}^1(\mathcal{M})$ and σ is a $\mathcal{C}^1$ section, then $\nabla_V(\phi \sigma) = d\phi(V)\sigma + \phi \nabla_V \sigma$.

(iii) ∇ is compatible with the action of the group $\mathfrak{G}$. This means that $\forall \gamma \in G$, $\nabla_V(\gamma \sigma) = \gamma \nabla_V \sigma$, where $s \longmapsto \gamma s$ is the action of γ on $\mathcal{F}$.

It suffices to know the action of ∇ on a family of sections $E = (E_1, ..., E_n)$, such that at every point M of $\mathcal{M}$, $(E_1(M), ..., E_n(M))$ is a basis of $\mathcal{F}_M$. It is determined by the *Maurer–Cartan* 1-forms ω_b^a on $\mathcal{M}$ ($1 \leq a, b \leq n$) using the relations

$$\nabla_V E_a = E_b \omega_a^b(V).$$

Thanks to the linearity of ∇ and to (ii), we have for any section $\sigma = E_a s^a$, $\nabla_V(\sigma) = E_a(\omega_b^a(V)s^b + ds^a(V))$.

A remarkable property is that there exists a 2-form Ω_∇ on $\mathcal{M}$, with coefficients in the endomorphisms of $\mathcal{F}_M$, called *curvature* form, such that $\forall V, W \in T_{P(M)}\mathcal{M}$, $\nabla_V \nabla_W \sigma - \nabla_W \nabla_V \sigma - \nabla_{[V,W]}\sigma = \Omega_\nabla(V,W).\sigma$.

The curvature is characterized by a family of 2-forms Ω_b^a on $\mathcal{M}$ such that, decomposing a section along $\sigma = E_a s^a$, we have $\Omega(V,W).\sigma = E_a \Omega_b^a(V,W)s^b$. We can check that $\Omega_b^a = d\omega_b^a + \omega_c^a \wedge \omega_b^c$.

The curvature is a barrier to the existence of parallel sections, i.e. of sections σ which are solutions of $\nabla \sigma = 0$. In fact, one can show, using the Frobenius theorem, that the local existence of n parallel sections $E_1, ..., E_n$ forming a moving frame is true if and only if $\Omega_\nabla = 0$.

As an example, we come back to one case that interests us: $\mathcal{M}$ is an open subset Ω of $\mathbb{R}^2$, F is $\mathbb{R}^n$ with the canonical Euclidean scalar product, $\mathfrak{G} = SO(n)$ and $\mathcal{F} = \Omega \times F$. Every orthonormal basis $\mathbf{E} = (\mathbf{E}_1, ..., \mathbf{E}_n)$ of $\mathbb{R}^n$ gives a family of ("constant") sections such that $\omega_b^a + \omega_a^b = 0$, i.e. ω is an $so(n)$-valued form. Conversely, to every 1-form on Ω

$$a = a_x dx + a_y dy \tag{2.30}$$

whose coefficients a_x and a_y take values in $so(n)$, we can associate a connection ∇ on $\mathcal{F}$ (for which $\omega = a$). The curvature of this connection is

$$\Omega_\nabla = da + a \wedge a = \left(\frac{\partial a_y}{\partial x} - \frac{\partial a_x}{\partial y} + [a_x, a_y] \right) dx \wedge dy. \tag{2.31}$$

Thus, condition (2.27) means that ∇ has vanishing curvature. This yields that there are sections ϕ_i of $\mathcal{F}$ such that $\nabla \phi_i = 0$ and that $(\phi_1, ..., \phi_n)$ is a basis of $\mathbb{R}^n$. Moreover, we can chose this basis to be orthonormal (because if it is orthonormal at a point (x, y), it will be so everywhere, since a has anti-symmetric coefficients). The matrix v whose columns are the components of the ϕ_i belongs to $O(n)$ and satisfies the relation $dv = va$. Thus, we have a solution of the system (2.26).

Similarly if $\mathfrak{G} = U(n)$, a_x and a_y take values in $u(n)$.

Exercises

We identify $\mathbb{R}^4$ with the set of quaternions $\mathbb{H}$, $\mathbb{R}^3$ with the set of imaginary quaternions $\Im(\mathbb{H})$ and S^3 with the set of quaternions of norm 1 in $\mathbb{H}$. We recall that for any $R \in SO(3)$, there exists some $p \in S^3$ such that, $\forall \zeta \in \mathbb{R}^3$,

$$R(\zeta) = p\zeta\bar{p},$$

and that p is unique up to sign.

We denote by Ω a simply connected domain of $\mathbb{C}$.

2.2 Let $X : \Omega \longrightarrow \mathbb{R}^3$ be a conformal immersion into a surface of constant mean curvature $\frac{1}{2}$. In particular, X is a solution of the

equation
$$\Delta X = \frac{\partial X}{\partial x}\frac{\partial X}{\partial y} \text{ in } \mathbb{H}.$$

Let us consider the map from Ω to $so(3)$ defined by
$$A := \begin{pmatrix} 0 & X^3 & -X^2 \\ -X^3 & 0 & X^1 \\ X^2 & -X^1 & 0 \end{pmatrix},$$
and consider the connection 1-form on Ω
$$a = -\frac{\partial A}{\partial y}dx + \frac{\partial A}{\partial x}dy.$$

Show that $da + a \wedge a = 0$ and deduce that there exists a map $U : \Omega \longrightarrow SO(3)$ such that $dU = Ua$ and that U is harmonic and conformal.

2.3 Prove that there exists a smooth map $p : \Omega \longrightarrow S^3$ such that $\forall \zeta \in \mathbb{R}^3 \simeq \Im(\mathbb{H})$,
$$U(\zeta) = p\zeta\bar{p} \text{ in } \mathbb{H}.$$
Show that
$$\bar{p}dp = \frac{1}{2}\left(\frac{\partial X}{\partial y}dx - \frac{\partial X}{\partial x}dy\right) \text{ in } \mathbb{H},$$
and that p is a harmonic conformal immersion in S^3 and hence minimal. Prove that conversely one may associate to each conformal minimal immersion of Ω in S^3 a conformal constant mean curvature immersion in $\mathbb{R}^3$ (see [104]).

2.4 Let u be a harmonic map from Ω into S^n (for $n \geq 2$). Let $M = \begin{pmatrix} 0 & {}^tu \\ -u & b \end{pmatrix}$ as in in the previous section. Consider the connection form
$$\theta = -\frac{\partial M}{\partial y}dx + \frac{\partial M}{\partial x}dy.$$
Prove that $d\theta + \theta \wedge \theta = 0$ and thus that there exists a map $g : \Omega \longrightarrow SO(n+2)$ such that $dg = g\theta$. Show that g is harmonic

2.2 Harmonic maps with values into Lie groups

and conformal. In the following we will assume that $n = 2$.

2.5 Prove that for any rotation $R \in SO(4)$ there exists a pair $(p, q) \in S^3 \times S^3$ such that $\forall \zeta \in \mathbb{H}$,

$$R(\zeta) = p\zeta\bar{q} \text{ in } \mathbb{H}$$

and that (p, q) is unique up to sign.

2.6 Let $g : \Omega \longrightarrow SO(4)$ be the harmonic map constructed in question 2.4. Prove that one can associate to g a pair of maps $(p, q) : \Omega \longrightarrow S^3 \times S^3$ such that

$$g(z)(\zeta) = p(z)\zeta\overline{q(z)}, \forall \zeta \in \mathbb{H}, \forall z \in \Omega,$$

and that

$$dp = \frac{1}{2}p\left(\frac{\partial X_+}{\partial y}dx - \frac{\partial X_+}{\partial x}dy\right) \text{ in } \mathbb{H},$$

$$dq = \frac{1}{2}q\left(\frac{\partial X_-}{\partial y}dx - \frac{\partial X_-}{\partial x}dy\right) \text{ in } \mathbb{H},$$

where X_+ and X_- are the conformal constant mean curvature immersions constructed in subsection 2.1.1 (outside degenerate points). Show that p and q are conformal harmonic maps into S^3. Conclude.

2.2.1 Families of curvature-free connections

After all this talk about different formulations, we will take a big step and introduce an auxiliary complex parameter λ (which we may call a "spectral parameter", see the comments at the end of this section). For $\lambda \in \mathbb{C}^\star$, we write

$$\begin{aligned} A_\lambda &= \frac{1}{2}[(1 - \lambda^{-1})fdz + (1 - \lambda)\bar{f}d\bar{z}] \\ &= \frac{1}{2}\left(1 - \frac{\lambda + \lambda^{-1}}{2}\right)a - \frac{1}{2}\frac{\lambda - \lambda^{-1}}{2i} \star a, \end{aligned} \quad (2.32)$$

where a is given by (2.30). We remark that A_λ is a deformation of a, in the sense that

(i) $A_{-1} = a$

(ii) $A_1 = 0$
(iii) for $|\lambda| = 1$, A_λ is a 1-form on Ω with coefficients in $\mathfrak{g}$.
(iv) In general, for $\lambda \in \mathbb{C}^*$, A_λ is a 1-form on Ω with coefficients in $\mathfrak{g} \otimes \mathbb{C}$.

Let us calculate the curvature of A_λ, i.e.

$dA_\lambda + A_\lambda \wedge A_\lambda$

$$= \left(\frac{\partial A_{\lambda,y}}{\partial x} - \frac{\partial A_{\lambda,x}}{\partial y} + [A_{\lambda,x}, A_{\lambda,y}] \right) dx \wedge dy$$

$$= \left(\frac{\partial}{\partial z}(A_{\lambda,\bar{z}}) - \frac{\partial}{\partial \bar{z}}(A_{\lambda,z}) + [A_{\lambda,z}, A_{\lambda,\bar{z}}] \right) dz \wedge d\bar{z}$$

$$= \left(\frac{\partial}{\partial z}\left(\frac{1-\lambda}{2}\bar{f}\right) - \frac{\partial}{\partial \bar{z}}\left(\frac{1-\lambda^{-1}}{2}f\right) + \frac{2-\lambda-\lambda^{-1}}{4}[f,\bar{f}] \right) dz \wedge d\bar{z}$$

$$= -\frac{1-\lambda^{-1}}{2}\left(\frac{\partial f}{\partial \bar{z}} + \frac{1}{2}[\bar{f},f]\right) dz \wedge d\bar{z} + \frac{1-\lambda}{2}\overline{\left(\frac{\partial f}{\partial \bar{z}} + \frac{1}{2}[\bar{f},f]\right)} dz \wedge d\bar{z}.$$
(2.33)

Comparing this last expression with (2.29), we can immediately obtain

Theorem 2.2.1 *[128], [185], [174] Let u be a map from Ω to $\mathfrak{G}$, and A_λ the connection form on Ω, with coefficients in $\mathfrak{g} \otimes \mathbb{C}$, defined by (2.32). Then u is harmonic if and only if the vanishing curvature relation*

$$dA_\lambda + A_\lambda \wedge A_\lambda = 0 \qquad (2.34)$$

is true for any value of $\lambda \in \mathbb{C}^*$.

Remark 2.2.2 *By relation (2.33), it is easy to see that it suffices to have relation (2.34) true for at least two distinct values of λ, different from 0 and 1, for u to be harmonic.*

Theorem 2.2.1 yields an "integrated" version of the characterization of harmonic maps, for simply connected domains.

Theorem 2.2.3 *Suppose that Ω is simply connected. Choose an arbitrary point $p \in \Omega$, and let u be a map from Ω to $\mathfrak{G}$ such that $u(p) = \mathbb{1}$. Then, the following properties are equivalent.*

2.2 Harmonic maps with values into Lie groups

(i) u is harmonic.

(ii) For any $\lambda \in S^1$, there exists a map E_λ from Ω to $\mathfrak{G}$ such that

$$E_\lambda(p) = \mathbb{1} \qquad (2.35)$$

$$E_\lambda^{-1} dE_\lambda = \lambda^{-1} \hat{\xi}_{-1} + \hat{\xi}_0 + \lambda \hat{\xi}_1, \qquad (2.36)$$

where $\hat{\xi}_{-1}$, $\hat{\xi}_0$ and $\hat{\xi}_1$ are smooth 1-forms on Ω with coefficients in $\mathfrak{g} \otimes \mathbb{C}$, and $\hat{\xi}_{-1}\left(\frac{\partial}{\partial \bar{z}}\right) = 0$

$$E_1 = \mathbb{1} \qquad (2.37)$$

$$E_{-1} = u. \qquad (2.38)$$

(iii) For every $\lambda \in \mathbb{C}^*$, there exists a map E_λ from Ω to $\mathfrak{G}^\mathbb{C}$ such that (2.35), (2.36), (2.37) and (2.38) are satisfied, and moreover,

$$E_{\overline{\lambda}^{-1}} = \overline{E_\lambda}. \qquad (2.39)$$

Furthermore, for any $z \in \Omega$, $\lambda \longmapsto E_\lambda$ is holomorphic in $\mathbb{C}^*$.

Remark 2.2.4 *Here we denote by $SO(n)^\mathbb{C} := \{R \in M(n, \mathbb{C})/{}^t RR = \mathbb{1}, \det R = 1\}$ and $U(n)^\mathbb{C} := GL(n, \mathbb{C})$ the complexifications of $SO(n)$ and $U(n)$. A map $h : \mathbb{C}^* \longrightarrow \mathfrak{G}^\mathbb{C}$ is holomorphic if and only if $\frac{\partial h}{\partial \bar{z}} = 0$.*

On the other hand, the point of view which turns out to be most convenient is to see E_λ not as a family of maps from Ω to $\mathfrak{G}^\mathbb{C}$ parametrized by λ, but rather as a map from Ω to the based loop group (based, since they are equal to $\mathbb{1}$ at 1)

$$\Omega \mathfrak{G} = \{g_\bullet : S^1 \longrightarrow \mathfrak{G}/g_1 = \mathbb{1}\}.$$

Here and in the following we systematically denote by $g_\bullet$ a map depending on a spectral parameter in the circle S^1 (or in $\mathbb{C}^*$) and its value at some $\lambda \in S^1$ (or $\mathbb{C}^*$) will be denoted g_λ. The product of $g_\bullet$ and $h_\bullet \in \Omega\mathfrak{G}$ is defined by $(g_\bullet h_\bullet)_\lambda = g_\lambda h_\lambda$. It will also be useful below to consider the complexification of $\Omega\mathfrak{G}$

$$\Omega\mathfrak{G}^\mathbb{C} = \{g_\bullet : S^1 \longrightarrow \mathfrak{G}^\mathbb{C} \mid g_1 = \mathbb{1}\},$$

and the subgroups

$$\Omega_{hol}\mathfrak{G}^{\mathbb{C}} := \{g_\bullet \in \Omega\mathfrak{G}^{\mathbb{C}} \mid g_\bullet \text{ extends in a holomorphic way to } \mathbb{C}^*\},$$

$$\Omega_{hol}\mathfrak{G} := \Omega\mathfrak{G} \cap \Omega_{hol}\mathfrak{G}^{\mathbb{C}}.$$

We denote the Lie algebras of these groups by $\Omega\mathfrak{g}$, $\Omega\mathfrak{g}^{\mathbb{C}}$, $\Omega_{hol}\mathfrak{g}^{\mathbb{C}}$, $\Omega_{hol}\mathfrak{g}$. For instance,

$$\Omega\mathfrak{g} = \{\xi_\bullet : S^1 \longrightarrow \mathfrak{g} \mid \xi_1 = 0\},$$

and $[\xi_\bullet, \eta_\bullet]_\lambda = [\xi_\lambda, \eta_\lambda]$.

We denote by $E_\bullet$ the map

$$\begin{aligned} E_\bullet : \quad \Omega &\longrightarrow \Omega_{hol}\mathfrak{G} \subset \Omega\mathfrak{G} \\ z &\longmapsto [\lambda \longmapsto E_\lambda(z)]. \end{aligned}$$

This map is called an "extended harmonic map", although it is not a harmonic map.

Proof of theorem 2.2.3
Proof of (i) $\Rightarrow$ (ii) This follows essentially from the Frobenius theorem and theorem 2.2.1. In fact, we know that if u is harmonic,

$$A_\lambda = \frac{1}{2}\left[(1-\lambda^{-1})fdz + (1-\lambda)\overline{f}d\overline{z}\right]$$

has zero curvature hence for every $A \in S^1$, there exists map $E_\bullet$ from Ω to $\mathfrak{G}$ such that

$$dE_\lambda = E_\lambda A_\lambda. \tag{2.40}$$

Choosing $E_\bullet$ such that $E_\bullet(p) = \mathbb{1}$, E_λ will be unique.

In particular, since $A_{-1} = u^{-1}du$, and $E_\lambda(p) = u(p) = \mathbb{1}$, we obtain that $E_{-1} = u$. Likewise, since $A_1 = 0$, $E_1 = \mathbb{1}$. Finally, relation (2.36) follows from (2.40). We remark that it follows from Frobenius' theorem and the uniqueness of E_λ that $(z, \lambda) \longmapsto E_\lambda(z)$ is continuous.

Proof of (ii) $\Rightarrow$ (iii) Consider

$$A_\lambda = \lambda^{-1}\hat{\xi}_{-1} + \hat{\xi}_0 + \lambda\hat{\xi}_1 = E_\lambda^{-1}dE_\lambda.$$

2.2 Harmonic maps with values into Lie groups

The fact that $dA_\lambda + A_\lambda \wedge A_\lambda = 0$ for $\lambda \in S^1$ implies that this relation is valid for all $\lambda \in \mathbb{C}^*$. Repeating the preceding arguments, we can thus show that we can define E_λ, for $\lambda \in \mathbb{C}^*$, which extends E_λ on S^1, and satisfies (2.35), (2.36), (2.37) and (2.38). And since A_λ is a holomorphic function of λ in $\mathbb{C}^*$, the same is true for E_λ.

Condition (2.39) is called the *reality condition*. It is automatically true for $\lambda \in S^1$, since $E_\lambda(z) \in \mathfrak{G}$ for $\lambda \in S^1$. Thus, we know from (ii) that for all z, $\lambda \longmapsto \left(\overline{E_{\overline{\lambda}^{-1}}}\right)^{-1}.E_\lambda$ is an analytic function in $\mathbb{C}^*$ which takes the value $1\!\!1$ over S^1. Hence, this function is necessarily constant and equal to $1\!\!1$ in $\mathbb{C}^*$ by the analytic extension principle. This proves (2.39).

Proof of (iii) $\Rightarrow$ (i) Suppose E_λ is as described in (iii) and write

$$A_\lambda = E_\lambda^{-1} dE_\lambda = \lambda^{-1}\hat{\xi}_{-1} + \hat{\xi}_0 + \lambda\hat{\xi}_1 .$$

Condition (2.37) implies that

$$A_1 = 0,$$

and condition (2.39) that

$$\hat{\xi}_{-1} = \overline{\hat{\xi}_1} \quad \text{and} \quad \hat{\xi}_0 = \overline{\hat{\xi}_0} .$$

Thus, if we write $\alpha = -2\hat{\xi}_{-1}$, then

$$A_\lambda = \frac{1}{2}\left[(1-\lambda^{-1})\alpha + (1-\lambda)\overline{\alpha}\right] .$$

But since $\hat{\xi}_{-1}\left(\frac{\partial}{\partial \overline{z}}\right) = 0$, we deduce that there is a map f from Ω to $\mathfrak{g} \otimes \mathbb{C}$ such that

$$A_\lambda = \frac{1}{2}\left[(1-\lambda^{-1})f dz + (1-\lambda)\overline{f}d\overline{z}\right] .$$

It is then easy to see that

$$f = u^{-1}\frac{\partial u}{\partial z},$$

and since A_λ must satisfy the vanishing curvature condition, it follows that

$$\frac{\partial f}{\partial \overline{z}} + \frac{1}{2}[\overline{f}, f] = 0,$$

and thus u is harmonic (see relation (2.29)). □

To present all the results that have been obtained using the setting of theorem 2.2.3 would take us too far. The interested reader should see [24] for a Hamiltonian approach, or [25], [26], [174] or the books [62], [76], [89].

JACOBI FIELDS

The results we have chosen to briefly present here are those that illustrate the existence of infinitesimal symmetries, i.e. the symmetries of the linearized problem.

Given a harmonic map $u : \Omega \longrightarrow \mathfrak{G}$, the "harmonic deformations" of u are represented, to the first order, by a map $v : \Omega \longrightarrow M(n, \mathbb{R})$ such that

$$(u + tv)^{-1}(u + tv) = \mathbb{1} + o(t) \tag{2.41}$$

and

$$d[\star(u + tv)^{-1}d(u + tv)] = 0 \tag{2.42}$$

where $\star(\alpha_x dx + \alpha_y dy) = -\alpha_y dx + \alpha_x dy$. It is more convenient to write $v = u\Lambda$. Condition (2.41) then becomes

$$\Lambda \in \mathfrak{g}, \tag{2.43}$$

and we can see that (2.42) is equivalent to

$$d[\star(d\Lambda + [A, \Lambda])] = 0, \tag{2.44}$$

where

$$A = u^{-1}du = fdz + \overline{f}d\overline{z}.$$

Definition 2.2.5 *Let u be a harmonic map and $A = u^{-1}du$. By a Jacobi field we mean any infinitesimal deformation of u, $u + t\delta u$, which is a harmonic map "up to order 1 in t", or equivalently, any map $\Lambda : \Omega \longrightarrow \mathfrak{g}$ which is a solution of (2.44).*

2.2 Harmonic maps with values into Lie groups

Equation (2.44) has trivial solutions: the constant solutions (because $d \star A = 0$). These solutions reflect the existence of symmetries for the harmonic map equation. Other solutions, much less trivial but remarkable, have been found by Louise Dolan.

Theorem 2.2.6 [45], [174] Suppose $a \in \mathfrak{g}$. Then, for every $\lambda \in \mathbb{C}^*$, the map

$$\Lambda_\lambda(a): \Omega \longrightarrow \mathfrak{g} \otimes \mathbb{C}$$
$$z \longmapsto E_\lambda^{-1}(z) a E_\lambda(z) \qquad (2.45)$$

is a solution of (2.44). Consequently, the real and imaginary parts of $\Lambda_\lambda(a)$ are Jacobi fields for u.

By a complexified Jacobi field we mean a map which takes values in $\mathfrak{g} \otimes \mathbb{C}$, and is a solution of (2.44). Notice that a complexified Jacobi field is not a Jacobi field in general, since it does not preserve the reality condition $u \in \mathfrak{G}$.

Remark 2.2.7 We can show that the vector space generated by

$$\{\Re(\Lambda_\lambda(a))/\lambda \in \mathbb{C}^*, a \in \mathfrak{g} \otimes \mathbb{C}\}$$

has a graded Kac–Moody algebra structure (a generalization of Lie algebras). More precisely, this space can be identified with a set of formal series in one complex variable, with coefficients in $\mathfrak{g} \otimes \mathbb{C}$ (see [45], [174]). Such symmetries have also been observed for the self-dual Yang–Mills equation in $\mathbb{R}^4$ or S^4 (see [34], [41]). This Lie algebra action on the "tangent space" to the set of harmonic maps u is the infinitesimal version of the action of a symmetry group on harmonic maps ("dressing"). This result, shown by [174], will be used below. An analogous result for self-dual Yang–Mills was proved in [41].

Proof of theorem 2.2.6 In order to lighten the notation let $\Lambda_\lambda = \Lambda_\lambda(a)$. We start by calculating

$$\begin{aligned} d\Lambda_\lambda &= -E_\lambda^{-1}(dE_\lambda)E_\lambda^{-1} a E_\lambda + E_\lambda^{-1} a dE_\lambda \\ &= [E_\lambda^{-1} a E_\lambda, E_\lambda^{-1} dE_\lambda] \end{aligned}$$

$$= -[A_\lambda, \Lambda_\lambda], \qquad (2.46)$$

and hence

$$d\Lambda_\lambda = -\frac{1}{2}(1-\lambda^{-1})[f, \Lambda_\lambda]dz - \frac{1}{2}(1-\lambda)[\overline{f}, \Lambda_\lambda]d\overline{z}.$$

Thus

$$\begin{aligned} d\Lambda_\lambda + [A, \Lambda_\lambda] &= \left(\frac{\partial \Lambda_\lambda}{\partial z} + [f, \Lambda_\lambda]\right) dz + \left(\frac{\partial \Lambda_\lambda}{\partial \overline{z}} + [\overline{f}, \Lambda_\lambda]\right) d\overline{z} \\ &= \frac{1}{2}(1+\lambda^{-1})[f, \Lambda_\lambda]dz + \frac{1}{2}(1+\lambda)[\overline{f}, \Lambda_\lambda]d\overline{z}. \end{aligned} \quad (2.47)$$

Now since $\star dz = -idz$ and $\star d\overline{z} = id\overline{z}$, (2.47) implies

$$\begin{aligned} \star (d\Lambda_\lambda + [A, \Lambda_\lambda]) &= i\left(-\frac{1}{2}(1+\lambda^{-1})[f, \Lambda_\lambda]dz + \frac{1}{2}(1+\lambda)[\overline{f}, \Lambda_\lambda]d\overline{z}\right) \\ &= i\frac{1+\lambda}{1-\lambda}\left(\frac{1}{2}(1-\lambda^{-1})[f, \Lambda_\lambda]dz + \frac{1}{2}(1-\lambda)[\overline{f}, \Lambda_\lambda]d\overline{z}\right) \\ &= -i\frac{1+\lambda}{1-\lambda}d\Lambda_\lambda, \end{aligned}$$

which proves (2.44). $\qquad \square$

2.2.2 The dressing

We will now see that there is an infinite-dimensional Lie group acting on the set of harmonic maps. This action is called the "dressing action". It is based on three facts.

(i) There exists a (loop) group $\mathcal{G}$ acting on $\Omega_{hol}\mathfrak{G}$. For every $g_\bullet \in \mathcal{G}$, and $e_\bullet \in \Omega_{hol}\mathfrak{G}$, we denote by $g_\bullet \sharp e_\bullet$ the image of $e_\bullet$ in $\Omega_{hol}\mathfrak{G}$ under the action of $g_\bullet$.

(ii) This induces an action of $\mathcal{G}$ on the set of maps from an open subset Ω of $\mathbb{C}$ to $\Omega_{hol}\mathfrak{G}$. For all $g_\bullet \in \mathcal{G}$, and for all maps $E_\bullet$ from Ω to $\Omega_{hol}\mathfrak{G}$, we write

$$\begin{aligned} g_\bullet \sharp E_\bullet : \Omega &\longrightarrow \Omega_{hol}\mathfrak{G} \\ z &\longmapsto g_\bullet \sharp (E_\bullet(z)). \end{aligned}$$

(iii) Miracle: if $E_\bullet$ is an extended harmonic map, i.e. if it satisfies (2.36), (2.37), (2.38) and (2.39), then the same is true for $g_\bullet \sharp E_\bullet$.

2.2 Harmonic maps with values into Lie groups

CONCLUSION: translating this last fact with the help of theorem 2.2.3, we deduce that $\mathcal{G}$ acts on the set of harmonic maps. The group $\mathcal{G}$ is the set of germs at 0 of holomorphic maps from a neighborhood of 0 in $\mathbb{C}$ to $\mathfrak{G}^{\mathbb{C}}$. In other words, if for $\epsilon \in (0,1]$, we write

$$B_\epsilon = \{\zeta \in \mathbb{C} \mid |\zeta| < \epsilon\}, \tag{2.48}$$

then we can define

$$\mathcal{G} = \bigcup_{\epsilon > 0} \{g_\bullet : B_\epsilon \longrightarrow \mathfrak{G}^{\mathbb{C}} \mid [\zeta \longmapsto g_\zeta] \text{ is a holomorphic function of } \zeta\}.$$

The group law on this set is given by the pointwise product, i.e. in general, if X is a domain, a set of maps from X to $\mathfrak{G}$ or $\mathfrak{G}^{\mathbb{C}}$ is equipped with a group structure given by the product

$$\forall \zeta \in X, \quad (g_\bullet h_\bullet)_\zeta = g_\zeta h_\zeta.$$

We present here two special cases of the dressing action, simpler to describe, and corresponding to certain subgroups of $\mathcal{G}$: first

$L^+\mathfrak{G}^{\mathbb{C}} = \{g_\bullet : S^1 \longrightarrow \mathfrak{G}^{\mathbb{C}} \mid g_\bullet \text{ has a holomorphic extension to } B_1 \text{ taking values in } \mathfrak{G}^{\mathbb{C}}\}$,

and, in the case where $\mathfrak{G} = U(n)$,

$\mathcal{A}(S^2, U(n)) := \{g_\bullet : \mathbb{C} \cup \{\infty\} \longrightarrow GL(n,\mathbb{C}) \mid g_\bullet \text{ is meromorphic on } \mathbb{C} \cup \{\infty\}, \text{ holomorphic at 0 and } +\infty, g_{\overline{\lambda}^{-1}} = \overline{g_\lambda}, \text{ and } g_1 = 1\}$.

We remark that the inclusion of $\mathcal{A}(S^2, U(n))$ in $\mathcal{G}$ can be conceived modulo the restriction of a map in $\mathcal{A}(S^2, U(n))$ to a neighborhood of 0.

ACTION OF $L^+\mathfrak{G}^{\mathbb{C}}$

Define

$$L\mathfrak{G}^{\mathbb{C}} = \{g_\bullet : S^1 \longrightarrow \mathfrak{G}^{\mathbb{C}}\}.$$

Theorem 2.2.8 *[130] Assume that $\mathfrak{G} = SO(n)$ or $U(n)$. Then the product*

$$\begin{aligned} \Omega\mathfrak{G} \times L^+\mathfrak{G}^{\mathbb{C}} &\longrightarrow L\mathfrak{G}^{\mathbb{C}} \\ (e_\bullet, g_\bullet) &\longmapsto e_\bullet g_\bullet \end{aligned}$$

is a diffeomorphism. In particular, for all $v_\bullet \in L\mathfrak{G}^{\mathbb{C}}$, $\exists! e_\bullet \in \Omega\mathfrak{G}$, $\exists! g_\bullet \in L^+\mathfrak{G}^{\mathbb{C}}$,

$$v_\bullet = e_\bullet g_\bullet. \tag{2.49}$$

Remark 2.2.9 *The proof of this result would be easy if we just needed to show the linearized version around the identity: we consider*

$$\psi_\bullet \in L\mathfrak{g}^{\mathbb{C}} := \{\psi_\bullet : S^1 \longrightarrow \mathfrak{g} \otimes \mathbb{C}\},$$

and we look for

$$\gamma_\bullet \in L^+\mathfrak{g}^{\mathbb{C}} := \{\gamma_\bullet \in L\mathfrak{g}^{\mathbb{C}} \mid \gamma_\bullet \text{ extends holomorphically to } B_1\}$$

and

$$\phi_\bullet \in \Omega\mathfrak{g} = \{\phi_\bullet : S^1 \longrightarrow \mathfrak{g} \mid \phi_1 = 0\},$$

such that

$$(\mathbb{1} + t\psi_\bullet) = (\mathbb{1} + t\phi_\bullet).(\mathbb{1} + t\gamma_\bullet) + o(t)$$

which is equivalent to

$$\psi_\bullet = \phi_\bullet + \gamma_\bullet. \tag{2.50}$$

The proof of (2.50) reduces to showing that

$$L^+\mathfrak{g}^{\mathbb{C}} \oplus \Omega\mathfrak{g} = L\mathfrak{g}^{\mathbb{C}},$$

a result that can easily be obtained by doing a series expansion. On the contrary, the proof of (2.49) cannot be obtained by such a simple argument, and is a deep result.

Remark 2.2.10 *In theorem 2.2.8, it is necessary to choose a topology on $L\mathfrak{G}^{\mathbb{C}}$: the result is true if we consider loops in $\mathcal{C}^\infty$, or in $H^s(S^1, \mathfrak{G}^{\mathbb{C}})$, for $s > \frac{1}{2}$ (see [130]).*

2.2 Harmonic maps with values into Lie groups

Let us describe the dressing action of $L^+\mathfrak{G}^{\mathbb{C}}$ on $\Omega\mathfrak{G}$: suppose $g_\bullet \in L^+\mathfrak{G}^{\mathbb{C}}$; for any $e_\bullet \in \Omega\mathfrak{G}$,

$$g_\bullet e_\bullet \in L\mathfrak{G}^{\mathbb{C}},$$

and thus, by theorem 2.2.8, $\exists ! \widetilde{g}_\bullet \in L^+\mathfrak{G}^{\mathbb{C}}$, $\exists ! \widetilde{e}_\bullet \in \Omega\mathfrak{G}$,

$$g_\bullet e_\bullet = \widetilde{e}_\bullet \widetilde{g}_\bullet. \tag{2.51}$$

Notice that $\widetilde{g}_\bullet$ depends on $e_\bullet$ and $g_\bullet$ in general (unless $\mathfrak{G}$ is Abelian). We denote

$$g_\bullet \sharp e_\bullet = \widetilde{e}_\bullet = g_\bullet e_\bullet \widetilde{g}_\bullet^{-1} \in \mathfrak{G}. \tag{2.52}$$

(The visual aspect of this expression is the origin of the term "dressing", an English translation of the original Russian word.) The expected result is the following.

Theorem 2.2.11 *Let $u : \Omega \longrightarrow \mathfrak{G}$ be a harmonic map, and $E_\bullet : \Omega \longrightarrow \Omega\mathfrak{G}$ the associated extended harmonic map. Suppose $g_\bullet \in L^+\mathfrak{G}^{\mathbb{C}}$ then the map*

$$g_\bullet \sharp E_\bullet : \Omega \longrightarrow \Omega\mathfrak{G}$$
$$z \longmapsto g_\bullet \sharp E_\bullet(z)$$

is also an extended harmonic map. This implies that

$$g_\bullet \sharp u := g_\bullet \sharp E_{-1} \tag{2.53}$$

is still a harmonic map (by theorem 2.2.3).

Proof We start by noticing that the diffeomorphism property stated in theorem 2.2.8 implies that $g_\bullet \sharp E_\bullet$ is smooth in Ω. By (2.52), we have for every z

$$(g_\bullet \sharp E_\bullet)(z) = g_\bullet E_\bullet(z) \widetilde{g}_\bullet^{-1}(z),$$

and thus,

$$\begin{aligned}(g_\bullet \sharp E_\bullet)^{-1} d(g_\bullet \sharp E_\bullet) &= \widetilde{g}_\bullet E_\bullet^{-1} g_\bullet^{-1}(g_\bullet dE_\bullet \widetilde{g}_\bullet^{-1} - g_\bullet E_\bullet \widetilde{g}_\bullet^{-1} d\widetilde{g}_\bullet \widetilde{g}_\bullet^{-1}) \\ &= \widetilde{g}_\bullet (E_\bullet^{-1} dE_\bullet - \widetilde{g}_\bullet^{-1} d\widetilde{g}_\bullet) \widetilde{g}_\bullet^{-1}\,. \end{aligned} \quad (2.54)$$

Recall first that

$$(E_\bullet^{-1} dE_\bullet)_\lambda = E_\lambda^{-1} dE_\lambda = A_\lambda$$

is a linear combination of $\lambda^{-1}, 1$ and λ, and second that since $\widetilde{g}_\bullet \in L^+ \mathfrak{G}^{\mathbb{C}}$, both $\widetilde{g}_\lambda$ and $\widetilde{g}_\lambda^{-1}$ may be expressed as a non-negative power series in λ. Then (2.54) yields

$$(g_\bullet \sharp E_\lambda)^{-1} d(g_\bullet \sharp E_\lambda) = \sum_{k \geq -1} \hat{\xi}_k \lambda^k, \quad (2.55)$$

where each $\hat{\xi}_{(k)}$ is a linear form on Ω, and

$$\hat{\xi}_{-1} = \widetilde{g}_0(-\frac{1}{2} f) \widetilde{g}_0^{-1} dz\,. \quad (2.56)$$

Now, notice that $g_\bullet \sharp E_\bullet \in \Omega \mathfrak{G}$, and hence it satisfies the reality condition

$$g_\bullet \sharp E_{\overline{\lambda}^{-1}} = \overline{g_\bullet \sharp E_\lambda}\,.$$

This property, together with (2.55), implies that

$$(g_\bullet \sharp E_\lambda)^{-1} d(g_\bullet \sharp E_\lambda) = \hat{\xi}_{-1} \lambda^{-1} + \hat{\xi}_0 + \hat{\xi}_1 \lambda\,. \quad (2.57)$$

Recall also that if $g_\bullet \sharp E_\bullet \in \Omega \mathfrak{G}$ then $g_\bullet \sharp E_1 = \mathbb{1}$, and hence

$$\hat{\xi}_{-1} + \hat{\xi}_0 + \hat{\xi}_1 = 0\,. \quad (2.58)$$

We conclude that (2.56), (2.57) and (2.58) imply $g_\bullet \sharp E_\bullet$ is an extended harmonic map, which proves our result. $\square$

2.2.3 Uhlenbeck factorization for maps with values in $U(n)$

A drawback of the dressing action is that it is difficult to picture its effect without using the loop group machinery. An alternative possibility was proposed in [174] for the case $\mathfrak{G} = U(n)$. The interest of considering the subgroup $\mathcal{A}(S^2, U(n))$ defined above is that it is possible to describe its action "more explicitly", in particular by decomposing each element $g_\bullet \in \mathcal{A}(S^2, U(n))$ as a product of elementary factors

$$g_\bullet = g_{\bullet,1} \cdots g_{\bullet,j},$$

where each $g_{\bullet,j} \in \mathcal{A}(S^2, U(n))$ has one pole and one zero. This is the approach used in [174] by Uhlenbeck, who obtains in this way a constructive proof of

Theorem 2.2.12 *Let $g_\bullet \in \mathcal{A}(S^2, U(n))$ and $e_\bullet \in \Omega_{hol} U(n)$ then $\exists! \tilde{g}_\bullet \in \mathcal{A}(S^2, U(n))$, $\exists! \tilde{e}_\bullet \in \Omega_{hol} U(n)$ such that*

$$g_\bullet e_\bullet = \tilde{e}_\bullet \tilde{g}_\bullet \, .$$

More precisely, we may decompose $g_\bullet$ as $g_\bullet = g_{\bullet,1} ... g_{\bullet,k}$, where each factor $g_{\bullet,j} \in \mathcal{A}(S^2, U(n))$ is a homographic function, and obtain $\tilde{g}_\bullet$ as $\tilde{g}_\bullet = \tilde{g}_{\bullet,1} ... \tilde{g}_{\bullet,k}$ where each $\tilde{g}_{\bullet,j}$ is also a homographic function which may be calculated explicitly. We write $g_\bullet \sharp e_\bullet = \tilde{e}_\bullet$.

Notice that replacing the group $L^+U(n)^{\mathbb{C}}$ by $\mathcal{A}(S^2, U(n))$, we have extended the dressing given by theorem 2.2.8 to certain elements which do not belong to $L^+U(n)^{\mathbb{C}}$. This is possible because we are acting on the subgroup $\Omega_{hol}U(n)$ of $\Omega U(n)$. But the fact of having an action defined only over $\Omega_{hol}U(n)$ does not pose any problem for us since, by theorem 2.2.3, every extended harmonic map $E_\bullet$ takes its values in $\Omega_{hol}U(n)$. We can thus proceed as in theorem 2.2.11, and show that if $g_\bullet \in \mathcal{A}(S^2, U(n))$ and $E_\bullet : \Omega \longrightarrow \Omega_{hol}U(n)$ is an extended harmonic map, then $g_\bullet \sharp E_\bullet$ is also an extended harmonic map (see [174]).

Notice that the dressing action of $\Omega_{hol}U(n)$ extends to the group $\mathcal{G}$, by adapting the arguments of [112] and [26].

In order to clarify the connections between the dressing and Dolan's Jacobi fields, constructed in theorem 2.2.6, we will study the effect of the dressing action described in theorem 2.2.12 for $g \in \mathcal{A}(S^2, U(n))$ close to $\mathbb{1}$. To do so, let

$$\mathcal{A}(S^2, u(n)) := \{\xi_\bullet : \mathbb{C} \cup \{\infty\} \longrightarrow u(n) \otimes \mathbb{C} |$$
$$\xi_\bullet \text{ is meromorphic on } \mathbb{C} \cup \{\infty\},$$
$$\text{holomorphic at } 0 \text{ and } \infty,$$
$$\xi_{\overline{\lambda}^{-1}} = \overline{\xi_\lambda}, \text{ and } \xi_1 = 0\}$$

be the Lie algebra of $\mathcal{A}(S^2, U(n))$. Let $\xi_\bullet \in \mathcal{A}(S^2, u(n))$, and $u : \Omega \longrightarrow U(n)$ be a harmonic map. For small t,

$$g_{\bullet,t} = \mathbb{1} + t\xi_\bullet + o(t) \in \mathcal{A}(S^2, U(n)),$$

and by theorem 2.2.12, $\exists! \widetilde{g}_{\bullet,t} \in \mathcal{A}(S^2, U(n)), \exists! \widetilde{E}_{\bullet,t} \in \Omega_{hol} U(n)$, such that

$$g_{\bullet,t} E_\bullet = \widetilde{E}_{\bullet,t} \widetilde{g}_{\bullet,t}.$$

Writing $\widetilde{E}_{\bullet,t} = E_\bullet + t\delta E_\bullet + o(t)$ and $\widetilde{g}_{\bullet,t} = \mathbb{1} + t\widetilde{\xi}_\bullet + o(t)$, we obtain

$$\delta E_\bullet = \xi_\bullet E_\bullet - E_\bullet \widetilde{\xi}_\bullet.$$

Exercise (solved)

Consider

$$\xi_\lambda = \frac{1}{2} \left[\frac{(\lambda-1)(\gamma+1)}{2(\lambda-\gamma)} a + \frac{(\lambda-1)(\overline{\gamma}+1)}{2(\overline{\gamma}\lambda-1)} \overline{a} \right], \quad (2.59)$$

where $a \in u(n) \otimes \mathbb{C}, \gamma \in \mathbb{C}^*$. It is easy to show that $\xi_\bullet \in \mathcal{A}(S^2, u(n))$, and $\xi_\bullet$ has poles at γ and $\overline{\gamma}^{-1}$.

Let us try to guess $\widetilde{\xi}_\bullet$. Since $\widetilde{\xi}_\bullet \in \mathcal{A}(S^2, U(n))$, it is a rational function of λ. Since $\delta E_\bullet$ is holomorphic in $\mathbb{C}^*$, it is necessary that the poles which show up in $\xi_\bullet E_\bullet$ are cancelled when we subtract $E_\bullet \widetilde{\xi}_\bullet$, and thus $\widetilde{\xi}_\bullet$ has poles at γ and $\overline{\gamma}^{-1}$. If we also use the reality condition $\xi_{\overline{\lambda}^{-1}} = \overline{\xi_\lambda}$ and $\xi_1 = 0$, we obtain

$$\widetilde{\xi}_\lambda = -\frac{1}{2} \left[\frac{(\lambda-1)(\gamma+1)}{2(\lambda-\gamma)} E_\gamma^{-1} a E_\gamma + \frac{(\lambda-1)(\overline{\gamma}+1)}{2(\overline{\gamma}\lambda-1)} E_{\overline{\gamma}^{-1}}^{-1} \overline{a} E_{\overline{\gamma}^{-1}} \right].$$

Thus,

$$\delta E_\lambda = \frac{1}{2} \left[\frac{(\lambda-1)(\gamma+1)}{2(\lambda-\gamma)} (aE_\lambda - E_\lambda E_\gamma^{-1} aE_\gamma) \right.$$
$$\left. + \frac{(\lambda-1)(\bar{\gamma}+1)}{2(\bar{\gamma}\lambda-1)} (\bar{a}E_\lambda - E_\lambda E_{\bar{\gamma}^{-1}}^{-1} \bar{a} E_{\bar{\gamma}^{-1}}) \right]. \quad (2.60)$$

The same reasoning with ξ_λ replaced by

$$\eta_\lambda = \frac{1}{2i} \left[\frac{(\lambda-1)(\gamma+1)}{2(\lambda-\gamma)} a - \frac{(\lambda-1)(\bar{\gamma}+1)}{2(\bar{\gamma}\lambda-1)} \bar{a} \right] \quad (2.61)$$

yields

$$\delta E_\lambda = \frac{1}{2i} \left[\frac{(\lambda-1)(\gamma+1)}{2(\lambda-\gamma)} (aE_\lambda - E_\lambda E_\gamma^{-1} aE_\gamma) \right.$$
$$\left. - \frac{(\lambda-1)(\bar{\gamma}+1)}{2(\bar{\gamma}\lambda-1)} (\bar{a}E_\lambda - E_\lambda E_{\bar{\gamma}^{-1}}^{-1} \bar{a} E_{\bar{\gamma}^{-1}}) \right]. \quad (2.62)$$

In equations (2.60) and (2.62), we constructed "extended Jacobi fields", i.e. infinitesimal deformations of an extended harmonic map $E_\bullet$. Specifying these relations for the case $\lambda = -1$, we obtain Jacobi fields for u, which are respectively

$$\delta u = \Re(au - uE_\gamma^{-1} aE_\gamma),$$

and

$$\delta u = \Im(au - uE_\gamma^{-1} aE_\gamma).$$

Thus, we recover Dolan's result (theorem 2.2.8).

2.2.4 S^1-action

Another consequence of the formulation using $\Omega_{hol}\mathfrak{G}$ is that it enables us to show that the group S^1 acts on the set of harmonic maps u from Ω to $\mathfrak{G}$, generating a family, parametrized by S^1, of deformations of a harmonic map u. For $\gamma \in S^1$, let

$$\widetilde{E_\lambda} := (\gamma \sharp E)_\lambda := E_{\lambda\gamma} E_\gamma^{-1}.$$

Theorem 2.2.13 *For any $\gamma \in S^1$, the map $\widetilde{f} : \Omega \longrightarrow \mathfrak{g} \otimes \mathbb{C}$ defined by*

$$\widetilde{f} = \gamma^{-1} E_\gamma f E_\gamma^{-1}$$

is such that

$$\widetilde{E}_\lambda^{-1} d\widetilde{E}_\lambda = \frac{1}{2}\left[(1-\lambda^{-1})\widetilde{f}dz + (1-\lambda)\overline{\widetilde{f}}d\bar{z}\right], \tag{2.63}$$

and hence, in particular, $\widetilde{E}_\lambda$ is an extended harmonic map, and $\gamma \sharp u = (\gamma \sharp E)_{-1}$ is a harmonic map from Ω to $\mathfrak{G}$.

Proof A direct calculation yields

$$\widetilde{E}_\lambda^{-1} \frac{\partial \widetilde{E}_\lambda}{\partial z} = \gamma^{-1} \frac{1-\lambda^{-1}}{2} E_\gamma f E_\gamma^{-1}$$

$$\widetilde{E}_\lambda^{-1} \frac{\partial \widetilde{E}_\lambda}{\partial \bar{z}} = \gamma \frac{1-\lambda}{2} E_\gamma \overline{f} E_\gamma^{-1},$$

which implies (2.63). □

CARTAN IMMERSION OF S^{n-1} IN $SO(n)$

Having in mind the study of S^{n-1}-valued harmonic maps, it is interesting to notice that there exists an immersion of S^{n-1} in $SO(n)$, given by

$$\begin{aligned} \mathcal{C}: S^{n-1} &\longrightarrow SO(n) \\ v &\longmapsto u_0(2v\,{}^tv - \mathbb{1}), \end{aligned}$$

such that for every map $v : \Omega \subset \mathbb{C} \longrightarrow S^{n-1}$, v is harmonic if and only if $\mathcal{C} \circ v$ is harmonic. Here u_0 is just a constant in $O(n)$, chosen so that $u \in SO(n)$ (det $u_0 = (-1)^{n-1}$). In fact, writing $u = \mathcal{C} \circ v$, we have

$$u^{-1}du = 2(v\,{}^tdv - dv\,{}^tv) = 2v \times dv, \tag{2.64}$$

where we use the notation of the previous section. It follows that

$$d(\star u^{-1}du) = 2d(\star v \times dv),$$

2.2 Harmonic maps with values into Lie groups

which gives the equivalence between the harmonicity of v and that of u.

We can interpret this using a general property characteristic of *totally geodesic maps*. A map $\phi : (\mathcal{N}, h) \longrightarrow (\mathcal{N}', h')$ is totally geodesic if and only if its second fundamental form

$$(\nabla d\phi)^i_{IJ} = \frac{\partial^2 \phi^i}{\partial y^I \partial y^J} - \Gamma^K_{IJ} \frac{\partial \phi^i}{\partial y^K} + \Gamma'^i_{jk} \frac{\partial \phi^j}{\partial y^I} \frac{\partial \phi^k}{\partial y^J}$$

vanishes. (Here Γ^K_{IJ} is the Christoffel symbol on $(\mathcal{N}, h)$ and Γ'^i_{jk} is the Christoffel symbol on $(\mathcal{N}', h')$.) But for every harmonic map $u : (\mathcal{M}, g) \longrightarrow (\mathcal{N}, h)$ and for every totally geodesic map $\phi : (\mathcal{N}, h) \longrightarrow (\mathcal{N}', h')$, the composition $\phi \circ u$ is still harmonic [51]. And here, $\mathcal{C}$ is precisely a totally geodesic map.

It follows that the preceding results concerning $SO(n)$-valued harmonic maps are still valid if the image is a sphere. In particular, the image of S^{n-1} by $\mathcal{C}$ in $SO(n)$ is preserved by the different groups which act on harmonic maps.

Nevertheless, in order to study harmonic maps taking values in the sphere (or any other homogeneous manifold), it is more convenient to use an alternative formulation which we will present in the following section.

A mystery remains: what is the meaning of the parameter λ? Although the importance of this parameter is obvious, its meaning is not clear. It is sometimes called the spectral parameter. We then refer to an analogous formulation of the Korteweg–de Vries or non-linear Schrödinger equations (see [105], [56]). For these equations, we associate to the unknown function an operator whose evolution is governed by the Lax equation. The spectrum of this operator has the property of being conserved in time, and λ then plays the role of an eigenvalue.

Other authors call λ a twistor parameter: this is a comparison with self-dual Yang–Mills connections on S^4 or $\mathbb{R}^4$. In fact, in this setting, we are led to similar equations by lifting the equation to the bundle of complex structures on S^4 or $\mathbb{R}^4$. This bundle has fiber $S^2 \simeq \mathbb{C} \cup \{\infty\}$, and in the case of S^4 it is identified with projective space $P\mathbb{C}^3$. We are then in a well understood situation, thanks to Penrose's twistor theory ([126]). By Ward's construction, we obtain a system of equations where an extra variable $\lambda \in \mathbb{C}$ appears: it is precisely a coordinate on the fiber of the complex structure bundle (see [175], [3], [41]).

But in the case of harmonic maps on a surface, I do not know of a precise "geometric" interpretation for λ.

2.3 Harmonic maps with values into homogeneous spaces

A formulation analogous to that we just saw, for maps taking values in a quotient of Lie groups $\mathfrak{G}/\mathfrak{K}$, where $\mathfrak{K}$ is a Lie subgroup of the Lie group $\mathfrak{G}$, was developed in [25], [26], [46]. Notice that it is always possible to embed $\mathfrak{G}/\mathfrak{K}$ in $\mathfrak{G}$, in such a way that the composition of a harmonic map taking values in $\mathfrak{G}/\mathfrak{K}$ with this embedding is a $\mathfrak{G}$-valued harmonic map (see [27]), and this enables us to use the preceding results. But the following new formulation turns out to be more convenient. We will present it for the case where

$$\mathfrak{G} := SO(n+1), \ \mathfrak{K} := SO(n) \text{ and } \mathfrak{G}/\mathfrak{K} \simeq S^n.$$

We start by defining the quotient $SO(n+1)/SO(n)$. The simplest way is to consider

$$SO(n) \simeq \mathfrak{K} := \left\{ \begin{pmatrix} 1 & 0 \\ 0 & R \end{pmatrix} \mid R \in SO(n) \right\} \subset SO(n+1).$$

But we can also define $\mathfrak{K}$ in a more "intrinsic" way by considering

$$P := \begin{pmatrix} 1 & 0 \\ 0 & -\mathbb{1}_n \end{pmatrix} \in O(n+1),$$

and letting

$$SO(n) \simeq \mathfrak{K} := \{ g \in \mathfrak{G} \mid Pg - gP = 0 \}.$$

An equivalent way to define $\mathfrak{K}$ is to introduce the canonical basis $(\mathbf{E}_0, ..., \mathbf{E}_n) = \mathbf{E}$ of $\mathbb{R}^{n+1}$, and to let

$$\mathfrak{K} := \{ g \in \mathfrak{G} \mid g(\mathbf{E}_0) = \mathbf{E}_0 \}.$$

In all cases the result is the same, and it is clear that $\mathfrak{K}$ is a subgroup of $\mathfrak{G}$. Then, we define the following equivalence relation in $\mathfrak{G}$:

$$g \mathcal{R} g' \iff g^{-1} g' \in \mathfrak{K}.$$

We define, for each $g \in \mathfrak{G}$,

$$[g] := g.\mathfrak{K},$$

the class of g, and

2.3 Harmonic maps with values into homogeneous spaces

$$\mathfrak{G}/\mathfrak{K} := \{[g] \mid g \in \mathfrak{G}\}.$$

Thanks to the introduction of the basis $\mathbf{E}$ on $\mathbb{R}^{n+1}$, we can easily see that this space is diffeomorphic to S^n: it suffices to realize that the map

$$\begin{aligned} \mathfrak{G} &\longrightarrow S^n \\ g &\longmapsto g(\mathbf{E}_0) \end{aligned}$$

yields, by passing to the quotient, a diffeomorphism between $\mathfrak{G}/\mathfrak{K}$ and S^n.

The idea is to represent a map u from $\Omega \subset \mathbb{C}$ to S^n (not necessarily harmonic) by a lifting into $\mathfrak{G}$, i.e. a map

$$F : \Omega \longrightarrow \mathfrak{G},$$

such that $\forall z \in \Omega$, $[F(z)] = u(z)$. If Ω is contractible, which we will assume in the sequel, there is no problem in constructing such a lifting. We remark that giving F corresponds to specifying

$$u = e_0 = \mathbf{E}_k F_0^k \in S^n,$$

and

$$e_j = \mathbf{E}_k F_j^k, \text{ for } k = 1, ..., n,$$

i.e. an orthonormal basis $(e_1, ..., e_n)$ of $T_u S^n$. Both preceding relations are condensed in

$$e = (e_0, e_1, ..., e_n) = (\mathbf{E}_0, ..., \mathbf{E}_n) F = \mathbf{E} F.$$

We should now study the "movement" of u as a function of the variable $z = x + iy$. For this, it is convenient to introduce the 1-form on Ω, with coefficients in $\mathfrak{g}$, given by

$$\theta = F^{-1} dF = {}^t F dF,$$

called the Maurer–Cartan form. We can think of θ as an anti-symmetric matrix whose elements are

$$\theta_j^i = F_i^k dF_j^k = \langle \mathbf{E}_k F_i^k, \mathbf{E}_l dF_j^l \rangle = \langle e_i, de_j \rangle.$$

For $j = 0$, we write

$$\theta_0^i = \langle e_i, du \rangle = \alpha^i,$$

and if $i, j \neq 0$,

$$\theta_j^i = \langle e_i, de_j \rangle = \omega_j^i.$$

We can also introduce the column vector of 1-forms

$$\alpha = \begin{pmatrix} \alpha^1 \\ \vdots \\ \alpha^n \end{pmatrix} = \begin{pmatrix} \theta_0^1 \\ \vdots \\ \theta_0^n \end{pmatrix},$$

and the anti-symmetric $n \times n$ matrix,

$$\omega = \begin{pmatrix} 0 & \omega_2^1 & \cdots & \omega_n^1 \\ \omega_1^2 & 0 & \cdots & \omega_n^2 \\ \vdots & \vdots & & \vdots \\ \omega_1^n & \omega_2^n & \cdots & 0 \end{pmatrix}.$$

In this way,

$$\theta = \begin{pmatrix} 0 & -{}^t\alpha \\ \alpha & \omega \end{pmatrix}.$$

We remark that the information contained in α, and that contained in ω, are of different kinds: α represents the projection of du in the basis $(e_1, ..., e_n)$, while ω describes the movement of this basis. This is why sometimes α is compared with a Higgs field and ω with a connection (see [94]). It is then natural to decompose θ along

$$\theta_1 = \begin{pmatrix} 0 & -{}^t\alpha \\ \alpha & 0 \end{pmatrix},$$

representing du in the moving frame $(e_1, ..., e_n)$, and

$$\theta_0 = \begin{pmatrix} 0 & 0 \\ 0 & \omega \end{pmatrix},$$

representing the connection.

Such a decomposition has a simple Lie algebra interpretation. The automorphism of $\mathfrak{G}$

$$\tau : g \longmapsto PgP^{-1}$$

induces a Lie algebra automorphism in $\mathfrak{g}$:

$$\tau : a \longmapsto PaP^{-1},$$

which also has the property of being a linear operator on $\mathfrak{g}$ with square equal to $\mathbb{1}$. Thus, τ diagonalizes over $\mathfrak{g}$, with eigenvalues 1 and -1. Let

$$\mathfrak{g}_0 = \{a \in \mathfrak{g} \mid \tau(a) = a\},$$

and

$$\mathfrak{g}_1 = \{a \in \mathfrak{g} \mid \tau(a) = -a\},$$

the corresponding eigenspaces. In fact, $\mathfrak{g}_0$ is the Lie algebra $\mathfrak{k} \simeq so(n)$ of $\mathfrak{K} \simeq SO(n)$. It is clear (once you think of it) that θ_0 is a 1-form with coefficients in $\mathfrak{g}_0$, and θ_1 is a 1-form with coefficients in $\mathfrak{g}_1$. Hence, the decomposition $\theta = \theta_0 + \theta_1$ follows from the decomposition

$$\mathfrak{g} = \mathfrak{g}_0 \oplus \mathfrak{g}_1.$$

Remark 2.3.1 *Suppose that a, b are two elements of $\mathfrak{g}_0$; then*

$$P[a, b]P^{-1} = [PaP^{-1}, PbP^{-1}] = [a, b],$$

and thus $[a, b]$ is also an element of $\mathfrak{g}_0$. By the same reasoning, we can also see easily that $[\mathfrak{g}_0, \mathfrak{g}_1] \subset \mathfrak{g}_1$, and $[\mathfrak{g}_1, \mathfrak{g}_1] = \mathfrak{g}_0$ (we have a $\mathbb{Z}_2$-graduation of $\mathfrak{g}$).

In order to write the structure equations for θ, expressing the fact that θ "derives" from F, we will introduce the following notation. If

$$\beta = \beta_x dx + \beta_y dy \text{ and } \gamma = \gamma_x dx + \gamma_y dy$$

are two 1-forms with matrix coefficients, we let

$$\begin{aligned}
[\beta \wedge \gamma] &= [\beta_x, \gamma_y] dx \wedge dy + [\beta_y, \gamma_x] dy \wedge dx \\
&= ([\beta_x, \gamma_y] + [\gamma_x, \beta_y]) dx \wedge dy \\
&= [\gamma \wedge \beta].
\end{aligned}$$

Now, using the fact that $\theta = F^{-1}dF$, we obtain

$$d\theta + \frac{1}{2}[\theta \wedge \theta] = 0. \tag{2.65}$$

Projecting this equation over $\mathfrak{g}_0$ and $\mathfrak{g}_1$, respectively, we obtain (using the $\mathbb{Z}_2$-graduation of $\mathfrak{g}$)

$$d\theta_0 + \frac{1}{2}[\theta_0 \wedge \theta_0] + \frac{1}{2}[\theta_1 \wedge \theta_1] = 0, \tag{2.66}$$

$$d\theta_1 + [\theta_0 \wedge \theta_1] = 0. \tag{2.67}$$

Remark 2.3.2 *Using the variables α and ω, we obtain an equivalent system*

$$d\alpha^i + \omega^i_j \wedge \alpha^j = 0, \tag{2.68}$$

$$d\omega^i_j + \omega^i_k \wedge \omega^k_j = \alpha^i \wedge \alpha^j, \tag{2.69}$$

or again

$$d\alpha + \omega \wedge \alpha = 0, \tag{2.70}$$

$$d\omega + \omega \wedge \omega = \alpha \wedge {}^t\alpha, \tag{2.71}$$

i.e. the Cartan system of equations.

Suppose that u is harmonic: what are the extra equations that we can obtain? Denote by $\star$ the Hodge star operator, whose action on 1-forms is described by

$$\star(\beta_x dx + \beta_y dy) = -\beta_y dx + \beta_x dy,$$

and observe that

$$\Delta u \, dx \wedge dy = d(\star du).$$

Since we have to impose on u the fact that Δu is everywhere orthogonal to $T_u S^n$, we obtain

2.3 Harmonic maps with values into homogeneous spaces

$$\langle d(\star du), e_i \rangle = 0, \quad \forall i = 1, ..., n.$$

This equation is equivalent to

$$d(\star \langle du, e_i \rangle) + \langle (\star du) \wedge de_i \rangle = 0, \quad \forall i = 1, ..., n. \tag{2.72}$$

We can formulate equation (2.72) in two ways:

$$d(\star \alpha^i) + \omega^i_j \wedge (\star \alpha^j) = 0, \quad \forall i = 1, ..., n, \tag{2.73}$$

or

$$d(\star \theta_1) + [\theta_0 \wedge (\star \theta_1)] = 0. \tag{2.74}$$

According to [25] and [46], we can condense equations (2.66), (2.67) and (2.74) by introducing the parameter $\lambda \in \mathbb{C}^*$ and writing

$$\Theta_\lambda := \lambda^{-1}\theta'_1 + \theta_0 + \lambda \theta_1'', \tag{2.75}$$

where

$$\begin{aligned}
\theta'_1 &= \frac{\theta_1 - i(\star\theta_1)}{2} \\
&= \frac{1}{2}\left(\theta_1\left(\frac{\partial}{\partial x}\right) - i\theta_1\left(\frac{\partial}{\partial y}\right)\right)(dx + idy) \\
&= \theta_1\left(\frac{\partial}{\partial z}\right) dz, \\
\theta_1'' &= \theta_1\left(\frac{\partial}{\partial \bar{z}}\right) d\bar{z} = \overline{\theta'_1}.
\end{aligned}$$

Theorem 2.3.3 *For any map u from Ω, an open subset of $\mathbb{C}$, to S^n, u is harmonic if and only if for every $\lambda \in \mathbb{C}^*$,*

$$d\Theta_\lambda + \frac{1}{2}[\Theta_\lambda \wedge \Theta_\lambda] = 0. \tag{2.76}$$

Proof By adding and subtracting equation (2.67) and i times equation (2.74), we obtain that (2.67) and (2.74) are equivalent to

$$d\theta_1' + [\theta_0 \wedge \theta_1'] = 0, \tag{2.77}$$

$$d\theta_1'' + [\theta_0 \wedge \theta_1''] = 0. \tag{2.78}$$

(Notice that these two equations are in fact equivalent by complex conjugation.) On the other hand, a direct calculation yields

$$\begin{aligned}d\Theta_\lambda + \frac{1}{2}[\Theta_\lambda \wedge \Theta_\lambda] &= \lambda^{-1}(d\theta_1' + [\theta_0 \wedge \theta_1']) \\ &\quad + d\theta_0 + \frac{1}{2}[\theta_0 \wedge \theta_0] + \frac{1}{2}[\theta_1 \wedge \theta_1] \\ &\quad + \lambda(d\theta_1'' + [\theta_0 \wedge \theta_1'']),\end{aligned} \tag{2.79}$$

which shows that (2.76) is true if and only if (2.66), (2.67) and (2.74) are all true. $\square$

As in the previous section, condition (2.76) allows us to integrate Θ_λ over any simply connected domain Ω. This leads to a new formulation of harmonic maps using extended harmonic liftings taking their values in twisted loop groups. The translation between the two problems is based on the following result.

Theorem 2.3.4 *[46] Let u be a map from Ω to S^n. Suppose that $u(p) = \mathbf{E}_0$, where p is a fixed point in Ω. Then,*

(i) *If, for any $\lambda \in S^1$, there exists a map F_λ from Ω to $\mathfrak{G}$ such that*

$$[F_1] = u, \tag{2.80}$$

$$F_\lambda^{-1} dF_\lambda = \lambda^{-1}\hat{a}_{-1} + \hat{a}_0 + \lambda\hat{a}_1, \tag{2.81}$$

where $\hat{a}_{-1} = \xi dz$, $\hat{a}_0$ is a 1-form with coefficients in $\mathfrak{g}_0$, $\hat{a}_1 = \bar\xi d\bar z$ and ξ is a map from Ω to $\mathfrak{g}_1 \otimes \mathbb{C}$, then u is harmonic.

(ii) *Conversely, if Ω is simply connected and u is harmonic, then for any $\lambda \in S^1$, there exists a map F_λ from Ω to $\mathfrak{G}$ such that (2.80) are (2.81) are satisfied. Moreover, we can choose F_λ such that*

$$F_\lambda(p) = \mathbb{1}, \quad \forall \lambda \in S^1, \tag{2.82}$$

2.3 Harmonic maps with values into homogeneous spaces   89

and F_λ will then be unique. F_λ is called an *extended harmonic lifting*.

Proof

(i) Let u and F_λ be such that (2.80) and (2.81) are true. Then, (2.80) expresses the fact that $F := F_1$ is a lifting of u and it follows from (2.81) (for $\lambda = 1$) that $\hat{a}_{-1} = \theta_1'$, $\hat{a}_0 = \theta_0$ and $\hat{a}_1 = \theta_1''$. This yields that $\Theta_\lambda = \lambda^{-1}\theta_1' + \theta_0 + \lambda\theta_1'' = F_\lambda^{-1}dF_\lambda$ satisfies equation (2.76), and hence that u is harmonic by theorem 2.3.3.

(ii) Conversely, if u is harmonic, we can construct a lifting F of u, such that $F(p) = \mathbb{1}$ (because $u(p) = \mathbf{E}_0$), and construct from F a 1-form Θ_λ, depending on the parameter $\lambda \in S^1$ as in (2.75), which is a solution of (2.76), thanks to theorem 2.3.3. Using Frobenius' theorem and the fact that Ω is simply connected, we deduce that there exists an unique map F_λ from Ω to $\mathfrak{G}$ such that

$$dF_\lambda = F_\lambda \Theta_\lambda, \qquad (2.83)$$

satisfying the initial condition (2.82). It is then easy to see that F_λ satisfies (2.81). □

Corollary 2.3.5 *Let $F_\lambda : \Omega \longrightarrow \mathfrak{G}$ be given by the preceding theorem. Then, for any value of $\lambda \in S^1$, the map $u_\lambda := [F_\lambda]$ is a harmonic map from Ω to S^n.*

Proof It suffices to apply case (i) of theorem 2.3.4 with $\widetilde{F_\lambda} = F_{\lambda/\gamma}$, for any $\gamma \in S^1$. □

In this way, we recover the fact that any harmonic map may be continuously deformed into a family of harmonic maps parametrized by S^1. The relation between theorem 2.2.13, seen earlier, and corollary 2.3.5 will be discussed in section 2.4 of this chapter.

There is an elegant way of characterizing the maps F_λ from Ω to $\mathfrak{G}$ depending on the parameter $\lambda \in S^1$ that satisfy (2.81). First of all, it is convenient to consider the map $F_\bullet$ from Ω to the loop group

$$L\mathfrak{G} := \{g_\bullet : S^1 \longrightarrow \mathfrak{G}\},$$

given by

$$F_\bullet : \Omega \longrightarrow L\mathfrak{G}$$
$$z \longmapsto [\lambda \longmapsto F_\lambda(z)].$$

If we impose on any map $F_\bullet$ from Ω o $L\mathfrak{G}$ the condition

for every fixed z, $\lambda \longmapsto \lambda F_\lambda(z)^{-1} dF_\lambda(z)$
extends holomorphically to the disk $|\lambda| < 1$, (2.84)

then we will have

$$F_\lambda^{-1} dF_\lambda = \sum_{k=-1}^{+\infty} \hat{a}_k \lambda^k, \qquad (2.85)$$

where each $\hat{a}_k$ is a 1-form on Ω with coefficients in $\mathfrak{g} \otimes \mathbb{C}$. But knowing that F_λ is $\mathfrak{G}$-valued for $|\lambda| = 1$, we have the reality condition

$$F_{\bar{\lambda}^{-1}} = \overline{F_\lambda},$$

which implies, using (2.85),

$$F_\lambda^{-1} dF_\lambda = \sum_{k=-1}^{1} \hat{a}_k \lambda^k. \qquad (2.86)$$

Let us impose on $F_\bullet$ the "twist" condition

$$\tau(F_\lambda) = P F_\lambda P = F_{-\lambda}. \qquad (2.87)$$

Then, we see that (2.86) and (2.87) are equivalent to (2.86) plus the condition that $\hat{a}_{-1}$ and $\hat{a}_1$ have coefficients in $\mathfrak{g}_1 \otimes \mathbb{C}$, and $\hat{a}_0$ in $\mathfrak{g}_0$.

Therefore, if we add to the hypotheses (2.84) and (2.87) the condition that

$$\hat{a}_{-1}\left(\frac{\partial}{\partial \bar{z}}\right) = 0, \qquad (2.88)$$

then $F_\bullet$ satisfies condition (2.81) of theorem 2.3.4, and thus is an extended harmonic lifting.

Hence, we are led to introduce the set

2.3 Harmonic maps with values into homogeneous spaces

$$L\mathfrak{G}_\tau := \{g_\bullet : S^1 \longrightarrow \mathfrak{G} \mid \tau(g_\lambda) = g_{-\lambda}\}.$$

Note that, since $\tau^2(g_\bullet) = g_\bullet$, it is obvious that $L\mathfrak{G}_\tau$ is a subgroup of $L\mathfrak{G}$, called the twisted loop group.

As a result, we have the following reformulation of theorem 2.3.4.

Theorem 2.3.6 *Let $F_\bullet$ be a map from Ω to $L\mathfrak{G}_\tau$ such that*

$$F_\lambda^{-1} dF_\lambda = \sum_{k=-1}^{+\infty} \hat{a}_k \lambda^k, \qquad (2.89)$$

where $\hat{a}_{-1}\left(\frac{\partial}{\partial \bar{z}}\right) = 0$. Then, $F_\bullet$ is an extended harmonic lifting, i.e. $u := [F_1]$ is an S^n-valued harmonic map. Conversely, if Ω is simply connected, any harmonic map from Ω to S^n can be obtained in this way.

We are no longer very far from being able to prove the following result: every harmonic lifting (and hence every harmonic map) may be obtained from a holomorphic map g from Ω to the complexified loop group $L\mathfrak{G}^\mathbb{C}$, such that

$$g_\lambda^{-1} dg_\lambda = \sum_{k=-1}^{+\infty} \hat{\xi}_k(z) \lambda^k dz,$$

where each $\hat{\xi}_k$ is a holomorphic function of z in Ω, taking values in $so(n+1) \otimes \mathbb{C}$. This means that there exists an (algebraic) algorithm which enables us to produce any harmonic map from holomorphic data, i.e. from solutions of a linear elliptic problem. Such a construction, due to Joseph Dorfmeister, Franz Pedit and HongYu Wu, makes true the dream of constructing a Weierstrass-type representation, analogous to that for minimal surfaces.

But the algorithm is clearly more complicated than for minimal surfaces: we will need a version of the decomposition theorem 2.2.8 for the complexified loop group, which is adapted to the case of twisted loop groups. An ingredient we need to add is the Iwasawa decomposition. Let $\mathfrak{K}^\mathbb{C}$ be the complexified group of $\mathfrak{K}$,

$$\mathfrak{K}^\mathbb{C} = \left\{ \begin{pmatrix} 1 & 0 \\ 0 & R \end{pmatrix} \mid R \in SO(n)^\mathbb{C} \right\}.$$

Lemma 2.3.7 (Iwasawa decomposition) *(see [90]). There exists a solvable subgroup $\mathfrak{B}$ of $\mathfrak{K}^{\mathbb{C}}$ such that*

$$\forall g \in \mathfrak{K}^{\mathbb{C}}, \ \exists! h \in \mathfrak{K}, \ \exists! b \in \mathfrak{B},$$

$$g = hb.$$

Remark 2.3.8 *In our case, we can choose as $\mathfrak{B}$ the subgroup of $\mathfrak{K}^{\mathbb{C}}$ that leaves invariant the subsets of $\mathbb{C}^{n+1}$*

$$V_k := (0, +\infty)(\mathbf{E}_{2k-1} - i\mathbf{E}_{2k}) + \mathbb{C}(\mathbf{E}_{2k-3} - i\mathbf{E}_{2k-2}) + \cdots + \mathbb{C}(\mathbf{E}_1 - i\mathbf{E}_2),$$

for $1 \leq 2k \leq n$.

We also define

$$L_{\mathfrak{B}}^+ \mathfrak{G}_\tau^{\mathbb{C}} := \{g_\bullet : S^1 \longrightarrow \mathfrak{G}^{\mathbb{C}} \mid \tau(g_\lambda) = g_{-\lambda},$$

$\lambda \longmapsto g_\lambda$ has a holomorphic extension inside B_1 and $g_0 \in \mathfrak{B}\}$.
(See formula (2.48) for a definition of B_1.)

The main tool is the following variant of theorem 2.2.8.

Theorem 2.3.9 *[46] The product map*

$$\begin{array}{rcl} L\mathfrak{G}_\tau \times L_{\mathfrak{B}}^+ \mathfrak{G}_\tau^{\mathbb{C}} & \longrightarrow & L\mathfrak{G}^{\mathbb{C}} \\ (\phi_\bullet, b_\bullet) & \longmapsto & \phi_\bullet b_\bullet \end{array}$$

is a diffeomorphism. In particular, for any $g_\bullet$ in $L\mathfrak{G}_\tau^{\mathbb{C}}$, $\exists! \phi_\bullet \in L\mathfrak{G}_\tau$, $\exists! b_\bullet \in L_{\mathfrak{B}}^+ \mathfrak{G}_\tau^{\mathbb{C}}$, such that

$$g_\bullet = \phi_\bullet b_\bullet \,.$$

Thanks to this result we have the following construction:

Theorem 2.3.10 *[46] Let $\xi_\bullet$ be a holomorphic map from Ω to the Lie algebra $L\mathfrak{g}_\tau^{\mathbb{C}}$ of $L\mathfrak{G}_\tau^{\mathbb{C}}$, of the form*

$$\xi_\lambda = \sum_{k=-1}^{+\infty} \hat{\xi}_k(z) \lambda^k dz, \qquad (2.90)$$

2.3 Harmonic maps with values into homogeneous spaces

where each $\hat{\xi}_k$ is holomorphic from Ω to $\mathfrak{g} \otimes \mathbb{C}$. Such a $\xi_\bullet$ is called a *holomorphic potential*. Then,

(i) *(Integration of $\xi_\bullet$ w.r.t. z).* There exists a unique map $g_\bullet$ from Ω to $L\mathfrak{G}_\tau^\mathbb{C}$ such that

$$g_\bullet(p) = \mathbb{1}, \tag{2.91}$$

$$dg_\bullet = g_\bullet \xi_\bullet. \tag{2.92}$$

(ii) *(Decomposition of $g_\bullet(z)$, for every fixed value of z).* The "component" of $g_\bullet$ along $L\mathfrak{G}_\tau$, i.e. the map $\phi_\bullet : \Omega \longrightarrow L\mathfrak{G}_\tau$ such that for each fixed z,

$$g_\bullet(z) = \phi_\bullet(z) b_\bullet(z), \tag{2.93}$$

where $b_\bullet \in L_\mathfrak{B}^+ \mathfrak{G}_\tau^\mathbb{C}$, is an extended harmonic lifting.

(iii) *Consequently, $z \longmapsto [\phi_1(z)]$ is a harmonic map from Ω to S^n.*

Proof

Step 1 Remark that, since $\frac{\partial \xi_{(k)}}{\partial \bar{z}} = 0$, $\forall k$, and $dz \wedge dz = 0$, we have that

$$d\xi_\bullet = 0$$

and

$$[\xi_\bullet \wedge \xi_\bullet] = 0.$$

Hence,

$$d\xi_\bullet + \frac{1}{2}[\xi_\bullet \wedge \xi_\bullet] = 0. \tag{2.94}$$

The existence of a solution $g_\bullet$ of (2.91) and (2.92) follows from the Frobenius theorem. Recall that

$$g_\lambda^{-1} dg_\lambda = \sum_{k=-1}^{+\infty} \hat{\xi}_k \lambda^k dz \tag{2.95}$$

(this will be used below).

Step 2 We use theorem 2.3.9. For any value of z, $\exists! b_\bullet(z) \in L_\mathfrak{B}^+ \mathfrak{G}_\tau^\mathbb{C}$,

$$g_\bullet(z) = \phi_\bullet(z)b_\bullet(z),$$

or, equivalently, $\phi_\bullet = g_\bullet b_\bullet^{-1}$. Moreover, theorem 2.3.9 tells us that $g_\bullet(z)$ and $b_\bullet(z)$ are smooth functions of z. It follows that

$$\phi_\bullet^{-1}d\phi_\bullet = b_\bullet[g_\bullet^{-1}dg_\bullet - b_\bullet^{-1}db_\bullet]b_\bullet^{-1}. \tag{2.96}$$

For the moment, let us fix z and analyze equation (2.96) w.r.t. λ, specifically the Laurent series expansions of both sides of this equation. Since $b_\bullet \in L_\mathfrak{B}^+ \mathfrak{G}_\tau^\mathbb{C}$, $b_\lambda, b_\lambda^{-1}$ and $b_\lambda^{-1}db_\lambda$ only contain non-negative powers of λ. Using (2.95), we deduce that

$$\phi_\lambda^{-1}d\phi_\lambda = \sum_{k=-1}^{+\infty} \hat{\alpha}_k \lambda^k, \tag{2.97}$$

where

$$\hat{\alpha}_{-1} = \hat{b}_0 \hat{\xi}_{-1} \hat{b}_0^{-1} dz.$$

Hence, thanks to theorem 2.3.6, we see that $\phi_\bullet$ is an extended harmonic lifting. This proves the second and third points of the theorem. $\square$

In [46] the authors show that, conversely, any harmonic map from Ω to S^n may be obtained in this way. Furthermore, they prove an even stronger version of this converse, more precisely, that it suffices to apply the preceding algorithms with potentials $\xi_\bullet$ having the special form

$$\xi_\lambda = \lambda^{-1}\hat{\xi}_{-1}(z)dz,$$

where $\hat{\xi}_{-1}(z)$ is a meromorphic function of z, taking values in $\mathfrak{g}_1 \otimes \mathbb{C}$. This last result uses the "twisted" version of the Birkhoff–Grothendieck theorem (see [130] and [46]).

To conclude this section (which just gives an overview of a developing subject), we mention that a study of the dressing action was carried out in this formalism in [26].

On the other hand, it is useful to remark that we can conveniently study the special case of S^2-valued harmonic maps, by replacing the group $SO(3)$ by its universal covering

$$SU(2) = \left\{ \begin{pmatrix} a & -\bar{b} \\ b & \bar{a} \end{pmatrix} \mid a, b \in \mathbb{C}, |a|^2 + |b|^2 = 1 \right\},$$

noticing that $S^2 \simeq SU(2)/SO(2)$, where

$$SO(2) \simeq \left\{ \begin{pmatrix} e^{i\theta} & 0 \\ 0 & e^{-i\theta} \end{pmatrix} \mid \theta \in \mathbb{R} \right\}.$$

2.4 Synthesis: relation between the different formulations

Let Ω be a simply-connected domain in $\mathbb{R}^2$, choose a point p in Ω, and consider

$$\mathcal{H} := \{ u : \Omega \longrightarrow S^n \mid u \text{ is harmonic}, u(p) = \mathbf{E}_0 \}.$$

By what we saw above, we have two approaches using loop groups to study $\mathcal{H}$.

(i) Thanks to the Cartan immersion $\mathcal{C}$ from S^n to $SO(n+1)$, we can associate an $SO(n+1)$-valued harmonic map

$$S = \mathcal{C} \circ u = S_0(2u\,{}^t u - \mathbb{1}),$$

where S_0 is any constant in $O(n)$ such that $\det S_0 = (-1)^n$. For instance

$$S_0 = P = \begin{pmatrix} 1 & 0 \\ 0 & -\mathbb{1} \end{pmatrix},$$

so that $S(p) = \mathbb{1}$. In this way we have a bijection between $\mathcal{H}$ and

$$\mathcal{S} := \{ S : \Omega \longrightarrow SO(n+1) \text{ harmonic } \mid \forall z \in \Omega, S_0 S(z) \text{ is a symmetry w.r.t. a straight line and } S(p) = \mathbb{1} \}.$$

Furthermore, it is possible for each $S \in \mathcal{S}$ to construct an extended harmonic map $E_\bullet$, from Ω to $\Omega SO(n+1)$, such that

$$E_{-1} = S, \qquad (2.98)$$

$$E_\bullet(p) = \mathbb{1}. \qquad (2.99)$$

This allows us to define an action of S^1 on $\mathcal{S}$, by letting, for any $\lambda \in S^1$ and $S \in \mathcal{S}$,

$$\sigma_\lambda(S) = E_{-\lambda} E_\lambda^{-1}.$$

In fact, we saw (theorem 2.2.13) that $\sigma_\lambda(S)$ was an $SO(n+1)$-valued harmonic map. The fact that σ_λ maps $\mathcal{S}$ to itself (i.e. preserves the condition that $S_0.S$ is a symmetry w.r.t. a straight line) will be shown below.

(ii) The other approach consists of associating to $u \in \mathcal{H}$ a lifting F of u, which we choose such that

$$F(p) = \mathbb{1}.$$

Then, we know how to construct an extended harmonic lifting $F_\bullet$, from Ω to $LSO(n+1)_\tau$. We define

$$\mathcal{F} := \{F : \Omega \longrightarrow SO(n+1) \mid \exists u \text{ harmonic map,} \\ [F] = u \text{ and } F(p) = \mathbb{1}\}.$$

In this way we can also identify $\mathcal{H}$ with $\mathcal{F}/\mathcal{G}$, where

$$\mathcal{G} := \{\gamma : \Omega \longrightarrow \mathfrak{K}/\gamma(p) = \mathbb{1}\}$$

can be interpreted as a gauge transformation group. The bundle on which this gauge group acts is the inverse image by u of the fiber bundle $SO(n+1) \longrightarrow S^n$.

The existence of an extended harmonic lifting $F_\bullet$ associated to F enables us to define an action of S^1 on $\mathcal{F}$ by letting, for all $\lambda \in S^1$ and $F \in \mathcal{F}$,

$$\rho_\lambda(F) = F_\lambda$$

(see corollary 2.3.5).

A natural question is to ask whether the two actions σ_λ and ρ_λ are in correspondence via the identification

$$\mathcal{H} \simeq \mathcal{S} \simeq \mathcal{F}/\mathcal{G}.$$

2.4 Synthesis: relation between the different formulations

This turns out to be the case. Before proving that σ_λ and ρ_λ are essentially the same, let us prove:

Lemma 2.4.1 *Suppose $F \in \mathcal{F}$ and $S \in \mathcal{S}$ describe the same harmonic map $u \in \mathcal{H}$. Then, for any $\lambda \in \mathbb{C}$,*

$$E_\lambda = F_\lambda F^{-1}. \tag{2.100}$$

Proof The first step consists of writing the relation that exists between S and F, knowing that $S = \mathcal{C} \circ u$ and $[F] = u$.

After thinking for a while, one sees that

$$S = PFPF^{-1}. \tag{2.101}$$

Thus, we define

$$\begin{aligned} T: \mathcal{F} &\longrightarrow \mathcal{S} \\ F &\longmapsto PFPF^{-1} \end{aligned}. \tag{2.102}$$

Next, we express $a := S^{-1}dS$ in terms of $\theta = F^{-1}dF$.

$$\begin{aligned} a &= (PFPF^{-1})^{-1}d(PFPF^{-1}) \\ &= F(PF^{-1}dFP - F^{-1}dF)F^{-1} \\ &= F(P\theta P - \theta)F^{-1} \\ &= -2F\theta_1 F^{-1}. \end{aligned} \tag{2.103}$$

Let us calculate

$$\begin{aligned} (F_\lambda F^{-1})^{-1}d(F_\lambda F^{-1}) &= F(F_\lambda^{-1}dF_\lambda - F^{-1}dF)F^{-1} \\ &= F(\Theta_\lambda - \Theta_1)F^{-1} \\ &= -F[(1-\lambda^{-1})\theta'_1 + (1-\lambda)\theta''_1]F^{-1} \end{aligned} \tag{2.104}$$

and using (2.103) in (2.104)

$$(F_\lambda F^{-1})^{-1}d(F_\lambda F^{-1}) = \frac{1}{2}[(1-\lambda^{-1})a' + (1-\lambda)a''] = A_\lambda. \tag{2.105}$$

Hence, we deduce from (2.105) that

$$d[(F_\lambda F^{-1})E_\lambda^{-1}] = 0,$$

and hence there exists a matrix $M \in SO(n+1)$ such that

$$F_\lambda F^{-1} = M E_\lambda.$$

But since $F_{\lambda(p)} = F(p) = E_\lambda = \mathbb{1}$, necessarily $M = \mathbb{1}$, which proves (2.100). □

It is now easy to show that the action of ρ_λ and that of σ_λ coincide through the map T defined in (2.102).

Theorem 2.4.2 *The following diagram commutes for all $\lambda \in S^1$.*

$$\begin{array}{ccc} \mathcal{F} & \xrightarrow{T} & \mathcal{S} \\ \rho_\lambda \downarrow & & \downarrow \sigma_\lambda \\ \mathcal{F} & \xrightarrow{T} & \mathcal{S} \end{array}$$

Proof Let $F \in \mathcal{F}$ and $S = T(F) = PFPF^{-1}$. Define

$$S_\lambda = \sigma_\lambda(S) = \sigma_\lambda \circ T(F),$$

and

$$\widetilde{S}_\lambda = T(F_\lambda) = T \circ \rho_\lambda(F).$$

We have that

$$\begin{aligned} \widetilde{S}_\lambda^{-1} d\widetilde{S}_\lambda &= (PF_\lambda PF_\lambda^{-1})^{-1} d(PF_\lambda PF_\lambda^{-1}) \\ &= F_\lambda (PF_\lambda^{-1} dF_\lambda P - F_\lambda^{-1} dF_\lambda) F_\lambda^{-1} \\ &= F_\lambda (P\Theta_\lambda P - \Theta_\lambda) - F_\lambda^{-1} \\ &= -2F_\lambda (\lambda^{-1} \theta'_1 + \lambda \theta''_1) F_\lambda^{-1}. \end{aligned} \qquad (2.106)$$

On the other hand, using lemma 2.4.1,

$$\begin{aligned} S_\lambda^{-1} dS_\lambda &= (E_{-\lambda} E_\lambda^{-1})^{-1} d(E_{-\lambda} E_\lambda^{-1}) \\ &= (F_{-\lambda} F_\lambda^{-1})^{-1} d(E_{-\lambda} E_\lambda^{-1}) \\ &= F_\lambda (\Theta_{-\lambda} - \Theta_\lambda) F_\lambda^{-1} \\ &= -2F_\lambda (\lambda^{-1} \theta'_1 + \lambda \theta''_1) F_\lambda^{-1}. \end{aligned} \qquad (2.107)$$

2.4 Synthesis: relation between the different formulations

We deduce from (2.106) and (2.107) that

$$\widetilde{S_\lambda}^{-1} d\widetilde{S_\lambda} = S_\lambda^{-1} dS_\lambda,$$

and hence that $S_\lambda = \widetilde{S_\lambda}$, since $\widetilde{S_\lambda}(p) = S_\lambda(p) = \mathbb{1}$.

□

Remark 2.4.3 *By proving this result, we have also shown that σ_λ really maps $\mathcal{S}$ to itself.*

To conclude, we come back to a construction already seen in section 2.1. There we saw that if Ω is simply connected, to each harmonic map u from Ω to S^n we can associate a map b from Ω to $so(n+1)$ (given by (2.17)). In fact, the existence of b reflects the action of S^1 on harmonic maps, which was the subject of the preceding theorem. More precisely, the following result says that b is the infinitesimal variation of u under the action of S^1.

Theorem 2.4.4 *Let $u \in \mathcal{H}$, and for any λ in S^1, let*

$$u_\lambda = [F_\lambda] = [\rho_\lambda(F)]$$

be the deformation of u along S^1. Let $b : \Omega \longrightarrow so(n+1)$ be a solution of

$$\begin{cases} db &= -\star(u \times du) \\ b(p) &= 0. \end{cases} \quad (2.108)$$

Then

$$\frac{d}{dt}(u_{e^{it}})|_{t=0} = bu. \quad (2.109)$$

Proof Write

$$F_{e^{it}} = (\mathbb{1} + t\xi)F + o(t),$$

where $\xi : \Omega \longrightarrow so(n+1)$, and develop the relation

$$dF_{e^{it}} = F_{e^{it}} \Theta_{e^{it}}.$$

It follows that

$$d[(\mathbb{1}+t\xi)F] = (\mathbb{1}+t\xi)F(\theta+t(\star\theta_1)) + o(t),$$

and hence

$$d\xi = \star(F\theta_1 F^{-1}). \qquad (2.110)$$

A simple calculation then yields

$$d\xi = \star(du \times u) = db,$$

and thus, since $\xi(p) = b(p) = 0$, we obtain our result. $\square$

A second way of understanding the meaning of b is to see that b "coincides" with the harmonic map $u_{\sqrt{-1}}$, in the following sense.

Problem 2.4.5 *Let $u : \Omega \longrightarrow S^n$ be a harmonic map, $F : \Omega \longrightarrow SO(n+1)$ a harmonic lifting of u and let, as in section 2.3,*

$$(e_0, ..., e_n) = (\mathbf{E}_0, ..., \mathbf{E}_n).F,$$

and

$$\alpha^i = \langle de_0, e_i \rangle = \langle du, e_i \rangle, \text{ for } i = 1, ..., n.$$

In a Euclidean space $\mathbb{R}^N$, let us look for an orthonormal family of n vectors $(f_1, ..., f_n)$ depending smoothly on $z \in \Omega$ (i.e. a smooth map from Ω to the family of n-tuples of orthonormal vectors in $\mathbb{R}^N$), such that

$$d[(\star\alpha^1)f_1 + \cdots + (\star\alpha^n)f_n] = 0. \qquad (2.111)$$

Here we have two solutions to this problem:

SOLUTION GIVEN BY NOETHER'S THEOREM

We choose $\mathbb{R}^N = \mathbb{R}^{\frac{n(n+1)}{2}} \simeq so(n+1)$ and

$$f_j = e_j \times e_0.$$

Then

$$(\star\alpha^1)f_1 + \cdots + (\star\alpha^n)f_n = db,$$

which implies (2.111).

SOLUTION GIVEN BY THE EXTENDED HARMONIC FRAME

We take $\mathbb{R}^N = \mathbb{R}^{n+1}$, and

$$f_j = F_{\sqrt{-1}}(\mathbf{E}_j).$$

Then

$$(\star\alpha^1)f_1 + \cdots + (\star\alpha^n)f_n = du_i.$$

In case $n = 2$, we can even push the comparison between B (defined in section 2.1 in (2.4)) and u_i a little further: the inverse images of the Euclidean metric on $\mathbb{R}^3$ by B and u_i coincide. Moreover, in case this metric is non-degenerate, B and u_i are harmonic maps taking values in their image.

Still in the case where $n = 2$, we can combine the analysis done in section 2.1 with the action of the circle described in the following sections. Thus, we obtain that there exists an action of S^1 on the constant Gauss curvature immersions of surfaces in $\mathbb{R}^3$. Then, we recover a result first proved by Ossian Bonnet (see [17], [113]).

2.5 Compactness of weak solutions in the weak topology

In this and the following sections we consider very different questions from above. We will be interested in weakly harmonic maps in $H^1(\Omega, S^n)$, where Ω is an open subset of $\mathbb{R}^m$. Nevertheless, the following results rest on the same observation we used in sections 2.1, 2.2 and 2.3: the existence of a conservation law thanks to Noether's theorem.

The general problem is the following: $\mathcal{M}$ is a manifold with boundary and $\mathcal{N}$ is compact without boundary, as in chapter 1. Let $(u_k)_{k \in \mathbb{N}}$ be a sequence of maps in $H^1(\mathcal{M}, \mathcal{N})$, and suppose that there exists $u \in H^1(\mathcal{M}, \mathcal{N})$ such that

$$u_k \rightharpoonup u \text{ weakly in } H^1(\mathcal{M}, \mathcal{N}), \tag{2.112}$$

or equivalently, for every function $\phi \in H^1(\mathcal{M}, \mathbb{R}^N)$,

$$\lim_{k \to +\infty} \int_\mathcal{M} [\langle \phi, u_k \rangle + \langle d\phi, du_k \rangle] d\text{vol}_g$$

$$= \int_\mathcal{M} [\langle \phi, u \rangle + \langle d\phi, du \rangle] d\text{vol}_g. \qquad (2.113)$$

We will then say that u_k converges weakly to u in the H^1 topology. Suppose that each u_k is weakly harmonic, may we deduce that u is also weakly harmonic?

We do not know the answer to this question, in general. We do not know of any example such that the limit u is not harmonic, but cannot prove the contrary except in certain special cases. But in fact, it suffices that there exists an isometry group acting transitively on $\mathcal{N}$, in order to conclude that the weak limit is weakly harmonic. As an example, consider the case where $\mathcal{N}$ is the sphere S^n.

Theorem 2.5.1 *Let u_k be a sequence in $H^1(\mathcal{M}, S^n)$ such that*

$$u_k \rightharpoonup u \text{ weakly in } H^1(\mathcal{M}, S^n), \qquad (2.114)$$

$$\Delta u_k + u_k |du_k|^2 = 0 \text{ in } \mathcal{D}'(\mathcal{M}). \qquad (2.115)$$

Then u is weakly harmonic, i.e.

$$\Delta u + u |du|^2 = 0 \text{ in } \mathcal{D}'(\mathcal{M}). \qquad (2.116)$$

Proof It seems very hard to pass to the limit in (2.115), because of the quadratic term in du_k.

Let us use Noether's theorem 1.3.1: we obtain that for every map $\alpha \in H^1_0(\mathcal{M}, so(n+1)) \cap L^\infty(\mathcal{M}, so(n+1))$,

$$\int_\mathcal{M} \langle d\alpha, u_k \times du_k \rangle d\text{vol}_g = 0. \qquad (2.117)$$

Let Ω be any open bounded subset of $\mathcal{M}$, and suppose that the support of α is contained in Ω. A consequence of the compact embedding of $H^1(\Omega, \mathbb{R}^{n+1})$ in $L^2(\Omega, \mathbb{R}^{n+1})$ (Rellich's theorem, see [19]), is that

$$u_k \to u \text{ in } L^2(\Omega, \mathbb{R}^{n+1}).$$

2.5 Compactness of weak solutions in the weak topology

This implies that

$$u_k \times du_k \rightharpoonup u \times du \quad \text{weakly in } L^1(\Omega, so(n+1)).$$

Hence, we can pass to the limit in (2.117) and obtain

$$\int_{\mathcal{M}} \langle d\alpha, u \times du \rangle dvol_g = 0. \tag{2.118}$$

This implies that u is weakly harmonic, i.e. a solution of (2.116). To check it, let $\phi \in H_0^1(\mathcal{M}, \mathbb{R}^{n+1}) \cap L^\infty(\mathcal{M}, \mathbb{R}^{n+1})$, and write

$$\psi = \phi - \langle u, \phi \rangle u.$$

We have the identity, true a.e.

$$\begin{aligned}
\langle d\phi, du \rangle - \langle u, \phi \rangle |du|^2 &= \langle du, d\psi \rangle \\
&= \langle u, u \rangle \langle du, d\psi \rangle - \langle u, d\psi \rangle \cdot \langle u, du \rangle \\
&= \langle u \times du, u \times d\psi \rangle \\
&= \langle u \times du, d(u \times \psi) \rangle \\
&= \langle u \times du, d(u \times \phi) \rangle. \tag{2.119}
\end{aligned}$$

(Where we used, among other things,

$$\langle u \times du, du \times \psi \rangle = \langle u, du \rangle \cdot \langle du, \psi \rangle - \langle u, \psi \rangle \langle du, du \rangle = 0.)$$

Using (2.119) and (2.118) with $\alpha = u \times \phi$, we obtain

$$\int_{\mathcal{M}} (\langle d\phi, du \rangle - \langle u, \phi \rangle |du|^2) dvol_g =$$

$$\int_{\mathcal{M}} \langle d\alpha, u \times du \rangle dvol_g = 0,$$

which yields our result. □

This type of result has been used independently by numerous authors ([99], [35], [153]), in order to construct weak solutions for sphere-valued evolution problems. We will examine, as an example, the result due to Jalal Shatah regarding the S^n-valued wave equation (in fact the work in [153] concerns $O(4)$-valued maps).

It consists of finding S^n-valued maps u, defined on the product $I \times \mathcal{M}$, where $I = [0, T)$, $T \in (0, +\infty]$, and $(\mathcal{M}, g)$ is a Riemannian manifold which we suppose to be complete and without boundary, and which are solutions of

$$\Box u + \lambda(u) u = 0, \tag{2.120}$$

where

$$\lambda(u) = |d_x u|_g^2 - \left|\frac{\partial u}{\partial t}\right|^2,$$

and

$$\Box u = \Delta_g u - \frac{\partial^2 u}{\partial t^2}.$$

We consider the initial conditions

$$u(0, x) = f(x), \tag{2.121}$$

$$\frac{\partial u}{\partial t}(0, x) = g(x), \tag{2.122}$$

where $f : \mathcal{M} \longrightarrow S^n$, and $g : \mathcal{M} \longrightarrow \mathbb{R}^{n+1}$ is a map such that

$$\langle f(x), g(x) \rangle = 0. \tag{2.123}$$

As is shown in [153], the Cauchy problem does not have, in general, a smooth solution when $m = \dim \mathcal{M} \geq 3$, even if the initial conditions are smooth. On the other hand, when $m = 1$, C.H. Gu [75], J. Ginibre and G. Vélo [72], and afterwards J. Shatah, proved the existence of a smooth solution for all times, if the initial conditions are smooth. The case $m \geq 2$ is still not well understood. The fact that the wave equation "develops" singularities in general does not exclude the possibility of constructing weak solutions. Following J. Shatah we define a weak solution of (2.120) as being a function u in $L^1_{loc}(I \times \mathcal{M}, \mathbb{R}^{n+1})$ such that

$$\Box u + \lambda(u) u = 0 \text{ in } \mathcal{D}'(I \times \mathcal{M}), \tag{2.124}$$

$$u(x, t) \in S^n \text{ a.e.}, \tag{2.125}$$

2.5 Compactness of weak solutions in the weak topology

$$
\begin{aligned}
I &\longrightarrow H^1(\mathcal{M}, S^n) \\
t &\longmapsto [x \longmapsto u(t,x)]
\end{aligned}
\tag{2.126}
$$

is continuous for the weak topology of H^1, and

$$
\begin{aligned}
I &\longrightarrow L^2(\mathcal{M}, \mathbb{R}^{n+1}) \\
t &\longmapsto \left[x \longmapsto \frac{\partial u}{\partial t}(t,x)\right]
\end{aligned}
\tag{2.127}
$$

is continuous for the weak topology of L^2.

Before stating an existence result, we can make the following observation, made by J. Shatah: if equations (2.125), (2.126) and (2.127) are satisfied, then the condition (2.124) is equivalent to

$$
-\frac{\partial}{\partial t}\left(u \times \frac{\partial u}{\partial t}\right) + \sum_{\alpha=1}^{m} \frac{\partial}{\partial x^\alpha}\left(g^{\alpha\beta}\sqrt{\det g}\, u \times \frac{\partial u}{\partial x^\beta}\right) = 0 \text{ in } \mathcal{D}'(I \times \mathcal{M}).
\tag{2.128}
$$

This follows from the following equality whose proof rests on an identity similar to equation (2.119) given above: for any function $\phi \in C_c^\infty(I \times \mathcal{M}, \mathbb{R}^{n+1})$,

$$
\int\int_{I\times\mathcal{M}}\left(-\left\langle\frac{\partial u}{\partial t},\frac{\partial \phi}{\partial t}\right\rangle + \langle d_x u, d_x \phi\rangle - \lambda(u)\langle u,\phi\rangle\right) dt\, dvol_g
$$
$$
= \int\int_{I\times\mathcal{M}}\left(\left\langle u \times \frac{\partial u}{\partial t},\frac{\partial}{\partial t}(u \times \phi)\right\rangle - \langle u \times d_x u, d_x(u \times \phi)\rangle\right) dt\, dvol_g.
\tag{2.129}
$$

Theorem 2.5.2 *[153] Let $f \in H^1(\mathcal{M}, S^n)$ and $g \in L^2(\mathcal{M}, \mathbb{R}^{n+1})$ satisfy (2.123). Then there exists a weak solution $u \in L^1_{loc}(I \times \mathcal{M}, S^n)$ to the wave equation (2.120), i.e. a function u satisfying (2.124), (2.125), (2.126) and (2.127), and such that*

$$
\begin{cases} u(0,x) &= f(x), \\ \dfrac{\partial u}{\partial t}(0,x) &= g(x). \end{cases}
$$

Proof
Step 1 We start by solving an approximation problem. Let W be a $\mathcal{C}^1$ function from $\mathbb{R}^{n+1}$ to $\mathbb{R}$ such that

(i) $W(y)$ is a function of $|y|$.
(ii) $W(y) \geq 0$.

(iii) $W(y) = 0 \iff |y| = 1$.

An example of such a function is $W(y) := (1 - |y|^2)^2$. However, it is better to use a modified version of this function, in order to also have

(iv) $\exists C > 0, \forall y \in \mathbb{R}^{n+1}, \ W(y) \leq C(1 + |y|^2)$.

For $\epsilon > 0$, we look for maps u from $I \times \mathcal{M}$ to $\mathbb{R}^{n+1}$ which are solutions of

$$\Box u + \frac{1}{\epsilon^2} \nabla W(u) = 0, \tag{2.130}$$

$$u(0, x) = f(x), \tag{2.131}$$

$$\frac{\partial u}{\partial t}(0, x) = g(x). \tag{2.132}$$

Using classical results (see [159]), we can show that problem (2.130), (2.131), (2.132), has a solution over any time interval $I = (0, T)$ ($T \in (0, +\infty)$). Let u_ϵ be such a solution, then

$$u_\epsilon \in \mathcal{C}(I, H^1_{loc}(\mathcal{M}, \mathbb{R}^{n+1})), \tag{2.133}$$

$$\frac{\partial u_\epsilon}{\partial t} \in \mathcal{C}(I, L^2(\mathcal{M}, \mathbb{R}^{n+1})), \tag{2.134}$$

and moreover, u_ϵ satisfies the following energy conservation identity. For every time $t \in I$, we have

$$\int_\mathcal{M} \left(\frac{1}{2} \left| \frac{\partial u_\epsilon}{\partial t} \right|^2 + \frac{1}{2} |d_x u_\epsilon|^2 + \frac{1}{4\epsilon^2} W(u_\epsilon) \right) \mathrm{dvol}_{g|t}$$

$$= \int_\mathcal{M} \left(\frac{1}{2} |g|^2 + \frac{1}{2} |d_x f|^2 \right) \mathrm{dvol}_g. \tag{2.135}$$

In particular, this implies that each term in the l.h.s. of (2.135) is uniformly bounded for all times. Thus, we deduce that there is a subsequence ϵ' of ϵ such that

$$u_{\epsilon'} \rightharpoonup u_0 \text{ weak } \star \text{ in } L^\infty(I, \text{weak } H^1(\mathcal{M}, \mathbb{R}^{n+1})), \tag{2.136}$$

2.5 Compactness of weak solutions in the weak topology

$$\frac{\partial u_{\epsilon'}}{\partial t} \rightharpoonup v_0 \quad \text{weak} \star \quad \text{in} \quad L^\infty(I, \text{weak } L^2(\mathcal{M}, \mathbb{R}^{n+1})). \tag{2.137}$$

Here, the convergences are weak $\star$ w.r.t. t, and weak w.r.t. x. Notice that (2.136) implies that

$$u_{\epsilon'} \rightharpoonup u_0 \quad \text{weakly in} \quad H^1_{loc}(I \times \mathcal{M}, \mathbb{R}^{n+1}). \tag{2.138}$$

Step 2 By Rellich's theorem (see [19]), the convergence (2.138) implies that for any compact K in $I \times \mathcal{M}$, we can extract a subsequence of ϵ' (which, for convenience, we still denote by ϵ') such that

$$u_{\epsilon'} \longrightarrow u_0 \quad \text{in} \quad L^2(K, \mathbb{R}^{n+1}), \tag{2.139}$$

and

$$u_{\epsilon'} \longrightarrow u_0 \quad \text{a.e. on } K. \tag{2.140}$$

Thus, applying Fatou's lemma and (2.140), we deduce that

$$\int\int_K W(u_0) dt \, dvol_g \leq \liminf_{\epsilon' \to 0} \int\int_K W(u_{\epsilon'}) dt \, dvol_g.$$

But since we know, thanks to (2.135), that the r.h.s. of this inequality vanishes, we obtain

$$\int\int_K W(u_0) dt \, dvol_g = 0,$$

and because of conditions (ii) and (iii) imposed on W, this yields

$$u_0(t, x) \in S^n \quad \text{a.e. on } K. \tag{2.141}$$

Finally, since (2.141) is valid for any compact K, this shows that

$$u_0 \in L^\infty_{loc}(I, H^1(\mathcal{M}, S^n)).$$

Step 3 We check that $v_0 = \frac{\partial u_0}{\partial t}$. To do this, it suffices to pass to the limit in the identity

$$\int\int_{I \times \mathcal{M}} \left(\left\langle u_\epsilon, \frac{\partial \phi}{\partial t} \right\rangle + \left\langle \frac{\partial u_\epsilon}{\partial t}, \phi \right\rangle \right) dt \, dvol_g = 0,$$

valid $\forall \phi \in \mathcal{C}_c^\infty(I \times \mathcal{M}, \mathbb{R}^{n+1})$.

Step 4 We show that $u_0 \in \mathcal{C}(I, \text{weak } H^1(\mathcal{M}, S^n))$ and $\frac{\partial u_0}{\partial t} \in \mathcal{C}(I, \text{weak } L^2(\mathcal{M}, \mathbb{R}^{n+1}))$. Proving the first statement consists in showing that, for any function $\psi \in \mathcal{C}_c^\infty(\mathcal{M}, \mathbb{R}^{n+1})$, the map $h_0 : I \longrightarrow \mathbb{R}$ defined by

$$h_0(t) = \int_{\mathcal{M}} \langle d\psi, d_x u_0(t,x) \rangle \, d\text{vol}_g$$
$$= -\int_{\mathcal{M}} \langle \Delta_g \psi, u_0(t,x) \rangle \, d\text{vol}_g$$

is continuous over I. But, by (2.138), we know that h_0 is the limit of $h_{\epsilon'}$, for the weak topology of $H^1(I)$, where

$$h_{\epsilon'}(t) = \int_{\mathcal{M}} \langle d\psi(x), d_x u_{\epsilon'}(t,x) \rangle \, d\text{vol}_g$$
$$= -\int_{\mathcal{M}} \langle \Delta_g \psi, u_{\epsilon'}(t,x) \rangle \, d\text{vol}_g.$$

Moreover, we have

$$h'_{\epsilon'}(t) = -\int_{\mathcal{M}} \langle \Delta_g \psi, \frac{\partial u_{\epsilon'}}{\partial t}(t,x) \rangle \, d\text{vol}_g,$$

and $\int_{\mathcal{M}} |\frac{\partial u_\epsilon}{\partial t}|^2 d\text{vol}_g$ is uniformly bounded in time, by (2.135). This yields, using the Cauchy–Schwarz inequality, that $||h'_{\epsilon'}||_{L^\infty}$ is uniformly bounded in ϵ' and in t. Hence, this implies that, up to passing to a new subsequence ϵ', $h_{\epsilon'}$ converges uniformly to a certain continuous function k_0. But we already know that $h_{\epsilon'}$ converges weakly to h_0 in $H^1(I)$; hence we conclude that $h_0 = k_0$. This proves that h_0 is continuous and therefore that $u_0 \in \mathcal{C}(I, \text{weak } H^1(\mathcal{M}, S^n))$. We proceed in the same way for $\frac{\partial u_0}{\partial t}$.

Step 5 It only remains to show that u_0 is a weak solution of the wave equation. In order to achieve this, we notice that (2.140) implies that, in the sense of distributions,

$$-\frac{\partial}{\partial t}\left(u_\epsilon \times \frac{\partial u_\epsilon}{\partial t}\right) + \sum_{\alpha=1}^m \frac{\partial}{\partial x^\alpha}\left(g^{\alpha\beta} \sqrt{\det g} \, u_\epsilon \times \frac{\partial u_\epsilon}{\partial x^\beta}\right) = 0 \quad (2.142)$$

(we have also used property (i) of W). As in theorem 2.5.1, we can pass to the limit in (2.142) and, using (2.136), (2.137) and (2.139), we show that u_0 satisfies equation (2.142) in the sense of distributions. In view

2.6 Regularity of weak solutions

of identity (2.129) this proves that u_0 is a solution of the wave equation. □

2.6 Regularity of weak solutions

We conclude this chapter by showing how the conservation law formulation for the sphere-valued harmonic map equation enables us to prove, in a rather simple way, the regularity of harmonic maps from a surface to a sphere.

But first, we need to introduce one more ingredient. It consists of the compensation phenomenon, which enables us to improve slightly the classical elliptic estimates, and which is of the same nature as the phenomenon that enables us to pass to the limit in equation (2.115) of the preceding section. Everything starts with a calculation trick invented by Henry Wente in 1969 ([177]), in a work about constant mean curvature surfaces.

We have already seen in the first section of this chapter the equation verified by a conformal parametrization X of a constant mean curvature surface in $\mathbb{R}^3$. It can be written as

$$\Delta X = 2H \frac{\partial X}{\partial x} \times \frac{\partial X}{\partial y}, \qquad (2.143)$$

where H is the (constant) value of the mean curvature. The result of H. Wente mentioned above is the following regularity theorem.

Theorem 2.6.1 *[177] Let Ω be an open domain in $\mathbb{R}^2$, and suppose X is a map in $H^1(\Omega, \mathbb{R}^3)$ that is a distributional solution of equation (2.143). Then X is C^∞ in Ω.*

The crucial point in the proof of this result is that each component in the r.h.s. of (2.143) is a Jacobian determinant of a map from Ω to $\mathbb{R}^2$ (obtained by choosing two distinct components of X). Following [21], consider a slightly more general problem: let $a, b \in H^1(\Omega, \mathbb{R})$, and define

$$\{a, b\} := \frac{\partial a}{\partial x}\frac{\partial b}{\partial y} - \frac{\partial a}{\partial y}\frac{\partial b}{\partial x}. \qquad (2.144)$$

Consider the following question: we suppose that the function $\phi \in L^1_{loc}(\Omega, \mathbb{R})$ is a solution of

$$\begin{cases} -\Delta\phi = \{a,b\} & \text{in } \Omega \\ \phi = 0 & \text{on } \partial\Omega, \end{cases} \qquad (2.145)$$

(the homogeneous Dirichlet condition has been chosen for simplicity). We want to know the regularity of ϕ. More precisely, to which Lebesgue, Sobolev or continuous function spaces does ϕ belong?

According to the classical theory, which is based on Calderón–Zygmund estimates (see [156]), the determinant $\{a,b\}$ is in $L^1(\Omega)$, with

$$\|\{a,b\}\|_{L^1} \leq \|da\|_{L^2}\|db\|_{L^2}.$$

And hence it follows that, if Ω is bounded,

$$\phi \in L^q(\Omega, \mathbb{R}), \ \forall 1 \leq q < +\infty, \qquad (2.146)$$

$$d\phi \in L^p(\Omega, \mathbb{R}^2), \ \forall 1 \leq p < 2. \qquad (2.147)$$

This means that ϕ is very close to belonging to $H^1(\Omega) \cap L^\infty(\Omega)$, but we cannot deduce from classical theory that ϕ belongs precisely to this space. We are in a limit case (this is the disadvantage of L^1 and L^∞ when compared to L^p, for $1 < p < +\infty$).

But if we use the very special algebraic structure of $\{a,b\}$, we can slightly improve estimates (2.146) and (2.147).

Theorem 2.6.2 *[177], [178], [21] Let $\Omega = B^2$ be the unit ball in $\mathbb{R}^2$. Let $\phi \in L^1(\Omega, \mathbb{R})$ be a solution of (2.145). Then, $\phi \in H^1(B^2, \mathbb{R}) \cap \mathcal{C}(B^2, \mathbb{R})$, and moreover,*

$$\|\phi\|_{L^\infty} + \|d\phi\|_{L^2} \leq C\|da\|_{L^2}\|db\|_{L^2}, \qquad (2.148)$$

where C is a universal constant.

Remark 2.6.3 *The fact of having restricted ourselves to the case $\Omega = B^2$ is not essential. The proof and some generalizations of this result are the subject of chapter 3 (see theorem 3.1.2).*

Proof of theorem 2.6.1 Ever since [103], we know that every continuous solution X of (2.143) is $\mathcal{C}^\infty$ in the interior of Ω. Thus, it suffices to show that X is continuous in Ω. Moreover, since we need to prove a local

2.6 Regularity of weak solutions

result, we can always reduce to the case where $\Omega = B^2$. We notice that equation (2.143) may be written as

$$\begin{cases} -\Delta X^1 &= 2H\{X^2, X^3\} \\ -\Delta X^2 &= 2H\{X^3, X^1\} \\ -\Delta X^3 &= 2H\{X^1, X^2\} \end{cases} \quad (2.149)$$

and by theorem 2.6.2, this immediately implies that X is continuous inside B^2 (it suffices to decompose each component of X^i into the sum $\phi^i + \psi^i$ where ψ^i is the harmonic extension of the trace of X^i over ∂B^2. In B^2, ψ^i is C^∞ since it is a real harmonic function, and ϕ^i is continuous by theorem 2.6.2). $\square$

We now consider the case of weakly harmonic maps.

Theorem 2.6.4 *[83] Let $(\mathcal{M}, g)$ be a Riemannian surface and $u \in H^1(\mathcal{M}, S^n)$. Then if u is weakly harmonic, u is C^∞ on $\mathcal{M}$.*

Proof It suffices to study the case where $\mathcal{M}$ is the unit ball B^2 of $\mathbb{R}^2$, with the standard metric. In fact, the study of the regularity of u in the neighborhood of a point in $\mathcal{M}$ always reduces to this case by using local conformal coordinates. Therefore we can write that

$$\frac{\partial}{\partial x}\left(u \times \frac{\partial u}{\partial x}\right) + \frac{\partial}{\partial y}\left(u \times \frac{\partial u}{\partial y}\right) = 0 \text{ in } \mathcal{D}'(B^2). \quad (2.150)$$

This implies, since $u \in L^2$, that there exists $b \in H^1(B^2, so(n+1))$ such that

$$\begin{cases} \dfrac{\partial b}{\partial x} &= u \times \dfrac{\partial u}{\partial y} \\ \dfrac{\partial b}{\partial y} &= -u \times \dfrac{\partial u}{\partial x}. \end{cases} \quad (2.151)$$

Using again the fact that u is weakly harmonic, and the condition $\langle u, du \rangle = 0$ (which is a consequence of $|u|^2 = 1$),

$$\begin{aligned}
-\Delta u &= u|du|^2 \\
&= u\left(\left\langle \frac{\partial u}{\partial x}, \frac{\partial u}{\partial x}\right\rangle + \left\langle \frac{\partial u}{\partial y}, \frac{\partial u}{\partial y}\right\rangle\right) \\
&= \left(u\frac{{}^t\partial u}{\partial x} - \frac{\partial u}{\partial x}{}^tu\right)\frac{\partial u}{\partial x} + \left(u\frac{{}^t\partial u}{\partial y} - \frac{\partial u}{\partial y}{}^tu\right)\frac{\partial u}{\partial y}.
\end{aligned}$$

Thus, we obtain

$$-\Delta u = \frac{\partial b}{\partial x}\frac{\partial u}{\partial y} - \frac{\partial b}{\partial y}\frac{\partial u}{\partial x}, \tag{2.152}$$

which, rewritten component by component, yields

$$-\Delta u^i = \sum_{j=1}^{n+1}\{b_j^i, u^j\}. \tag{2.153}$$

Applying Wente's theorem 2.6.2, as in theorem 2.6.1, we deduce that u is continuous. Using the results of [103], we conclude that u is $\mathcal{C}^\infty$. □

This result can be extended without major difficulties to the case where the image manifold is a symmetric space, thanks to Noether's theorem [84]. But if the isometry group acting on the image manifold is too small (which is what happens in the general case), we need to use the methods presented in chapter 4. In a different direction, we may try to generalize this theorem to the case where the dimension of the domain manifold is larger than 2. As announced in chapter 1, there are some results, but both the hypotheses and the conclusions are quite different. The case where the image manifold is a sphere was studied by L.C. Evans, and will be presented at the end of chapter 3, and the case of an arbitrary image manifold is due to F. Bethuel, and will be presented in chapter 4. In all cases, the compensation phenomena indicated in theorem 2.6.2 play an important role, but other ingredients will also be used: the ϵ-regularity techniques of Morrey, and the monotonicity formula. We must also use deeper versions of theorem 2.6.2 involving Hardy spaces and apply the Fefferman–Stein theorem concerning the duality between Hardy and BMO spaces. All these concepts will be introduced in the next chapter.

2.6 Regularity of weak solutions

To conclude, we mention that the technique used in theorem 2.6.4 has been used by several authors for studying the heat equation between a surface and a sphere, i.e. the weak solutions of

$$-\frac{\partial u}{\partial t} + \Delta u + u|du|^2 = 0. \tag{2.154}$$

Among the topics studied we mention the behavior of a solution in the neighborhood of a singularity, by Jie Qing [133], and the uniqueness of the solution for the Cauchy problem, done by Tristan Rivière [137] under small energy hypothesis, and by Alexandre Freire [60] in the general case.

We recall that Michael Struwe [160] had proved an existence theorem for weak solutions to the Cauchy problem for equation (2.154). He showed that the solutions are $\mathcal{C}^\infty$, except maybe at a finite number of points in space–time. This result is optimal in the sense that Kung Ching Chang, Wei Yue Ding and Rugang Ye [29] have constructed examples of solutions to equation (2.154) having point singularities.

3
Compensations and exotic function spaces

In the previous chapter we saw some regularity results for weak solutions of non-linear partial differential equations, whose proofs are based on the fact that, among the quadratic combinations of derivatives of functions, the Jacobian determinants have a subtle extra regularity. To our knowledge, this phenomenon was first used by Henry Wente [177]. It is analogous to the compensations that allow one to pass to the limit in certain non-linear combinations of weakly convergent sequences, and which were used by François Murat and Luc Tartar in their compensated compactness theory [122], [166].

Below, we follow the clear presentation, due to Haïm Brezis and Jean-Michel Coron [21], of H. Wente's discovery, and we also present some of its improvements. In particular, twenty years after H. Wente's result, works by Stefan Müller and also by Ronald Coifman, Pierre-Louis Lions, Yves Meyer and Stephen Semmes resulted in a finer study of these compensation phenomena, in particular using Hardy spaces.

Other function spaces, less known to specialists in partial differential equations, such as Lorentz spaces, have also proven to be useful. These developments are presented in sections 3.2 (Hardy spaces) and 3.3 (Lorentz spaces) of this chapter. As an example of the use of Hardy spaces we present the regularity theorem for weakly stationary maps, due to Lawrence C. Evans, in the last section of this chapter. Most of the analytic results of this chapter will be applied in chapters 4 and 5, to other geometrical settings.

3.1 Wente's inequality

3.1.1 The inequality on a plane domain

Let a and b be two functions ins $H^1(\Omega, \mathbb{R})$, where Ω is an open subset of $\mathbb{R}^2$, and define, as in the previous chapter,

$$\{a,b\} := \frac{\partial a}{\partial x}\frac{\partial b}{\partial y} - \frac{\partial a}{\partial y}\frac{\partial b}{\partial x}.$$

It is clear that $\{a,b\}$ belongs to $L^1(\Omega, \mathbb{R})$, and that

$$\|\{a,b\}\|_{L^1(\Omega)} \leq \|da\|_{L^2(\Omega)}\|db\|_{L^2(\Omega)}.$$

This implies, by classical elliptic theory, that if ϕ is a solution in $W^{1,p}(\Omega, \mathbb{R})$, for $p > 1$, of the equation

$$\begin{cases} -\Delta\phi &= \{a,b\} \quad \text{in } \Omega \\ \phi &= 0 \quad \text{on } \partial\Omega, \end{cases} \qquad (3.1)$$

then $\phi \in \bigcap_{q<+\infty} L^q(\Omega, \mathbb{R})$, and $\phi \in \bigcap_{q<2} W^{1,q}(\Omega, \mathbb{R})$, and it is possible to obtain the estimates

$$\begin{aligned} \|\phi\|_{L^q(\Omega)} &\leq C_0(q,\Omega)\|\{a,b\}\|_{L^1(\Omega)}, \quad \text{for } q < +\infty, \\ \|d\phi\|_{L^q(\Omega)} &\leq C_1(q,\Omega)\|\{a,b\}\|_{L^1(\Omega)}, \quad \text{for } q < 2. \end{aligned}$$

Below, we will write

$$\widetilde{ab}_\Omega := \phi,$$

the solution of (3.1), or sometimes $\widetilde{ab} = \widetilde{ab}_\Omega$ if there is no ambiguity.

Remark 3.1.1 *The bilinear operator* $(a,b) \longmapsto \widetilde{ab}_\Omega$ *is invariant under conformal transformations in the sense that if* $T: \Omega_1 \longrightarrow \Omega_2$ *is a conformal transformation, and if* $a, b \in H^1(\Omega_2, \mathbb{R})$, *then the equation*

$$-\Delta(\widetilde{ab}_{\Omega_2}) = \{a,b\} \text{ on } \Omega_2$$

implies

$$-\Delta\left[(\widetilde{ab}_{\Omega_2}) \circ T\right] = \{a \circ T, b \circ T\} \text{ on } \Omega_1.$$

Thus, we conclude that

$$\widetilde{ab}_{\Omega_2} \circ T = \widetilde{(a \circ T)(b \circ T)}_{\Omega_1}. \tag{3.2}$$

We will see that thanks to the special algebraic structure of $\{a,b\}$, it is possible to estimate $\widetilde{ab}$ in $H^1(\Omega, \mathbb{R})$ and $L^\infty(\Omega, \mathbb{R})$, which is better than what we would expect from (3.1). In what follows, B^2 is the unit ball in $\mathbb{R}^2$.

Theorem 3.1.2 [177], [178], [21] Let $a, b \in H^1(B^2, \mathbb{R})$ then
$$\widetilde{ab} = \widetilde{ab}_{B^2} \in C^0(B^2, \mathbb{R}) \cap H^1(B^2, \mathbb{R}),$$
and
$$\|\widetilde{ab}\|_{L^\infty(B^2)} + \|d\widetilde{ab}\|_{L^2(B^2)} \leq C \|da\|_{L^2(B^2)} \|db\|_{L^2(B^2)}. \tag{3.3}$$

Proof We reproduce here the proof given in [6], improving slightly the proof of [21] which is based on the ideas in [177]. The difference is that we take advantage of the conformal invariance property, which allows us to obtain an optimal estimate for $\|\widetilde{ab}\|_{L^\infty(B^2)}$.

Step 1 ESTIMATE AT 0
In this step, as well as in steps 2 and 3, we consider $a, b \in H^1(B^2, \mathbb{R}) \cap C^\infty(B^2, \mathbb{R})$.

Let us use the integral representation
$$\widehat{ab}(0) = \frac{1}{2\pi} \int_{B^2} \{a, b\}(x) \left(\log \frac{1}{r} \right) dx^1 dx^2,$$
where $r = |x|$. The trick consists in writing $\{a, b\}$ and the integral in polar coordinates r and $\theta = \arctan\left(\frac{y}{x}\right)$.

$$\begin{aligned}
\widehat{ab}(0) &= \frac{1}{2\pi} \int_0^1 dr \int_0^{2\pi} \left(\log \frac{1}{r} \right) \frac{1}{r} \left(\frac{\partial a}{\partial r} \frac{\partial b}{\partial \theta} - \frac{\partial a}{\partial \theta} \frac{\partial b}{\partial r} \right) r d\theta \\
&= \frac{1}{2\pi} \int_0^1 dr \int_0^{2\pi} \left(\log \frac{1}{r} \right) \left[\frac{\partial}{\partial r}\left(a \frac{\partial b}{\partial \theta} \right) - \frac{\partial}{\partial \theta}\left(a \frac{\partial b}{\partial r} \right) \right] d\theta \\
&= \frac{1}{2\pi} \int_0^1 \frac{dr}{r} \int_0^{2\pi} a \frac{\partial b}{\partial \theta} d\theta.
\end{aligned}$$

Let us define $\overline{a}^r = \frac{1}{2\pi} \int_0^{2\pi} a(r, \theta) d\theta$. Then we can write

3.1 Wente's inequality

$$\left|\int_0^{2\pi} a \frac{\partial b}{\partial \theta} d\theta\right| = \left|\int_0^{2\pi} (a - \overline{a}^r) \frac{\partial b}{\partial \theta} d\theta\right|$$
$$\leq \|a - \overline{a}^r\|_{L^2(0,2\pi)} \left\|\frac{\partial b}{\partial \theta}\right\|_{L^2(0,2\pi)}$$
$$\leq \left\|\frac{\partial a}{\partial \theta}\right\|_{L^2(0,2\pi)} \left\|\frac{\partial b}{\partial \theta}\right\|_{L^2(0,2\pi)}.$$

Hence,

$$|\widetilde{ab}(0)| \leq \frac{1}{2\pi} \int_0^1 \left\|\frac{\partial a}{\partial \theta}\right\|_{L^2(0,2\pi)} \left\|\frac{\partial b}{\partial \theta}\right\|_{L^2(0,2\pi)} \frac{dr}{r}$$
$$\leq \frac{1}{2\pi} \|da\|_{L^2(B)} \|db\|_{L^2(B)}. \qquad (3.4)$$

Step 2 ESTIMATE FOR $\|\widetilde{ab}\|_{L^\infty(B^2)}$
For each $x \in B^2$, choose a complex homographic transformation of B^2, T, such that $T(0) = x$ †, and apply inequality (3.4) to $a \circ T$ and $b \circ T$. We obtain

$$|\widetilde{(a \circ T)(b \circ T)}(0)| \leq \frac{1}{2\pi} \|d(a \circ T)\|_{L^2(B^2)} \|d(b \circ T)\|_{L^2(B^2)}.$$

But, by (3.2) (see remark 3.1.1) and the conformal invariance of the Dirichlet integral, this implies

$$|\widetilde{ab}(x)| = |\widetilde{ab} \circ T(0)|$$
$$\leq \frac{1}{2\pi} \|da\|_{L^2(B^2)} \|db\|_{L^2(B^2)},$$

and hence

$$\|\widetilde{ab}\|_{L^\infty(B^2)} \leq \frac{1}{2\pi} \|da\|_{L^2(B^2)} \|db\|_{L^2(B^2)}. \qquad (3.5)$$

Step 3 ESTIMATE FOR $\|d\widetilde{ab}\|_{L^2(B^2)}$
Using the Dirichlet boundary condition we have

† for instance $T(z) = \frac{z+x}{1+z\overline{x}}$

$$\|d\widetilde{ab}\|^2_{L^2(B^2)} = \int_{B^2} -\widetilde{ab}\Delta\widetilde{ab}\,dx^1 dx^2$$

$$= \int_{B^2} \widetilde{ab}\{a,b\}\,dx^1 dx^2$$

$$\leq \|\widetilde{ab}\|_{L^\infty(B^2)}\|\{a,b\}\|_{L^1(B^2)}$$

$$\leq \frac{1}{2\pi}\|da\|^2_{L^2}\|db\|^2_{L^2},$$

which implies inequality (3.3) with $C = \frac{1}{2\pi} + \frac{1}{\sqrt{2\pi}}$, for $a, b \in H^1(B^2, \mathbb{R}) \cap \mathcal{C}^\infty(B^2, \mathbb{R})$.

Step 4 CONCLUSION
Let $a, b \in H^1(B^2, \mathbb{R})$, and consider two sequences $a_n, b_n \in H^1(B^2, \mathbb{R}) \cap \mathcal{C}^\infty(B^2, \mathbb{R})$ such that

$$a_n \to a \text{ in } H^1(B^2),$$
$$b_n \to b \text{ in } H^1(B^2).$$

Using (3.3), we easily show that $\widetilde{a_n b_n}$ is a Cauchy sequence in $H^1(B^2, \mathbb{R}) \cap L^\infty(B^2, \mathbb{R})$. Thus, there exists $\phi \in H^1(B^2, \mathbb{R}) \cap L^\infty(B^2, \mathbb{R})$ such that $\widetilde{a_n b_n} \to \phi$ in $H^1(B^2)$ and $L^\infty(B^2)$, and

$$\|d\phi\|_{L^2(B^2)} + \|\phi\|_{L^\infty(B^2)} \leq C\|da\|_{L^2(B)}\|db\|_{L^2(B^2)}.$$

But since each $\widetilde{a_n b_n}$ belongs to $\mathcal{C}^\infty(B^2, \mathbb{R})$, this implies that ϕ is continuous. Moreover, we can independently show that $\widetilde{a_n b_n}$ converges to $\widetilde{ab}$ in a bigger space than $H^1(B^2, \mathbb{R})$, like $W^{1,p}(B^2, \mathbb{R})$, for $p < 2$. This implies that $\phi = \widetilde{ab}$, which concludes the proof. □

An important question is whether this result extends to the case where the unit ball B^2 is replaced by any open set Ω of $\mathbb{R}^2$. In case Ω is conformally equivalent to B^2 the answer is quite simple, if we keep in mind remark 3.1.1 and the fact that the norms $\|a\|_{L^\infty(\Omega)}$ and $\|da\|_{L^2(\Omega)}$ are invariant under conformal transformations: theorem 3.1.2 extends easily to the case of an open subset conformally equivalent to B^2, and inequality (3.3) remains valid with the same constant. However, if Ω has a non-trivial topology, this will demand more work and a first answer is:

3.1 Wente's inequality

Theorem 3.1.3 *[14] There is a positive constant C, such that for every open subset Ω of $\mathbb{R}^2$, and for any functions $a, b \in H^1(\Omega, \mathbb{R})$,*

$$\|\widetilde{ab}\|_{L^\infty(\Omega)} + \|d\widetilde{ab}\|_{L^2(\Omega)} \leq C\|da\|_{L^2(\Omega)}\|db\|_{L^2(\Omega)}. \quad (3.6)$$

The same authors (Fabrice Bethuel and Jean-Michel Ghidaglia) have also generalized this result to the case where the operator $-\Delta$ is replaced by an elliptic operator of the form

$$L = -\sum_{i,j=1}^{2} \frac{\partial}{\partial x^i}\left(a^{ij}(x)\frac{\partial}{\partial x^j}\right),$$

where the a^{ij} are L^∞ functions on Ω such that $\exists \alpha \in \mathbb{R}_+^*, \forall \xi \in \mathbb{R}^2$,

$$\alpha\|\xi\|^2 \leq \sum_{i,j=1}^{2} a^{ij}\xi^i\xi^j \leq \frac{1}{\alpha}\|\xi\|^2.$$

It is natural to wonder if the constant C of theorem 3.1.3 is greater than or equal to that in theorem 3.1.2. More precisely, we may consider the following two problems and we shall enlarge our setting to Riemann surfaces.

3.1.2 The inequality on a Riemann surface

Let $(\mathcal{M}, g)$ be a Riemannian surface, with or without boundary. For $a, b \in H^1(\mathcal{M}, \mathbb{R})$, let $\widetilde{ab}_\mathcal{M}$ be the solution of the equation

$$\begin{cases} -\Delta\phi = \{a,b\} & \text{on } \mathcal{M} \\ \phi = 0 & \text{on } \partial\mathcal{M}, \text{ if } \partial\mathcal{M} \neq \emptyset \\ \phi(m_0) = 0 & \text{at a certain } m_0 \in \mathcal{M} \text{ if } \partial\mathcal{M} = \emptyset. \end{cases} \quad (3.7)$$

Notice that $\widetilde{ab}_\mathcal{M}$ exists even in the case where $\partial\mathcal{M} = \emptyset$, since then $\int_\mathcal{M}\{a,b\}dxdy = 0$. Let

$$C_\infty(\mathcal{M}, g) := \sup_{a,b \neq 0} \frac{\sup(\widetilde{ab}_\mathcal{M}) - \inf(\widetilde{ab}_\mathcal{M})}{\|da\|_{L^2(\mathcal{M})}\|db\|_{L^2(\mathcal{M})}},$$

$$C_2(\mathcal{M}, g) := \sup_{a,b \neq 0} \frac{\|d(\widetilde{ab}_\mathcal{M})\|_{L^2(\mathcal{M})}}{\|da\|_{L^2(\mathcal{M})}\|db\|_{L^2(\mathcal{M})}}.$$

We remark that the norms used in this definition are conformally invariant. Therefore, by (3.2), $C_\infty(\mathcal{M},g)$ and $C_2(\mathcal{M},g)$ depend only on the conformal class of $(\mathcal{M},g)$. So these quantities actually make sense on a Riemann surface.

Calculation of $C_2(\mathcal{M})$ and $C_\infty(\mathcal{M})$

3.1.2.1 Study of $C_2(\mathcal{M})$

We start by noticing that because of the symmetry

$$\widetilde{(ta)\left(\frac{b}{t}\right)} = \widetilde{ab}, \quad \forall t \in \mathbb{R}^*,$$

and the inequality

$$\|da\|_{L^2}\|db\|_{L^2} \leq \frac{1}{2}(\|da\|_{L^2}^2 + \|db\|_{L^2}^2)$$

$$=: \frac{1}{2}\|d(a,b)\|_{L^2}^2,$$

an equivalent definition for $C_2(\mathcal{M})$ is

$$C_2(\mathcal{M}) = \sup_{a,b \in H^1(\mathcal{M}),\, (a,b) \neq 0} \frac{2\|\widetilde{dab}\|_{L^2(\mathcal{M})}}{\|d(a,b)\|_{L^2(\mathcal{M})}^2}.$$

Determining $C_2(\mathcal{M})$ is equivalent to determining the norm of the operator

$$H^1(\mathcal{M}, \mathbb{R}^2) \longrightarrow H^1(\mathcal{M}, \mathbb{R})$$
$$(a,b) \longmapsto \widetilde{ab},$$

where $H^1(\mathcal{M}, \mathbb{R})$ is equipped with the semi-norm $\|d\phi\|_{L^2(\mathcal{M})}$. This problem is analogous to the study of the continuous injection of $H^1(\mathbb{R}^n)$ into $L^{\frac{2n}{n-2}}(\mathbb{R}^n)$. This suggests that the variational problem

$$\inf_{\|\widetilde{dab}\|_{L^2}=1} \|d(a,b)\|_{L^2}^2 \tag{3.8}$$

is probably of interest. In fact, this variational problem is connected in a surprising way to constant mean curvature surfaces in $\mathbb{R}^3$.

Theorem 3.1.4 *Let $(a,b) \in H^1(\mathcal{M}, \mathbb{R}^2)$ be a critical point of $\|d(a,b)\|_{L^2(\mathcal{M})}^2$,*

3.1 Wente's inequality

satisfying the constraint $\|d\widetilde{ab}\|_{L^2(\mathcal{M})} = 1$. Then a, b and $\widetilde{ab}$ satisfy the equation

$$\begin{cases} -\Delta a &= \lambda \{b, \widetilde{ab}\} \\ -\Delta b &= \lambda \{\widetilde{ab}, a\} \\ -\Delta \widetilde{ab} &= \{a, b\} \end{cases} \tag{3.9}$$

on $\mathcal{M}$, with the boundary conditions

$$\begin{cases} \dfrac{\partial a}{\partial n} = \dfrac{\partial b}{\partial n} = 0 & \text{on } \partial \mathcal{M} \\ \widetilde{ab} = 0 & \text{on } \partial \mathcal{M}, \end{cases} \tag{3.10}$$

where $\lambda = \|da\|^2_{L^2(\mathcal{M})} = \|db\|^2_{L^2(\mathcal{M})}$. Equivalently, if we define u from $\mathcal{M}$ to $\mathbb{R}^3$ by

$$u = \frac{1}{2} \begin{pmatrix} \sqrt{\lambda} a \\ \sqrt{\lambda} b \\ \lambda \widetilde{ab} \end{pmatrix},$$

then

$$-\Delta u = 2 \frac{\partial u}{\partial x} \times \frac{\partial u}{\partial y},$$

in local conformal coordinates. Moreover, a, b and $\widetilde{ab}$ are $\mathcal{C}^\infty$ on $\mathcal{M}$.

Remark 3.1.5 *We may be led to think that the map u constructed above parametrizes a surface of constant mean curvature equal to 1. However, this is not so unless u is conformal, which supposes knowing that*

$$\left| \frac{\partial u}{\partial x} \right|^2 - \left| \frac{\partial u}{\partial y} \right|^2 - 2i \left\langle \frac{\partial u}{\partial x}, \frac{\partial u}{\partial y} \right\rangle = 0.$$

Furthermore, in case $\partial \mathcal{M} \neq \emptyset$, due to the boundary conditions (3.10), the image of $\partial \mathcal{M}$ by u is contained in the plane $u^3 = 0$, and if u is an immersion, then $u(\mathcal{M})$ intersects the plane $u^3 = 0$ orthogonally along $u(\partial \mathcal{M})$. Actually, whether u is an immersion or not, the above implies that we can always construct a C^∞ map from $\mathcal{M} \cup_{\partial \mathcal{M}} \overline{\mathcal{M}}$ to $\mathbb{R}^3$, extending u by reflection across $\partial \mathcal{M}$. Here $\mathcal{M} \cup_{\partial \mathcal{M}} \overline{\mathcal{M}}$ is the Riemann surface obtained by gluing along $\partial \mathcal{M}$, $\mathcal{M}$ and $\overline{\mathcal{M}}$, a copy of $\mathcal{M}$ with the opposite orientation.

The extension v is constructed by letting

$$v = \begin{pmatrix} u^1 \\ u^2 \\ u^3 \end{pmatrix} \text{ on } \mathcal{M} \text{ and } v = \begin{pmatrix} u^1 \\ u^2 \\ -u^3 \end{pmatrix} \text{ on } \overline{\mathcal{M}}.$$

In the proof of theorem 3.1.4, we will need the following lemma, whose proof is left to the reader.

Lemma 3.1.6 *Let a, b, c be three functions in $H^1(\mathcal{M}, \mathbb{R})$. Suppose that either $\partial \mathcal{M} \neq \emptyset$ and at least one of the three functions has vanishing trace over $\partial \mathcal{M}$, or $\partial \mathcal{M} = \emptyset$. Then*

$$\int_{\mathcal{M}} a\{b,c\} dx dy = \int_{\mathcal{M}} c\{a,b\} dx dy = \int_{\mathcal{M}} b\{c,a\} dx dy. \qquad (3.11)$$

Proof of theorem 3.1.4 Let (a,b) be a critical point in $H^1(\mathcal{M})$ of the variational problem (3.8). Then, for any functions $\alpha, \beta \in H^1(\mathcal{M})$, the first variation formula of $\|\widetilde{dab}\|_{L^2}^2$ under the change $(a,b) \longmapsto (a+t\alpha, b+t\beta)$ is

$$\begin{aligned}
\delta \|\widetilde{dab}\|_{L^2}^2 &= \int_{\mathcal{M}} 2\langle \widetilde{dab}, d(\widetilde{\alpha b} + \widetilde{a\beta})\rangle dx dy \\
&= -2 \int_{\mathcal{M}} \widetilde{ab} \Delta(\widetilde{\alpha b} + \widetilde{a\beta}) dx dy \\
&= -2 \int_{\mathcal{M}} \widetilde{ab}(\{\alpha, b\} + \{a, \beta\}) dx dy \\
&= -2 \int_{\mathcal{M}} (\{b, \widetilde{ab}\}\alpha + \{\widetilde{ab}, a\}\beta) dx dy,
\end{aligned}$$

where we used lemma 3.1.6 in the last equality. Likewise,

$$\begin{aligned}
\delta \|d(a,b)\|_{L^2}^2(\alpha, \beta) &= 2 \int_{\partial \mathcal{M}} \left(\alpha \frac{\partial a}{\partial n} + \beta \frac{\partial b}{\partial n} \right) ds \\
&\quad -2 \int_{\mathcal{M}} (\Delta a\, \alpha + \Delta b\, \beta) dx dy.
\end{aligned}$$

3.1 Wente's inequality

The assumption that (a,b) is a critical point implies that for any $\alpha, \beta \in H^1(\mathcal{M})$ such that $\delta \|d(a,b)\|_{L^2}^2 (\alpha, \beta) = 0$, we have $\delta \|d\widetilde{ab}\|_{L^2}^2 (\alpha, \beta) = 0$.

This yields that there exists a Lagrange multiplier $\lambda \in \mathbb{R}$ such that

$$\begin{cases} -\Delta a = \lambda \{b, \widetilde{ab}\} \\ -\Delta b = \lambda \{\widetilde{ab}, a\}, \end{cases} \tag{3.12}$$

and that

$$\frac{\partial a}{\partial n} = \frac{\partial b}{\partial n} = 0$$

on $\partial \mathcal{M}$.

Multiplying the first equation of (3.12) by a, and integrating over $\mathcal{M}$, we obtain

$$\int_{\mathcal{M}} -\Delta a \, a \, dx dy = \int_{\mathcal{M}} \lambda a \{b, \widetilde{ab}\} dx dy. \tag{3.13}$$

But since $\frac{\partial a}{\partial n} = 0$ on $\partial \mathcal{M}$,

$$\int_{\mathcal{M}} -\Delta a \, a \, dx dy = \int_{\mathcal{M}} |da|^2 dx dy \tag{3.14}$$

and using lemma 3.1.6 again,

$$\begin{aligned} \int_{\mathcal{M}} a\{b, \widetilde{ab}\} dx dy &= \int_{\mathcal{M}} \widetilde{ab} \{a, b\} dx dy \\ &= \int_{\mathcal{M}} -\Delta(\widetilde{ab}) \widetilde{ab} \, dx dy \\ &= \int_{\mathcal{M}} |d\widetilde{ab}|^2 dx dy. \end{aligned} \tag{3.15}$$

Since by hypothesis $\|d\widetilde{ab}\|_{L^2(\mathcal{M})} = 1$, it follows from (3.13), (3.14) and (3.15) that

$$\lambda = \|da\|_{L^2(\mathcal{M})}^2.$$

A similar calculation, starting from the second equation in (3.12), leads to

$$\lambda = \|db\|_{L^2(\mathcal{M})}^2.$$

Knowing that λ is a strictly positive real, we can define u by

$$u = \frac{1}{2}\begin{pmatrix} \sqrt{\tilde{\lambda}}a \\ \sqrt{\tilde{\lambda}}b \\ \tilde{\lambda}ab \end{pmatrix},$$

and a simple calculation, using (3.12), yields

$$-\Delta u = 2\frac{\partial u}{\partial x} \times \frac{\partial u}{\partial y}.$$

Wente's theorem 3.1.2 then gives us that u, and consequently a, b and $\widetilde{ab}$, are $\mathcal{C}^\infty$ on $\mathcal{M}$. $\square$

Yuxin Ge determined the best constants C_2 for every Riemann surface. We reproduce here part of his proofs, which emphasize the link between the variational problem (3.8) and the isoperimetric inequality in $\mathbb{R}^3$.

Theorem 3.1.7 [64]

(i) If $\mathcal{M} = B^2$, then $C_2(\mathcal{M}) = \sqrt{\frac{3}{16\pi}}$, and moreover, the minimum in (3.8) is attained for

$$(a,b): B^2 \longrightarrow \mathbb{R}^2$$
$$(x,y) \longmapsto \frac{2(x,y)}{1+x^2+y^2}.$$

(ii) If $\mathcal{M} = S^2$, then $C_2(\mathcal{M}) = \sqrt{\frac{3}{32\pi}}$, and moreover, the minimum in (3.8) is attained for

$$(a,b): S^2 \longrightarrow \mathbb{R}^2$$
$$\begin{pmatrix} x \\ y \\ z \end{pmatrix} \longmapsto (x,y).$$

(iii) If $\mathcal{M}$ is a surface with non-empty boundary and is not diffeomorphic to B^2, then $C_2(\mathcal{M}) = \sqrt{\frac{3}{16\pi}}$ and the infimum in (3.8) is not attained.

(iv) If $\mathcal{M}$ is a surface without boundary and is not diffeomorphic to S^2, then $C_2(\mathcal{M}) = \sqrt{\frac{3}{32\pi}}$ and the infimum in (3.8) is not attained.

Proof of (i) *and* (ii)
(i) Let $(a,b) \in C^\infty(B^2, \mathbb{R}^2) \cap H^1(B^2, \mathbb{R}^2)$. We construct a map u from S^2 to $\mathbb{R}^3$ as follows. We identify S^2 with $\mathbb{C} \cup \{\infty\}$ via the stereographic projection, and let

$$u(z) = \begin{pmatrix} a(z) \\ b(z) \\ t\,\widetilde{ab}(z) \end{pmatrix} \quad \text{if } |z| \leq 1,$$

and

$$u(z) = \begin{pmatrix} a(z|z|^{-2}) \\ b(z|z|^{-2}) \\ -t\,\widetilde{ab}(z|z|^{-2}) \end{pmatrix} \quad \text{if } |z| \geq 1,$$

where t is a real parameter. Let V be the algebraic volume enclosed by the image of u in $\mathbb{R}^3$, and A the area of the image of u.

On the one hand,

$$\begin{aligned} V &= 2\int_{B^2} u^3\{u^1, u^2\}\,dxdy \\ &= 2t\int_{B^2} \widetilde{ab}\{a,b\}\,dxdy \\ &= -2t\int_{B^2} \widetilde{ab}\Delta(\widetilde{ab})\,dxdy \\ &= 2t\int_{B^2} |d\widetilde{ab}|^2\,dxdy\,. \end{aligned} \qquad (3.16)$$

On the other hand,

$$\begin{aligned} A &= 2\int_{B^2} \left|\frac{\partial u}{\partial x} \times \frac{\partial u}{\partial y}\right| \leq \int_{B^2} |du|^2\,dxdy \\ &= \int_{B^2} (|da|^2 + |db|^2 + t^2|d\widetilde{ab}|^2)\,dxdy\,. \end{aligned} \qquad (3.17)$$

Using (3.16), (3.17) and the isoperimetric inequality

$$36\pi V^2 \leq |A|^3,$$

we obtain

$$144\pi t^2 \|d\widetilde{ab}\|_{L^2}^4 \leq \left(\|d(a,b)\|_{L^2}^2 + t^2\|d\widetilde{ab}\|_{L^2}^2\right)^3.$$

Choosing

$$t^2 = \frac{\|d(a,b)\|_{L^2}^2}{2\|d\widetilde{ab}\|_{L^2}^2}$$

in this inequality, it follows that

$$\|d\widetilde{ab}\|_{L^2}^2 \leq \frac{3}{16\pi}\left(\frac{\|d(a,b)\|_{L^2}^2}{2}\right)^2, \qquad (3.18)$$

which yields that $C_2(B) \leq \sqrt{\frac{3}{16\pi}}$, since $\mathcal{C}^\infty(B,\mathbb{R}^2) \cap H^1(B,\mathbb{R}^2)$ is dense in $H^1(B,\mathbb{R}^2)$.

This value is attained for $(a,b) = \frac{(2x,2y)}{1+x^2+y^2}$ since in this case $\widetilde{ab} = \frac{1-x^2-y^2}{2(1+x^2+y^2)}$, and we have equality in inequality (3.18).

(ii) The approach is the same for $C_2(S^2)$. We start from $(a,b) \in \mathcal{C}^\infty(S^2,\mathbb{R}^2)$, and we consider the map $u : S^2 \longrightarrow \mathbb{R}^3$,

$$u = \begin{pmatrix} a \\ b \\ t\widetilde{ab} \end{pmatrix},$$

where t is a real parameter. We estimate the area, A, of the image of u:

$$A = \int_{S^2} \left|\frac{\partial u}{\partial x} \times \frac{\partial u}{\partial y}\right| dx dy \leq \frac{1}{2}\|d(a,b)\|_{L^2}^2 + \frac{t^2}{2}\|d\widetilde{ab}\|_{L^2}^2$$

and the algebraic volume enclosed by the image of u,

$$V = \int_{S^2} u^3\{u^1,u^2\} dx dy = t\|d\widetilde{ab}\|_{L^2}^2.$$

Then, using the isoperimetric inequality

$$36\pi V^2 \leq |A|^3,$$

we obtain

$$36\pi t^2 \|\widetilde{dab}\|_{L^2}^4 \leq \left(\frac{\|d(a,b)\|_{L^2}^2 + t^2\|\widetilde{dab}\|_{L^2}^2}{2}\right)^3.$$

We then choose $t^2 = \frac{\|d(a,b)\|_{L^2}^2}{2\|\widetilde{dab}\|_{L^2}^2}$ which yields $C_2(S^2) \leq \sqrt{\frac{3}{32\pi}}$. Finally, we check that this value is attained for

$$(a, b, \widetilde{ab}) : S^2 \longrightarrow \mathbb{R}^3$$
$$\begin{pmatrix} x \\ y \\ z \end{pmatrix} \longmapsto (x, y, \frac{z}{2})$$

□

Problem 3.1.8 *For a long time it was believed that there was no constant mean curvature immersion of the torus in* $\mathbb{R}^3$. *In 1984, Henry Wente gave an example of such an immersion, thus contradicting what used to be called the Hopf conjecture [179]. Two years later, U. Abresch gave a quasi-explicit expression of such an immersion, using elliptic integrals [1]. The surfaces constructed by Wente and Abresch have the common feature of being invariant under symmetry with respect to a plane.*

We can then obtain them by starting from a constant mean curvature immersion of an annulus of the plane into $\mathbb{R}^3$, *such that the image of the boundary of the annulus is contained in a plane* P *of* $\mathbb{R}^3$, *and the image of the annulus intersects the plane* P *orthogonally along the image of the boundary. In fact, it suffices to complete this immersion by gluing the immersion of a copy of the annulus, which is the mirror image in* P *of the first immersion, to obtain an immersion of a torus.*

We recognize here the construction described in remark 3.1.5. We thus conclude that the immersions of the annulus constructed by Wente and Abresch correspond to critical points of the variational problem (3.8).

A natural question will then be: can we find these immersions using a variational method? (see [64], [65] and [66]).

3.1.2.2 Study of $C_\infty(\mathcal{M})$

A study similar to the one done above for $C_2(\mathcal{M})$ was partially carried out by Sami Baraket and, very recently, completed by Peter Topping.

Theorem 3.1.9 *[178], [6], [170] For every Riemann surface $\mathcal{M}$, with or without boundary, we have*

$$C_\infty(\mathcal{M}) = \frac{1}{2\pi}.$$

We will not prove this result. The inequality $C_\infty(\mathcal{M}) \leq \frac{1}{2\pi}$ was proved in [178] and [6], in case Ω is conformally equivalent to the ball B^2, and in [170] in the general case. The inverse inequality $C_\infty(\mathcal{M}) \geq \frac{1}{2\pi}$ is due to [6].

3.2 Hardy spaces

In 1989, Stefan Müller noticed the following property. Let Ω be an open subset of $\mathbb{R}^m$ and denote by $L^1 \log L^1(\Omega)$ the set of measurable functions f from Ω to $\mathbb{R}$ (or $\mathbb{C}$) such that

$$\int_\Omega |f|(1 + \log(1 + |f|))dx < +\infty. \tag{3.19}$$

Theorem 3.2.1 *[120] Let $u \in W^{1,m}(\Omega, \mathbb{R}^m)$, and*

$$f(x) = \det(du(x)) \quad a.e.$$

Then, if $f(x) \geq 0$ a.e., $f \in L^1 \log L^1(\Omega)$.

A special case of this result is that if $\Omega \subset \mathbb{R}^2$ and $u = (a,b) \in H^1(\Omega, \mathbb{R}^2)$, then $\{a,b\} \in L^1 \log L^1(\Omega)$, as long as $\{a,b\} \geq 0$ a.e. Hence, we see that the algebraic structure of $\det(du(x))$ yields a slight improvement of the integrability of this function which, a priori, we would just expect to be in $L^1(\Omega)$.

We know a space close to $L^1(\Omega)$, the Hardy space $\mathcal{H}^1(\Omega)$, which has an entirely analogous property to the one described in theorem 3.2.1: $\mathcal{H}^1(\Omega)$ is a subspace of $L^1(\Omega)$, and any function f in $\mathcal{H}^1(\Omega)$ which is positive a.e. is necessarily in $L^1 \log L^1$ (see [156]).

This somehow suggests the following result, due to Ronald Coifman, Pierre-Louis Lions, Yves Meyer and Stephen Semmes.

Theorem 3.2.2 *[39] Let u be a map in $W^{1,m}(\mathbb{R}^m)$. Then $f := \det(du)$ belongs to $\mathcal{H}^1(\mathbb{R}^m)$, and moreover, there exists a constant C_m, depending only on m, such that*

$$\|f\|_{\mathcal{H}^1(\mathbb{R}^m)} \leq C_m \|u\|_{W^{1,m}(\mathbb{R}^m)} . \tag{3.20}$$

Another result, which coincides with theorem 3.2.2 in case $m = 2$, is the following.

Theorem 3.2.3 *[39] Let ϕ be a map in $H^1(\mathbb{R}^m, \mathbb{R})$, and E a divergence-free vector field in $L^2(\mathbb{R}^m, \mathbb{R}^m)$. Then the quantity*

$$f := \langle \mathrm{grad}\phi, E \rangle = \sum_{\alpha=1}^{m} \frac{\partial \phi}{\partial x^\alpha} E^\alpha$$

belongs to $\mathcal{H}^1(\mathbb{R}^m)$. Moreover, there is a constant C_m, depending only on m, such that

$$\|f\|_{\mathcal{H}^1(\mathbb{R}^m)} \leq C_m \|\phi\|_{H^1(\mathbb{R}^m)} \|E\|_{L^2(\mathbb{R}^m)} . \tag{3.21}$$

Before proceeding, we need to define what $\mathcal{H}^1(\mathbb{R}^m)$ is (the definition of $\mathcal{H}^1(\Omega)$ being considerably harder). We will give several definitions. The equivalence between these different definitions is far from obvious, and relies on a difficult theorem by Charles Fefferman ([57], [58]).

Definition 3.2.4 (Hardy space 1) *Let Ψ be a $\mathcal{C}_c^\infty(\mathbb{R}^m)$ function such that*

$$\int_{\mathbb{R}^m} \Psi = 1 .$$

For each $t > 0$, we let $\Psi_t(x) = t^{-m} \Psi\left(\frac{x}{t}\right)$. Finally, for each function $f \in L^1(\mathbb{R}^m)$, we define

$$f^*(x) = \sup_{t>0} |(\Psi_t * f)(x)| .$$

Then, f belongs to Hardy space $\mathcal{H}^1(\mathbb{R}^m)$ if and only if $f^* \in L^1(\mathbb{R}^m)$. Hardy space is equipped with the norm

$$\|f\|_{\mathcal{H}^1(\mathbb{R}^m)} = \|f\|_{L^1(\mathbb{R}^m)} + \|f^*\|_{L^1(\mathbb{R}^m)} . \tag{3.22}$$

Definition 3.2.5 (Hardy space 2) *For any function f in $L^1(\mathbb{R}^m)$, we denote by $R_\alpha f$ the function defined by*

$$\mathcal{F}(R_\alpha f)(\xi) = \frac{\xi_\alpha}{|\xi|} \mathcal{F}(f)(\xi),$$

where

$$\mathcal{F}(f)(\xi) = \frac{1}{(2\pi)^{m/2}} \int_{\mathbb{R}^m} e^{-i\langle x,\xi\rangle} f(x) dx$$

is the Fourier transform of f. $R_\alpha f$ is the Riesz transform of f. Then f belongs to $\mathcal{H}^1(\mathbb{R}^m)$ if and only if

$$\forall \alpha = 1, ..., m, \quad R_\alpha f \in L^1(\mathbb{R}^m).$$

We define a second norm on $\mathcal{H}^1(\mathbb{R}^m)$ by

$$\|f\|_{\mathcal{H}^1(\mathbb{R}^m)} = \|f\|_{L^1(\mathbb{R}^m)} + \sum_{\alpha=1}^m \|R_\alpha f\|_{L^1(\mathbb{R}^m)}. \tag{3.23}$$

Remark 3.2.6 *Although we have used the same notation, the norms on $\mathcal{H}^1(\mathbb{R}^m)$ defined by (3.22) and (3.23) are not the same. Nevertheless, they are equivalent.*

Definition 3.2.7 (*VMO* **and** *BMO* **space**) *For any locally integrable function f from $\mathbb{R}^m$ to $\mathbb{R}$, we define, for each $x \in \mathbb{R}^m, r > 0$,*

$$f_{x,r} = \frac{1}{|B(x,r)|} \int_{B(x,r)} f(y) dy,$$

and we let

$$\|f\|_{BMO(\mathbb{R}^m)} = \sup_{x\in\mathbb{R}^m} \sup_{r>0} \frac{1}{|B(x,r)|} \int_{B(x,r)} |f(y) - f_{x,r}| dy.$$

The space $BMO(\mathbb{R}^m)$ (Bounded Mean Oscillations) is the space of functions f such that $\|f\|_{BMO(\mathbb{R}^m)} < +\infty$. The quantity $\|f\|_{BMO(\mathbb{R}^m)}$ is a semi-norm ($\|f\|_{BMO(\mathbb{R}^m)} = 0$ if and only if f is equal to a constant a.e.). The quotient $BMO(\mathbb{R}^m)/\mathbb{R}$ is a Banach space. The space $VMO(\mathbb{R}^m)$ (Vanishing Mean Oscillations) is the subspace of functions f in $BMO(\mathbb{R}^m)$ such that $\forall x \in \mathbb{R}^m$,

$$\lim_{r\to 0} \frac{1}{|B(x,r)|} \int_{B(x,r)} |f(y) - f_{x,r}| dy = 0.$$

Definition 3.2.8 (Hardy space 3) $\mathcal{H}^1(\mathbb{R}^m)$ is the dual space of $VMO(\mathbb{R}^m)$. The dual space of $\mathcal{H}^1(\mathbb{R}^m)$ is $BMO(\mathbb{R}^m)$.

Notice that any of the definitions given enables us to easily show that if $f \in \mathcal{H}^1(\mathbb{R}^m)$, then $\int_{\mathbb{R}^m} f = 0$.

There are several ways of apprehending this strange space. A first way is to look at a function f in $\mathcal{H}^1(\mathbb{R}^m)$ through a microscope, and to see that even if f oscillates violently (but not too much, since $f \in L^1(\mathbb{R}^m)$), these oscillations are balanced at all scales, i.e. calculating the mean value of f in a ball of arbitrary position and size, the oscillations cancel each other out.

Such a point of view is illustrated by other definitions: the atomic one [40], or that using wavelets [114].

A second way is to think of Hardy space as the biggest subspace of $L^1(\mathbb{R}^m)$ for which we can use the recipes valid in $L^p(\mathbb{R}^m)$ for $1 < p < +\infty$, but which are no longer valid for $L^1(\mathbb{R}^m)$. An example of such a property was already mentioned in the previous chapter in relation to Wente's lemma. If ϕ is a solution of

$$-\Delta \phi = f, \quad \text{with } f \in L^p(\Omega),$$

and $1 < p < +\infty$, then, if the boundary conditions for ϕ are sufficiently smooth, $\phi \in W^{2,p}(\Omega)$. This follows from classical Calderón-Zygmund theory (see [156]). Such a result is false for $p = 1$, but we have:

Theorem 3.2.9 *[157]* Let $\phi \in L^1(\mathbb{R}^m)$ be a solution in $\mathbb{R}^m$ of

$$-\Delta \phi = f \in \mathcal{H}^1(\mathbb{R}^m).$$

Then all the second derivatives of ϕ belong to $L^1(\mathbb{R}^m)$, and

$$\left\| \frac{\partial^2 \phi}{\partial x^\alpha \partial x^\beta} \right\|_{L^1(\mathbb{R}^m)} \leq C \, \| f \|_{\mathcal{H}^1(\mathbb{R}^m)}.$$

Likewise, if $1 < p < +\infty$, every function $f \in L^p(\mathbb{R}^m)$ satisfies

$$f^* \in L^p(\mathbb{R}^m) \text{ and } R_\alpha f \in L^p(\mathbb{R}^m),$$

but this is no longer true for $p = 1$, unless we "replace" $L^1(\mathbb{R}^m)$ by $\mathcal{H}^1(\mathbb{R}^m)$.

Another property is that in dimension 2, if ϕ is a solution of

$$-\Delta \phi = f \in \mathcal{H}^1(\mathbb{R}^2),$$

then ϕ is continuous. We can deduce this from the fact that the kernel of $-\Delta$ on $\mathbb{R}^2$, $\frac{1}{2\pi} \log\left(\frac{1}{r}\right)$, belongs to $BMO(\mathbb{R}^2)$, using definition 3.2.8 of $\mathcal{H}^1(\mathbb{R}^2)$. In fact, we will see more precise results in the following section.

Now that we have quickly described $\mathcal{H}^1(\mathbb{R}^m)$, we see that theorems 3.2.2 and 3.2.3 establish subtle integrability improvements, and suggest the existence of a compensated integrability theory, analogous to that of Murat and Tartar. Theorem 3.2.3 is the prototype of a "div–curl" lemma for such a theory.

What is the interest of this space in the setting of harmonic maps? or, more generally, of non-linear partial differential equations? A first example is that of weakly harmonic maps from an open set Ω of $\mathbb{R}^m$ to S^2. Then it is clear that Wente's lemma is no longer very useful for studying the equation

$$-\Delta u = u|du|^2 \text{ in } \Omega, \qquad (3.24)$$

if $m \geq 3$.

Using $\mathcal{H}^1(\mathbb{R}^m)$, we can draw conclusions from this equation (by using the conservation laws that are hidden in (3.24)).

Lemma 3.2.10 *Let $u \in H^1(\Omega, S^n)$ be a weakly harmonic map. Then for any ball $B(x_0, r)$ whose closure is contained in Ω, there exists $f \in \mathcal{H}^1(\mathbb{R}^m, \mathbb{R}^{n+1})$ such that*

$$-\Delta u = f \quad \text{in } B(x_0, r). \qquad (3.25)$$

Remark 3.2.11 *This result seems to be just a slight improvement, but it enabled Lawrence C. Evans to prove that every stationary weakly harmonic map taking its values in a sphere is smooth outside a closed set whose Hausdorff $(m-2)$-dimensional measure is zero [54]. His proof uses in a crucial way the duality between $\mathcal{H}^1(\mathbb{R}^m)$ and $BMO(\mathbb{R}^m)$ (see theorem 3.5.1 below).*

Proof of lemma 3.2.10 We know that every solution u of (3.24) satisfies the conservation law

$$\operatorname{div}(u^i \nabla u^j - u^j \nabla u^i) = \sum_{\alpha=1}^{m} \frac{\partial}{\partial x^\alpha}\left(u^i \frac{\partial u^j}{\partial x^\alpha} - u^j \frac{\partial u^i}{\partial x^\alpha}\right) = 0. \quad (3.26)$$

It is convenient, when we work with an arbitrary number m of variables, to use the language of differential forms. Let $\star$ be the Hodge star operator acting on the set of differential 1-forms on Ω, $\Lambda^1 \Omega$. The action of $\star$ on 1-forms is given by

$$\star dx^\alpha = (-1)^{\alpha-1} dx^1 \wedge \ldots \wedge dx^{\alpha-1} \wedge dx^{\alpha+1} \wedge \ldots \wedge dx^m.$$

Define

$$\beta^{ij} = \star(u^i du^j - u^j du^i).$$

Then (3.26) may be written as

$$d\beta^{ij} = 0 \text{ in } \Omega. \quad (3.27)$$

Choose a ball B' contained in Ω, and containing $\overline{B(x_0, r)}$. Since B' is simply connected, there is an $(m-2)$-form α^{ij} on B', with coefficients in $H^1(B')$, such that

$$d\alpha^{ij} = \beta^{ij} \text{ in } B'. \quad (3.28)$$

(see lemma 4.3.6 in the next chapter).

Let χ be a function in $\mathcal{C}_c^\infty(B', \mathbb{R})$ taking the value 1 over $B(x_0, r)$, and let $f^i \in L^1(\mathbb{R})$ be defined by

$$\begin{cases} f^i dx^1 \wedge \ldots \wedge dx^m = \sum_{j=1}^{n+1} d(\chi u^j) \wedge d(\chi \alpha^{ij}) \text{ in } B', \\ f^i = 0 \text{ on } \mathbb{R}^m \setminus B'. \end{cases} \quad (3.29)$$

It is then easy to see that f^i is a sum of terms of the form $\langle \operatorname{grad}\phi, E \rangle$, where $\phi \in H^1(\mathbb{R}^m, \mathbb{R})$ (in fact, $\phi = \chi u^j$) and E is a divergence-free $L^2(\mathbb{R}^m, \mathbb{R}^m)$ vector field. Hence, it follows from theorem 3.2.3 that $f^i \in \mathcal{H}^1(\mathbb{R}^m)$.

Moreover, we can rewrite (3.24) as

$$-\Delta u^i = \sum_{j=1}^{n+1} \sum_{\alpha=1}^{m} \frac{\partial u^j}{\partial x^\alpha} \left(u^i \frac{\partial u^j}{\partial x^\alpha} - u^j \frac{\partial u^i}{\partial x^\alpha} \right) \text{ in } \Omega,$$

or also

$$-\Delta u^i dx^1 \wedge ... \wedge dx^m = \sum_{j=1}^{n+1} du^j \wedge \beta^{ij} \text{ in } \Omega.$$

We deduce from (3.28), (3.29) and (3.2) that $-\Delta u^i = f^i$ in B'. □

Remark 3.2.12 *Compensated compactness and regularity phenomena, of a quite different nature from those above, have recently been observed. Let $\phi \in H^1(\Omega, \mathbb{R})$ where Ω is an open subset of $\mathbb{R}^2$, and suppose that*

$$-\Delta \phi = f \quad in \, \Omega,$$

where f is either a positive Radon measure or an $L^1(\Omega)$ function. Then the quantity (which we write in complex notation for simplicity)

$$\omega = \left(\frac{\partial \phi}{\partial x} \right)^2 - \left(\frac{\partial \phi}{\partial y} \right)^2 - 2i \left(\frac{\partial \phi}{\partial x} \right) \left(\frac{\partial \phi}{\partial y} \right)$$

has compensation properties. More precisely, in the proof of an existence result for solutions of a 2-dimensional Euler equation for an incompressible fluid whose vorticity is a positive measure, Jean-Marc Delort proved that for any sequence (ϕ_n) converging weakly to ϕ in $H^1(\Omega, \mathbb{R})$ such that $(-\Delta \phi_n)$ is a bounded sequence of positive measures, the quantity ω_n associated to ϕ_n converges to ω in the sense of distributions [44].

In this result, the sign condition on $-\Delta f_n$ may be dropped, provided that the negative part of $-\Delta f_n$ belongs to $L^1(\Omega)$ [67]. Following this result, L.C. Evans and S. Müller showed that ω is in local Hardy space $\mathcal{H}^1_{loc}(\Omega)$ [55] (see also [152]).

The surprise is that ω does not have the same structure as the classical quantities of compensated compactness theory, and the geometric meaning of this result is unclear. We notice however that ω looks like the Hopf differential (see definition 1.3.10) and is holomorphic if $f = 0$.

3.3 Lorentz spaces

The use of Hardy spaces, and the theorems of the previous section, will enable us to state improved integrability results that are more precise than Wente's lemma. In order to do this, we need to define some new spaces, the Lorentz spaces, a sort of refinement of L^p spaces.

These spaces were first used in harmonic analysis (see, for instance, [111] or [158]). It was only later that people realized their impact on Sobolev embeddings and partial differential equations (see, for instance, [20], [23] and [165]). Here we will see that these spaces appear naturally in estimates for solutions of some differential equations.

For an $\mathbb{R}$-valued function f, on an open subset Ω of $\mathbb{R}^m$, the fact of belonging to a Lorentz space is determined by a condition on the non-decreasing rearrangement of $|f|$ on $[0, |\Omega|)$.

Definition 3.3.1 *Let $f : \Omega \longrightarrow \mathbb{R}$ be a measurable function. The non-increasing rearrangement of $|f|$ on $[0, |\Omega|)$ is the unique function, denoted by f^*, from $[0, |\Omega|)$ to $\mathbb{R}$ which is non-increasing and such that*

$$\text{measure } \{x \in \Omega \mid |f(x)| \geq s\} = \text{measure } \{t \in (0, |\Omega|) \mid f^*(t) \geq s\}.$$

Definition 3.3.2 *Let Ω be an open subset in $\mathbb{R}^m$, $p \in (1, +\infty)$, $q \in [1, +\infty]$. The Lorentz space $L^{(p,q)}(\Omega, \mathbb{R})$ is the set of measurable functions $f : \Omega \longrightarrow \mathbb{R}$ such that*

$$|f|_{(p,q)} = \left[\int_0^{+\infty} (t^{\frac{1}{p}} f^*(t))^q \frac{dt}{t} \right]^{\frac{1}{q}} < +\infty, \text{ if } q < +\infty,$$

or

$$|f|_{(p,\infty)} = \sup_{t>0} t^{\frac{1}{p}} f^*(t) < +\infty, \text{ if } q = +\infty.$$

These quantities $|f|_{(p,q)}$ are not norms since they do not satisfy the triangular inequality. However, if we define

$$f^{**}(t) = \frac{1}{t} \int_0^t f^*(s) ds$$

and

$$\|f\|_{(p,q)} = \left[\int_0^{+\infty} (t^{\frac{1}{p}} f^{**}(t))^q \frac{dt}{t}\right]^{\frac{1}{q}}, \text{ if } q < +\infty,$$

$$\|f\|_{(p,\infty)} = \sup_{t>0} t^{\frac{1}{p}} f^{**}(t),$$

we will have available in the Lorentz spaces norms which satisfy

$$\frac{1}{c}|f|_{p,q} \leq \|f\|_{p,q} \leq c|f|_{p,q}$$

(see [186]).

Hence the Lorentz spaces become Banach spaces. Classically these spaces show up in interpolation theory for linear operators between L^p spaces, as we will see below.

Each $L^{(p,q)}$ may be seen as a deformation of L^p. For instance, we have the strict inclusions

$$L^{(p,1)} \subset L^{(p,q')} \subset L^{(p,q'')} \subset L^{(p,\infty)},$$

if $1 < q' < q''$. Moreover,

$$L^{(p,p)} = L^p.$$

Furthermore, if $|\Omega|$ is finite, we have that for all q and q',

$$p < p' \Rightarrow L^{(p',q')} \subset L^{(p,q)}.$$

Finally, for $1 < p < +\infty$ and $1 \leq q \leq +\infty$, $L^{\left(\frac{p}{p-1}, \frac{q}{q-1}\right)}$ is the dual of $L^{(p,q)}$. For more details on Lorentz spaces, see [186], [158], [95] and [7].

The following interpolation result will be repeatedly used and will enable us to deduce estimates on Lorentz spaces from the corresponding estimates on L^p spaces.

Theorem 3.3.3 *[158], [95] Let Ω be an open subset of $\mathbb{R}^m$ and U an open subset of $\mathbb{R}^n$. Let r_0, r_1, p_0, p_1 be real numbers such that*

$$1 \leq r_0 < r_1 \leq \infty,$$

and

$$1 \leq p_0 \neq p_1 \leq \infty.$$

Let T be a linear operator whose domain D contains

$$\bigcup_{r_0 \leq r \leq r_1} L^r(\Omega),$$

and which maps continuously $L^{r_0}(\Omega)$ to $L^{p_0}(U)$, and $L^{r_1}(\Omega)$ to $L^{p_1}(U)$, with the norms

$$\forall f \in L^{r_0}(\Omega), \quad \|Tf\|_{L^{p_0}(U)} \leq k_0 \|f\|_{L^{r_0}(\Omega)},$$
$$\forall f \in L^{r_1}(\Omega), \quad \|Tf\|_{L^{p_1}(U)} \leq k_1 \|f\|_{L^{r_1}(\Omega)}.$$

Then, for each $1 \leq q \leq \infty$, and for every pair (p, r) such that $\exists \theta \in (0, 1)$,

$$\frac{1}{p} = \frac{1-\theta}{p_0} + \frac{\theta}{p_1} \quad \text{and} \quad \frac{1}{r} = \frac{1-\theta}{r_0} + \frac{\theta}{r_1},$$

f maps continuously $L^{(r,q)}(\Omega)$ to $L^{(p,q)}(U)$, and moreover,

$$\forall f \in L^{(r,q)}(\Omega), \quad \|Tf\|_{L^{(p,q)}(U)} \leq B_\theta \|f\|_{L^{(r,q)}(\Omega)},$$

where

$$B_\theta = \left(\frac{r}{|\gamma|p}\right)^{\frac{1}{q}} 2^{\frac{1}{p}} \left(\frac{rk_0}{r-r_0} + \frac{r_1 k_1}{r_1 - r}\right), \qquad (3.30)$$

and

$$\gamma = \left(\frac{1}{p_0} - \frac{1}{p}\right)\left(\frac{1}{r_0} - \frac{1}{r}\right)^{-1} = \left(\frac{1}{p} - \frac{1}{p_1}\right)\left(\frac{1}{r} - \frac{1}{r_1}\right)^{-1}.$$

Below, we will be essentially interested in the family L^2, i.e. the spaces $L^{(2,q)}$, for $1 \leq q \leq +\infty$. Three members of this family are particularly interesting: the mildest ($L^{(2,1)}$), the most violent ($L^{(2,\infty)}$), and the best-known ($L^{(2,2)} = L^2$).

Here are some results which illustrate these spaces (for $\mathbb{R}^2$).

Theorem 3.3.4 *Let Ω be an open subset of $\mathbb{R}^2$, with $\mathcal{C}^1$ boundary. Let $f \in H^1(\Omega)$ and suppose that $\frac{\partial f}{\partial x}$ and $\frac{\partial f}{\partial y}$ are in $L^{(2,1)}(\Omega)$. Then f is continuous and uniformly bounded in Ω.*

Remark 3.3.5 It is well-known that in general a function in $H^1(\Omega)$ is neither bounded nor continuous.

Theorem 3.3.6 Let Ω be an open subset of $\mathbb{R}^2$, with $\mathcal{C}^1$ boundary. Let $f \in L^1(\Omega)$, and ϕ be a solution of

$$\begin{cases} -\Delta \phi = f & \text{in } \Omega \\ \phi = 0 & \text{on } \partial\Omega. \end{cases} \quad (3.31)$$

Then $\frac{\partial \phi}{\partial x}, \frac{\partial \phi}{\partial y} \in L^{(2,\infty)}(\Omega)$, and there is a constant depending only on Ω, $C(\Omega)$, such that

$$\|d\phi\|_{L^{(2,\infty)}(\Omega)} \leq C(\Omega) \|f\|_{L^1(\Omega)}. \quad (3.32)$$

Remark 3.3.7 It is also well-known that in general a solution ϕ of (3.31) does not belong to $H^1(\Omega)$.

Theorem 3.3.8 Let Ω be an open subset of $\mathbb{R}^2$, with $\mathcal{C}^1$ boundary. Let $f \in \mathcal{H}^1(\mathbb{R}^2)$, and ϕ be a solution of

$$\begin{cases} -\Delta \phi = f & \text{in } \Omega \\ \phi = 0 & \text{on } \partial\Omega. \end{cases} \quad (3.33)$$

Then $\frac{\partial \phi}{\partial x}, \frac{\partial \phi}{\partial y} \in L^{(2,1)}(\Omega)$, and there is a constant depending only on Ω, $C(\Omega)$, such that

$$\|d\phi\|_{L^{(2,1)}(\Omega)} \leq C(\Omega) \|f\|_{\mathcal{H}^1(\mathbb{R}^2)}. \quad (3.34)$$

Before proving these results, we will present the deep reason that makes them be true. The kernel of the Laplacian in $\mathbb{R}^2$, i.e. the distribution $K \in \mathcal{D}'(\mathbb{R}^2)$ such that

$$-\Delta K = \delta_0 \text{ in } \mathbb{R}^2$$

is given by

$$K = \frac{1}{2\pi} \log\left(\frac{1}{r}\right),$$

and its derivative is

$$dK = -\frac{1}{2\pi r^2}(xdx + ydy).$$

We may check that K is in $BMO(\mathbb{R}^2)$, and that dK has coefficients in $L^{(2,\infty)}(\mathbb{R}^2)$. Thus, it is the duality between $BMO(\mathbb{R}^2)$ and $\mathcal{H}^1(\mathbb{R}^2)$, and that between $L^{(2,\infty)}$ and $L^{(2,1)}$, that are at the origin of these results.

Proof of theorem 3.3.4 Let f be a function in $\mathcal{C}^\infty(\overline{\Omega})$ with compact support. Consequently, its derivatives belong to $L^{(2,1)}(\Omega)$. By Whitney's method (see [19]), we may construct a compactly supported extension of f to $\mathbb{R}^2$, $\widehat{f}$. Since the boundary of Ω is $\mathcal{C}^1$, the extension operator $f \longmapsto \widehat{f}$ is continuous in all spaces $W^{1,p}$ for $1 \leq p \leq +\infty$ and hence, by the interpolation theorem 3.3.3, we may deduce that the derivatives of $\widehat{f}$ are in $L^{(2,1)}(\mathbb{R}^2)$, and that

$$\|d\widehat{f}\|_{L^{(2,1)}(\mathbb{R}^2)} \leq C\|df\|_{L^{(2,1)}(\Omega)} \tag{3.35}$$

where C is a constant depending only on Ω. Since $\widehat{f}$ is compactly supported we have, for all $z \in \mathbb{R}^2$,

$$\begin{aligned}
\widehat{f}(z) &= \int_{\mathbb{R}^2} \delta_0(z-\zeta)\widehat{f}(\zeta)d\zeta \\
&= \int_{\mathbb{R}^2} -\Delta K(z-\zeta)\widehat{f}(\zeta)d\zeta \\
&= \int_{\mathbb{R}^2} -\nabla K(z-\zeta).\nabla \widehat{f}(\zeta)d\zeta, \tag{3.36}
\end{aligned}$$

which yields that

$$\|\widehat{f}(z)\| \leq \|dK\|_{L^{(2,\infty)}(\mathbb{R}^2)}\|d\widehat{f}\|_{L^{(2,1)}(\mathbb{R}^2)}. \tag{3.37}$$

From (3.35) and (3.37) it follows that

$$\|f\|_{L^\infty(\Omega)} \leq \|\widehat{f}\|_{L^\infty(\mathbb{R}^2)} \leq C\|dK\|_{L^{(2,\infty)}(\mathbb{R}^2)}\|df\|_{L^{(2,1)}(\mathbb{R}^2)}. \tag{3.38}$$

This estimate implies, by the density of $\mathcal{C}_c^\infty(\overline{\Omega})$ in $\{f \in H^1(\Omega,\mathbb{R}) \mid df \in L^{(2,1)}(\Omega)\}$, that every function in $H^1(\Omega,\mathbb{R})$ with derivatives in $L^{(2,1)}(\Omega)$ is the uniform limit of a sequence of $\mathcal{C}^\infty$ functions and therefore is continuous. □

In the proofs of theorems 3.3.6 and 3.3.8, we will use the interpolation theorem 3.3.3 with a well chosen operator. This operator is obtained by a construction which we will use below and which may in general be useful for obtaining estimates with Lorentz norms. This is why we chose to present it separately.

Hodge decomposition of a vector field

Proposition 3.3.9 *We suppose that Ω is a bounded open subset of $\mathbb{R}^2$, with $\mathcal{C}^1$ boundary. To each vector field $g = (g_1, g_2) \in L^1(\Omega, \mathbb{R}^2)$, we associate the functions α, β, v on Ω, which are solutions of*

$$\begin{cases} -\Delta \alpha = -\left(\dfrac{\partial g_1}{\partial x} + \dfrac{\partial g_2}{\partial y}\right) & \text{in } \Omega \\ \alpha = 0 & \text{on } \partial\Omega, \end{cases}$$

$$\begin{cases} -\Delta \beta = \dfrac{\partial g_1}{\partial y} - \dfrac{\partial g_2}{\partial x} & \text{in } \Omega \\ \beta = 0 & \text{on } \partial\Omega, \end{cases}$$

and

$$\begin{cases} g_1 = \dfrac{\partial \alpha}{\partial x} + \dfrac{\partial \beta}{\partial y} + \dfrac{\partial v}{\partial x} \\ g_2 = \dfrac{\partial \alpha}{\partial y} - \dfrac{\partial \beta}{\partial x} + \dfrac{\partial v}{\partial y}, \end{cases}$$

so that

$$-\Delta v = 0 \ \text{in } \Omega.$$

We let

$$P(g) = d\alpha = \left(\frac{\partial \alpha}{\partial x}, \frac{\partial \alpha}{\partial y}\right)$$

and

$$H(g) = dv = \left(\frac{\partial v}{\partial x}, \frac{\partial v}{\partial y}\right).$$

Elliptic theory gives us that for any $1 < p < +\infty$, P and H map

3.3 Lorentz spaces

continuously $L^p(\Omega, \mathbb{R}^2)$ to itself. Therefore, by theorem 3.3.3, these operators are also continuous between the $L^{(p,q)}(\Omega, \mathbb{R}^2)$, for $1 < p < +\infty$, $1 \leq q \leq +\infty$.

Proof of theorem 3.3.6 Let $f \in L^1(\Omega)$, and consider the function $\widehat{f} \in L^1(\mathbb{R}^2)$, obtained by extending f by 0 outside Ω. Let Ψ be the function defined on $\mathbb{R}^2$ by

$$\Psi(z) = \int_{\mathbb{R}^2} K(z - \zeta)\widehat{f}(\zeta)d\zeta,$$

where $K(z) = \frac{1}{2\pi}\log\left(\frac{1}{|z|}\right)$. We know that

$$-\Delta\Psi = \widehat{f} \quad \text{in } \mathbb{R}^2,$$

and that

$$d\Psi(z) = \int_{\mathbb{R}^2} dK(z - \zeta)\widehat{f}(\zeta)d\zeta.$$

Since $\widehat{f} \in L^1(\mathbb{R}^2)$ and $dK \in L^{(2,\infty)}(\mathbb{R}^2)$, we have that $d\Psi \in L^{(2,\infty)}(\mathbb{R}^2)$ and

$$\|d\Psi\|_{L^{(2,\infty)}(\mathbb{R}^2)} \leq \|dK\|_{L^{(2,\infty)}(\mathbb{R}^2)}\|f\|_{L^1(\mathbb{R}^2)}. \tag{3.39}$$

A way of seeing equation (3.31) is that $d\phi$ is also given by

$$d\phi = P(d\Psi_{|\Omega}),$$

where P is the operator defined in proposition 3.3.9. This operator is continuous from $L^{(2,\infty)}(\Omega)$ to itself, and hence we deduce that $d\phi \in L^{(2,\infty)}(\Omega)$, and estimate (3.39) gives (3.32). $\square$

Proof of theorem 3.3.8 We will need an extra result: the space $W^{1,1}(\mathbb{R}^2)$ is continuously embedded in $L^{(2,1)}(\mathbb{R}^2)$. This result is the purpose of theorem 3.3.10, which we will prove below. Let $f \in \mathcal{H}^1(\mathbb{R}^2)$. As in the previous proof, we consider Ψ defined by

$$\Psi(z) = \int_{\mathbb{R}^2} K(z - \zeta)f(\zeta)d\zeta.$$

Since

$$-\Delta \Psi = f \text{ on } \mathbb{R}^2,$$

by theorem 3.2.9, all second derivatives of Ψ belong to $L^1(\mathbb{R}^2)$ and $\Psi \in W^{2,1}(\mathbb{R}^2)$. This implies, using the continuous embedding of $W^{1,1}(\mathbb{R}^2)$ into $L^{(2,1)}(\mathbb{R}^2)$, that the derivatives of Ψ belong to $L^{(2,1)}(\mathbb{R}^2)$, and since all the embeddings are continuous, there is a constant C such that

$$\|d\Psi\|_{L^{(2,1)}(\mathbb{R}^2)} \leq C \|f\|_{\mathcal{H}^1(\mathbb{R}^2)}.$$

Using, as in the previous proof, the fact that (3.33) implies

$$d\phi = P(d\Psi_{|\Omega}),$$

we deduce that $d\phi \in L^{(2,1)}(\Omega)$ and

$$\|d\phi\|_{L^{(2,1)}(\Omega)} \leq C \|f\|_{\mathcal{H}^1(\mathbb{R}^2)},$$

thanks to the continuity of the operator P, constructed in 3.3.9. □

Theorem 3.3.10 *For each $m \geq 2$, the space $W^{1,1}(\mathbb{R}^m)$ is continuously embedded in $L^{(\frac{m}{m-1},1)}(\mathbb{R}^m)$.*

Remark 3.3.11 *The following proof was communicated to us independently by Haïm Brezis and Pierre-Louis Lions. Other proofs are also possible; in particular a method of Luc Tartar leads to more general results [167].*

Proof We consider first the case of a function in $C_c^\infty(\mathbb{R}^m)$. The general case then follows by the density of $C_c^\infty(\mathbb{R}^m)$ in $W^{1,1}(\mathbb{R}^m)$. Suppose $f \in C_c^\infty(\mathbb{R}^m)$. Let $\widetilde{f}$ be the non-increasing rearrangement of $|f|$, i.e. the unique function in $C_c^\infty(\mathbb{R}^m)$, depending only on $r = |x|$, non-increasing as a function of r, and such that

$$\text{measure } \{x \in \mathbb{R}^m \mid \widetilde{f}(x) \geq s\} = \text{measure } \{x \in \mathbb{R}^m \mid |f(x)| \geq s\}.$$

Denote by f^* the non-increasing rearrangement of f over $[0, +\infty)$, and notice that if we denote by $\alpha_m = \frac{|S^{m-1}|}{m}$ the measure of the unit ball in $\mathbb{R}^m$, we have the relation

$$f^*(\alpha_m r^m) = \tilde{f}(r).$$

Using Pólya–Szegö's inequality (see, for instance, [118]), we have

$$\begin{aligned}
\int_{\mathbb{R}^m} |df(x)| dx &\geq \int_{\mathbb{R}^m} |d\tilde{f}(x)| dx \\
&= \int_0^{+\infty} \left(\int_{S^{m-1}} -\frac{d\tilde{f}}{dr}(r) r^{m-1} d\sigma(x) \right) dr \\
&= -|S^{m-1}| \int_0^{+\infty} \frac{d\tilde{f}}{dr} r^{m-1} dr \\
&= |S^{m-1}| \int_0^{+\infty} \tilde{f}(r)(m-1) r^{m-2} dr.
\end{aligned}$$

Performing the change of variable $t = \alpha_m r^m$, we obtain

$$\begin{aligned}
\int_{\mathbb{R}^m} |df(x)| dx &\geq (m-1) \alpha_m^{\frac{1}{m}} \int_0^{+\infty} t^{\frac{m-1}{m}} f^*(t) \frac{dt}{t} \\
&\geq C(m) \|f\|_{L^{(\frac{m}{m-1},1)}(\mathbb{R}^m)},
\end{aligned}$$

which proves our result. $\square$

Before ending this section we will show that local L^p estimates on the gradient of a real harmonic map, of Morrey type, generalize to Lorentz spaces. We start by recalling a classical result.

Lemma 3.3.12 *Suppose that $1 \leq p < +\infty$, and let B^m be the unit ball in $\mathbb{R}^m$. For any function $f \in L^p(B^m)$, which is a solution of*

$$-\Delta f = 0 \quad \text{on } B^m,$$

the function

$$r \longmapsto \frac{1}{r^m} \int_{B_r^m} |f|^p dx \qquad (3.40)$$

is increasing (here $B_r^m = B^m(0,r)$).

Proof We remark that for every smooth convex function K from $\mathbb{R}$ to $\mathbb{R}$, and for every harmonic function f,

$$\Delta(K(f)) = K''(f)|df|^2 \geq 0.$$

Hence, integrating this inequality over the ball B_r^m, and using Stokes' formula, if $v = K(f)$,

$$0 \leq \int_{\partial B_r^m} \frac{\partial v}{\partial r} d\sigma = r^{m-1} \frac{d}{dr} \int_{\partial B_1^m} v(r\theta) d\sigma(\theta).$$

Therefore, the function $r \longmapsto \int_{\partial B_1^m} v(r\theta) d\theta$ is non-decreasing. It follows that

$$\begin{aligned}
\int_{B_r^m} v(x) dx &= \int_0^r \rho^{m-1} d\rho \int_{\partial B_1^m} v(\rho\theta) d\theta \\
&\leq \int_0^r \rho^{m-1} d\rho \int_{\partial B_1^m} v(r\theta) d\theta \\
&= \frac{r^m}{m} \int_{\partial B_r^m} v(r\theta) d\theta = \frac{r}{m} \frac{d}{dr} \int_{B_r^m} v(x) dx.
\end{aligned}$$

Integrating this differential inequality, we see that

$$r \longmapsto \frac{1}{r^m} \int_{B_r^m} v(x) dx$$

is non-decreasing. This result applies to $K_\epsilon(y) = (\epsilon^2 + y^2)^{p/2}$, for all $\epsilon > 0$ and $1 \leq p < \infty$. By writing that $r^{-m} \int_{B_r^m} K_\epsilon(f) dx$ is non-decreasing, and passing to the limit when ϵ goes to 0, we obtain our result. □

I was not able to obtain the same result for the norm $L^{(p,q)}$, but I found the following:

Lemma 3.3.13 *Suppose that $1 < p < +\infty$, $1 < q < +\infty$, and let $0 < \gamma < \dfrac{m}{p}$.*

Then, there exists a constant $C(p, q, \gamma)$ depending on p, q and γ, such that, for any function $f \in L^1(B^m)$ such that $df \in L^{(p,q)}(B^m)$, the solution F of

$$\begin{cases} -\Delta F &= 0 \quad \text{in } B^m \\ F &= f \quad \text{on } \partial B^m \end{cases}$$

satisfies

$$\|dF\|_{L^{(p,q)}(B_r^m)} \leq C(p,q,\gamma)r^\gamma \|df\|_{L^{(p,q)}(B^m)} . \qquad (3.41)$$

Proof We will just give the proof of this result for $m = 2$ (we leave to the reader the task of generalizing this result for all m – for this purpose, use the Hodge decomposition as in section 4.3). We remark that $dF = H(df)$, where H was defined in proposition 3.3.9. For any $r \in [0,1]$, define the restriction operator to B_r^m, acting on any function g defined on B^m, by

$$R_r : g \longmapsto g_{|B_r^m} .$$

This operator is obviously continuous in all L^p spaces. The conclusion of the lemma is equivalent to saying that the operator $R_r \circ H$ acts on $L^{(p,q)}(B)$, is continuous and its norm satisfies

$$\|R_r \circ H\|_{L^{(p,q)}(B)} \leq C(p,q,\gamma)r^\gamma .$$

Choose $p_1 < p < p_2$, where p_2 is such that $\gamma = \frac{2}{p_2}$, and apply lemma 3.3.12, in $L^{p_1}(B)$ and $L^{p_2}(B)$. This implies, in particular, that $R_r \circ H$ is bounded in $L^{p_1}(B)$ and $L^{p_2}(B)$ with the estimates

$$\|R_r \circ H\|_{L^{p_1}(B)} \leq Cr^{\frac{2}{p_1}} , \qquad (3.42)$$

$$\|R_r \circ H\|_{L^{p_2}(B)} \leq Cr^{\frac{2}{p_2}} . \qquad (3.43)$$

By theorem 3.3.3, we have that $R_r \circ H$ maps $L^{(p,q)}(B)$ to $L^{(p,q)}(B_r)$, and that

$$\|R_r \circ H\|_{L^{(p,q)}(B)} \leq C_1 r^{\frac{2}{p_1}} + C_2 r^{\frac{2}{p_2}} ,$$

where C_1 and C_2 are constants depending on p_1 and p_2. This gives (3.41). □

3.4 Back to Wente's inequality

By applying Hardy and Lorentz spaces, we will be able to prove finer versions of theorem 3.1.2.

Theorem 3.4.1 *Let Ω be an open subset of $\mathbb{R}^2$, with $\mathcal{C}^1$ boundary, $a, b \in H^1(\phi)$ and ϕ be a solution of*

$$\begin{cases} -\Delta \phi = \{a, b\} & \text{in } \Omega \\ \phi = 0 & \text{on } \partial \Omega. \end{cases} \quad (3.44)$$

Then $\frac{\partial \phi}{\partial x}, \frac{\partial \phi}{\partial y} \in L^{(2,1)}(\Omega)$, and

$$\|d\phi\|_{L^{(2,1)}(\Omega)} \leq C(\Omega) \|da\|_{L^2(\Omega)} \|db\|_{L^2(\Omega)}, \quad (3.45)$$

where $C(\Omega)$ is a constant depending only on Ω.

Proof Let $\widehat{a}, \widehat{b} \in H^1(\mathbb{R}^2)$ be two functions which extend a and b, respectively, to $\mathbb{R}$. We may build these extensions so that the maps $a \longmapsto \widehat{a}$ and $b \longmapsto \widehat{b}$ are continuous from $H^1(\Omega)$ to $H^1(\mathbb{R}^2)$. By theorem 3.2.2 (or 3.2.3), we know that $\{\widehat{a}, \widehat{b}\} \in \mathcal{H}^1(\mathbb{R}^2)$, and that

$$\|\{\widehat{a}, \widehat{b}\}\|_{\mathcal{H}^1(\mathbb{R}^2)} \leq C_2 \|d\widehat{a}\|_{L^2(\mathbb{R}^2)} \|d\widehat{b}\|_{L^2(\mathbb{R}^2)}$$

$$\leq C' \|da\|_{L^2(\mathbb{R}^2)} \|db\|_{L^2(\mathbb{R}^2)}. \quad (3.46)$$

This shows that $-\Delta \phi$ coincides with an $\mathcal{H}^1(\mathbb{R}^2)$ function in Ω. Hence theorem 3.3.8 implies that $d\phi \in L^{(2,1)}(\Omega)$, and then inequality (3.46) yields estimate (3.45). $\square$

Remark 3.4.2 *Lorentz space estimates had previously been proven by Luc Tartar. In fact, in [165], he shows that the Fourier transform of $d\phi$ (where ϕ is defined by (3.44) in the previous theorem, i.e. $\phi = \widetilde{ab}_\Omega$) belongs to $L^{(2,1)}(\mathbb{R}^2)$.*

Remark 3.4.3 *Using theorems 3.3.4 and 3.4.1, we can also see that $\widetilde{ab}$ is continuous.*

Remark 3.4.4 *Another proof of theorem 3.4.1 was communicated to me by Luc Tartar. It uses in a simple way the interpolation between bilinear operators acting on H^s spaces, and the continuous embedding of $H^{\frac{1}{2}}(\mathbb{R}^2)$ in $L^{(4,2)}(\mathbb{R}^2)$. We explain the idea in the case of $\mathbb{R}^2$.*

We consider the operator

3.4 Back to Wente's inequality

$$B: \begin{array}{c} H^1(\mathbb{R}^2) \times H^1(\mathbb{R}^2) \\ (a,b) \end{array} \begin{array}{c} \longrightarrow \\ \longmapsto \end{array} \begin{array}{c} L^2(\mathbb{R}^2, \mathbb{R}^2) \\ d(\widetilde{ab}). \end{array}$$

We will show that B maps $H^{\frac{1}{2}}(\mathbb{R}^2) \times H^{\frac{3}{2}}(\mathbb{R}^2)$ to $L^{(2,1)}(\mathbb{R}^2, \mathbb{R}^2)$ (and similarly, that B maps $H^{\frac{3}{2}}(\mathbb{R}^2) \times H^{\frac{1}{2}}(\mathbb{R}^2)$ to $L^{(2,1)}(\mathbb{R}^2, \mathbb{R}^2)$). Take $a, b \in \mathcal{C}_C^\infty(\mathbb{R}^2)$, and notice that

$$\{a,b\} = \frac{\partial}{\partial x}\left(a\frac{\partial b}{\partial y}\right) - \frac{\partial}{\partial y}\left(a\frac{\partial b}{\partial x}\right).$$

It follows that

$$\|\widetilde{dab}\|_{L^{(2,1)}} \leq C \left(\left\|a\frac{\partial b}{\partial y}\right\|_{L^{(2,1)}} + \left\|a\frac{\partial b}{\partial x}\right\|_{L^{(2,1)}} \right)$$

$$\leq C \|a\|_{L^{(4,2)}} \|db\|_{L^{(4,2)}}$$

$$\leq C \|a\|_{H^{\frac{1}{2}}} \|db\|_{H^{\frac{1}{2}}}$$

$$\leq C \|a\|_{H^{\frac{1}{2}}} \|b\|_{H^{\frac{3}{2}}}.$$

Here C is a universal constant. In the first inequality we used the fact that $\widetilde{dab} = H(-a\frac{\partial b}{\partial y}, b\frac{\partial a}{\partial x})$ (see proposition 3.3.9) and the interpolation theorem 3.3.3, and in the second inequality, the continuity of the product $L^{(4,2)} \times L^{(4,2)} \longrightarrow L^{(2,1)}$, and the continuous embedding $H^{\frac{1}{2}} \longrightarrow L^{(4,2)}$.

Doing the same for $H^{\frac{3}{2}}(\mathbb{R}^2) \times H^{\frac{1}{2}}(\mathbb{R}^2)$, and then interpolating between the two, we obtain theorem 3.4.1 for $\Omega = \mathbb{R}^2$. The proof for an arbitrary $\mathcal{C}^2$ bounded domain can be obtained using techniques similar to those used in the previous section.

To conclude, we present a version of theorem 3.4.1 under weaker hypotheses. It is due to Fabrice Bethuel [9], who used it to study the regularity of weak solutions of the prescribed mean curvature equation.

We start by noticing that if a and b are two functions such that $da \in L^p$ for $1 \leq p < 2$ and $db \in L^2$, we can still give a meaning to the quantity $\{a,b\}$ by letting

$$\{a,b\} := \frac{\partial}{\partial x}\left(a\frac{\partial b}{\partial y}\right) - \frac{\partial}{\partial y}\left(a\frac{\partial b}{\partial x}\right).$$

(See also [2].) In fact, in this case $da \in L^p$ and we deduce from the Sobolev embedding theorems that $a \in L^2$. With this convention we have:

Theorem 3.4.5 *Let Ω be a bounded domain of $\mathbb{R}^2$, with C^2 boundary. Suppose a and b are two functions on Ω such that $da \in L^{(2,\infty)}(\Omega)$, and $db \in L^2(\Omega)$. Let ϕ be the solution of*

$$\begin{cases} -\Delta \phi &= \{a,b\} \quad \text{in } \Omega \\ \phi &= 0 \quad \text{on } \partial\Omega. \end{cases}$$

Then, ϕ is in $H^1(\Omega)$, and there is a constant C, depending only on Ω, such that

$$\|d\phi\|_{L^2(\Omega)} \leq C \|da\|_{L^{(2,\infty)}(\Omega)} \|db\|_{L^2(\Omega)}. \tag{3.47}$$

Proof Let U be a smooth bounded open subset of $\mathbb{R}^2$, containing Ω, such that we know how to construct a continuous extension operator for functions $f \in H^1(\Omega)$ giving functions $\widehat{f} \in H_0^1(U)$.

We first show inequality (3.47) for $a, b \in H^1(\Omega)$, and later extend it to the case where $da \in L^{(2,\infty)}$.

For any $a, b \in H^1(\Omega)$, let $\widehat{a}, \widehat{b} \in H_0^1(U)$ be their extensions. Integrating by parts we have (temporarily writing $\widehat{a} = a$ and $\widehat{b} = b$)

$$\begin{aligned} \|\widetilde{dab}_U\|^2_{L^2(U)} &= -\int_U \widetilde{ab}\,\Delta\widetilde{ab} = \int_U \widetilde{ab}\,\{a,b\} \\ &= \int_U a\,\{b, \widetilde{ab}\} = \int_U a\,\Delta(\widetilde{b(\widetilde{ab})}) \\ &= \int_U da \cdot d(\widetilde{b(\widetilde{ab})}) \\ &\leq C\,\|da\|_{L^{(2,\infty)}(U)} \|d(\widetilde{b(\widetilde{ab})})\|_{L^{(2,1)}(U)} \\ &\leq C^2\,\|da\|_{L^{(2,\infty)}(U)} \|db\|_{L^2(U)} \|\widetilde{dab}\|_{L^2(U)}. \end{aligned}$$

We used the fact that $\widetilde{ab} = 0$ on ∂U, then lemma 3.1.6, the fact that $\widetilde{b(\widetilde{ab})} = 0$ on ∂U, and the hypothesis $a = 0$ on ∂U. Finally, we applied theorem 3.4.1 twice. Next, since

3.4 Back to Wente's inequality

$$\begin{cases} -\Delta \widetilde{ab}_\Omega = -\Delta \widetilde{ab}_U & \text{in } \Omega \\ \widetilde{ab}_\Omega = 0 & \text{on } \partial\Omega, \end{cases}$$

we deduce that $\|d\widetilde{ab}_\Omega\|_{L^2(\Omega)} \leq \|d\widetilde{ab}_U\|_{L^2(\Omega)}$. The previous inequality then yields

$$\|d\widetilde{ab}_\Omega\|_{L^2(\Omega)} \leq C^2 \|da\|_{L^{(2,\infty)}(\Omega)} \|db\|_{L^2(\Omega)}.$$

It remains to prove the theorem in the general case. Let a be such that $da \in L^{(2,\infty)}$. Notice that in general there is no sequence $(a_k)_{k \in \mathbb{N}}$ in $H^1(\Omega)$ such that $\lim_{k \to +\infty} \|d(a - a_k)\|_{L^{(2,\infty)}} = 0$. (This lack of density is analogous to the fact that $\mathcal{C}^\infty$ is not dense in L^∞.) However, it is possible to construct a sequence $(a_k)_{k \in \mathbb{N}}$ in $\bigcap_{1 \leq p < 2} W^{1,p}(\Omega)$ such that

$$a_k \to a \text{ in } W^{1,p}(\Omega), \text{ for } p < 2, \tag{3.48}$$

$$\lim_{k \to +\infty} \|da_k\|_{L^{(2,\infty)}(\Omega)} = \|da\|_{L^{(2,\infty)}(\Omega)}. \tag{3.49}$$

By (3.47) and (3.49), it is clear that $\widetilde{a_k b}_\Omega$ is bounded in $H^1(\Omega)$ and thus, up to passing to a subsequence of k,

$$\widetilde{a_k b}_\Omega \rightharpoonup \phi \text{ weakly in } H^1(\Omega). \tag{3.50}$$

Moreover,

$$\|d\phi\|_{L^2(\Omega)} \leq C^2 \|da\|_{L^{(2,\infty)}(\Omega)} \|db\|_{L^2(\Omega)}.$$

Thus, in order to conclude we just need to check that $\phi = \widetilde{ab}_\Omega$. The convergence (3.48) implies that

$$\{a_k, b\} \longrightarrow \{a, b\} \text{ in } W^{-1,p}(\Omega), \text{ for } 1 \leq p < 2,$$

(where we used $\{a_k, b\} = \frac{\partial}{\partial x}(a_k \frac{\partial b}{\partial y}) - \frac{\partial}{\partial y}(a_k \frac{\partial b}{\partial x})$). And hence,

$$\widetilde{a_k b}_\Omega \longrightarrow \{a, b\} \text{ in } W^{1,p}(\Omega) \text{ for } 1 \leq p < 2.$$

It follows from (3.50) that $\phi = \widetilde{ab}_\Omega$. $\square$

Remark 3.4.6 *For a more general version of this theorem see [65].*

3.5 Weakly stationary maps with values into a sphere

We end this chapter with a result where Hardy space and BMO play an essential role: Lawrence Craig Evans' theorem [54] on partial regularity of weakly stationary maps (see definition 1.4.17). Its proof brings together the results presented above and more classical arguments due essentially to Charles B. Morrey and Ennio de Giorgi, as well as the writing of the harmonic map equation in the form of a conservation law, as in theorem 2.6.4.

Theorem 3.5.1 *Let Ω be an open subset of an m-dimensional Riemannian manifold $(\mathcal{M}, g)$, and $u \in H^1(\Omega, S^n)$. Suppose that $g \in \mathcal{C}^{k,\alpha}$, where $k \geq 0$ and $0 < \alpha < 1$. Then, if u is weakly stationary, there exists a closed subset $\mathcal{S}$ of Ω (which we call the singular set of u) whose $(m-2)$-dimensional Hausdorff measure is zero, and such that $u \in \mathcal{C}^{k+1,\alpha}$ in $\Omega \setminus \mathcal{S}$.*

To simplify, in the proof we will suppose that the metric g on $\mathcal{M}$ is Euclidean, and hence that Ω is an open subset of $\mathbb{R}^m$. We start by explaining what Hausdorff measure is. For each measurable subset $\mathcal{S}$ of $(\mathcal{M}, g)$, and for any $s \in [0, m]$, is is possible to define a measure (called s-dimensional Hausdorff measure) of $\mathcal{S}$, which we denote by

$$\mathcal{H}^s(\mathcal{S}),$$

taking values in $[0, +\infty]$. If, for instance, $\mathcal{S}$ is a k-dimensional $\mathcal{C}^1$ submanifold of $(\mathcal{M}, g)$, then

$$\begin{aligned}
\mathcal{H}^s(\mathcal{S}) &= +\infty & &\text{if } 0 \leq s < k \\
\mathcal{H}^k(\mathcal{S}) &= \text{Lebesgue } k\text{-dimensional measure of } \mathcal{S} \\
\mathcal{H}^s(\mathcal{S}) &= 0 & &\text{if } k < s \leq m.
\end{aligned}$$

The idea consists of covering $\mathcal{S}$ by a countable union of balls C_j of radii r_j, and considering

$$\alpha(s) \sum_{j=1}^{\infty} r_j^s, \tag{3.51}$$

which measures approximately the s-dimensional volume of $\mathcal{S}$. The right definition will be obtained by choosing coverings that are at the same time the finest and most economical possible. Concern for economy leads us to define, for $\delta > 0$,

3.5 Weakly stationary maps with values into a sphere

$$\mathcal{H}^s_\delta(\mathcal{S}) = \inf \left\{ \alpha(s) \sum_{j=1}^\infty r_j^s \mid \mathcal{S} \subset \bigcup_{j=1}^\infty C_j \text{ and } r_j \leq \delta \right\}, \qquad (3.52)$$

and to optimize the refinement we will adopt:

Definition 3.5.2 *The s-dimensional Hausdorff measure of $\mathcal{S}$ is given by*

$$\mathcal{H}^s(\mathcal{S}) = \lim_{\delta \to 0} \mathcal{H}^s_\delta(\mathcal{S}) = \sup_{\delta > 0} \mathcal{H}^s_\delta(\mathcal{S}).$$

For details see [186]. Notice that the constant $\alpha(s)$ in (3.51) is just a normalization coefficient which is present so that when s is an integer, $\mathcal{H}^s$ will coincide with the s-dimensional Lesbesgue measure. We will now see how the Hausdorff measure plays a role in theorem 3.5.1. We introduce the following notation: for each $x \in \Omega$ and $r \in (0, +\infty)$, such that $B(x, r)$ is contained in Ω, we let

$$E_{x,r}(u) = \frac{1}{r^{m-2}} \int_{B(x,r)} |du|^2 dx. \qquad (3.53)$$

Theorem 3.5.1 is a consequence of the following two independent results.

Theorem 3.5.3 *(ϵ-regularity) Let $u \in H^1(\Omega, S^n)$ be a weakly stationary map. Then, there is a constant $\epsilon > 0$ such that if for a point x in Ω, there exists $r > 0$ such that*

$$E_{x,r}(u) \leq \epsilon^2, \qquad (3.54)$$

then u is Hölder continuous in a neighborhood of x.

Lemma 3.5.4 *Let $u \in H^1(\Omega, S^n)$. Define, for any $\epsilon > 0$,*

$$\mathcal{S}' = \left\{ x \in \Omega \mid \sup_{r > 0} E_{x,r}(u) > \epsilon^2 \right\}. \qquad (3.55)$$

Then, we have

$$\mathcal{H}^{m-2}(\mathcal{S}') = 0. \qquad (3.56)$$

Proposition 3.5.5 *Theorem 3.5.1 is a consequence of theorem 3.5.3 and lemma 3.5.4.*

Proof In fact, choosing for ϵ in lemma 3.5.4 the value given by theorem 3.5.3, if u is weakly stationary, then by theorem 3.5.3, u is Hölder continuous in a neighborhood of each point in $\Omega \setminus \mathcal{S}'$. Thus there is an open set $\omega \subset \Omega$ (the union of the neighborhoods given by theorem 3.5.3) such that $\Omega \setminus \mathcal{S}' \subset \omega$ and $u_{|\omega}$ is Hölder continuous. Let $\mathcal{S} := \Omega \setminus \omega$. It is closed and contained in $\mathcal{S}'$. To conclude the proof of theorem 3.5.1, we just need to show that:

(i) If u is weakly stationary and Hölder continuous in $\omega = \Omega \setminus \mathcal{S}$, then u is $\mathcal{C}^{k+1,\alpha}$ in ω. This follows from theorem 1.5.1.

(ii) $\mathcal{H}^{m-2}(\mathcal{S}) = 0$. Since $\mathcal{S} \subset \mathcal{S}'$, this follows from lemma 3.5.4.

□

We shall first prove lemma 3.5.4 and terminate this section by the proof of theorem 3.5.3.

Proof of lemma 3.5.4 We will proceed in several steps.

Step 1 CONSTRUCTION OF A COVERING OF $\mathcal{S}'$ OF RADIUS $\delta > 0$
Here it will be important that every point in Ω is covered at most N times, where N is a certain constant depending only on the dimension. We consider a covering of $\mathcal{S}'$ by k balls $B(x_i, \frac{\delta}{2})$ such that the balls $B(x_i, \frac{\delta}{4})$ are pairwise disjoint: we may for instance construct a periodic lattice of balls of radius $\frac{\delta}{2}$ which cover Ω and then select those which intersect $\mathcal{S}'$. For each i, we choose

$$y_i \in \mathcal{S}' \subset B\left(x_i, \frac{\delta}{2}\right),$$

and we consider the family of balls $B(y_i, \delta)$. We can easily see that this family covers $\mathcal{S}'$ and that each point in Ω is covered at most a certain number N times by these balls. This yields

$$\sum_{i=1}^{k} \int_{B(y_i,\delta)} |du|^2 dx \leq N \int_{\cup_{i=1}^{k} B(y_i,\delta)} |du|^2 dx. \quad (3.57)$$

Step 2 $\mathcal{H}^{m-2}(\mathcal{S}')$ IS BOUNDED
By definition of $\mathcal{S}'$, we have that for any y_i,

$$\frac{1}{\delta^{m-2}} \int_{B(y_i,\delta)} |du|^2 dx > \epsilon^2. \tag{3.58}$$

This inequality and (3.57) imply that

$$k\epsilon^2 \delta^{m-2} < N \int_\Omega |du|^2 dx, \tag{3.59}$$

proving that $\mathcal{H}^{m-2}_\delta(\mathcal{S}')$ is bounded by $N\epsilon^{-2} \| du \|^2_{L^2(\Omega)}$, independently of δ, and hence that $\mathcal{H}^{m-2}(\mathcal{S}')$ is bounded.

Step 3 $\mathcal{H}^m(\mathcal{S}')$ IS ZERO
This follows from

$$\mathcal{H}^m \left(\bigcup_{i=1}^k B(y_i,\delta) \right) \leq \alpha(m) k \delta^m$$
$$\leq \left(\alpha(m) \tfrac{N}{\epsilon^2} \int_\Omega |du|^2 dx \right) \delta^2,$$

where we used (3.59). Since the coefficient of δ^2 in the last inequality is bounded, this yields that $\mathcal{H}^m(\mathcal{S}') = 0$.

Step 4 $\mathcal{H}^{m-2}(\mathcal{S}') = 0$
We write the following inequality which, like (3.59), follows from (3.57) and (3.58), but is slightly more precise:

$$k\epsilon^2 \delta^{m-2} < N \int_{\bigcup_{i=1}^k B(y_i,\delta)} |du|^2 dx. \tag{3.60}$$

We already know that the Lebesgue measure of the integration domain on the r.h.s. tends to 0 as δ tends to 0. Hence, it follows from the Lebesgue dominated convergence theorem that this integral converges to 0. Passing to the limit in (3.60) will then yield that $\mathcal{H}^{m-2}(\mathcal{S}') = 0$. □

It remains to prove the ϵ-regularity theorem 3.5.3. First we will show a crucial inequality for $E_{x,r}(u)$. This property, called the "monotonicity formula", is a consequence of the fact that u is Noether harmonic. We remark that theorems 3.5.1 and 3.5.3 would still be valid if instead of supposing that u is weakly stationary, we supposed only that u is weakly harmonic and satisfies the monotonicity formula.

Lemma 3.5.6 Let $\mathcal{N}$ be a compact n-dimensional manifold without boundary, C^2 embedded in $\mathbb{R}^N$. Let $u \in H^1(\Omega, \mathcal{N})$ be a weakly Noether harmonic map. Then, for every $x \in \Omega$ and $0 < r \leq r_1$ such that $B(x, r_1) \subset \Omega$, we have the "monotonicity formula"

$$E_{x,r}(u) \leq E_{x,r_1}(u).$$

Proof This result has already been mentioned in chapter 1 (example 1.3.7) for the smooth case. In the present case the proof is analogous. Recall that any Noether harmonic map has a divergence-free (in the sense of distributions) stress–energy tensor (theorem 1.4.15). Define

$$S_{\alpha\beta} = \frac{1}{2}|du|^2 \delta_{\alpha\beta} - \left\langle \frac{\partial u}{\partial x^\alpha}, \frac{\partial u}{\partial x^\beta} \right\rangle,$$

the stress–energy tensor. For each sufficiently small $\epsilon > 0$, we construct a C^1 function, χ_ϵ, from $(0, +\infty)$ to $[0, 1]$, such that $\chi_\epsilon = 1$ in $(0, 1-\epsilon)$, $\chi_\epsilon = 0$ in $(1+\epsilon, +\infty)$ and $\chi'_\epsilon \leq 0$ everywhere. We fix a point $x_0 \in \Omega$, and let

$$F_\epsilon(r) = \frac{1}{r^{m-2}} \int_\Omega \chi_\epsilon\left(\frac{\rho}{r}\right) |du|^2 \, dx,$$

where $\rho = |x - x_0|$. We remark that $F_0(r) = E_{x,r}(u)$, and that F_ϵ converges to F_0 in $W^{1,1}((0, \text{dist}\,(x_0, \partial\Omega)))$. Moreover, using the fact that $S_{\alpha\beta}$ is divergence-free (theorem 1.4.15), we obtain

$$\begin{aligned} 0 &= \int_\Omega \frac{\partial}{\partial x^\alpha}\left(\chi_\epsilon\left(\frac{\rho}{r}\right) x^\beta S_{\alpha\beta}\right) dx \\ &= \frac{1}{2} \int_\Omega \chi'_\epsilon\left(\frac{\rho}{r}\right) \frac{\rho}{r} \left(|du|^2 - \left|\frac{\partial u}{\partial \rho}\right|^2\right) dx \\ &\quad + \frac{1}{2} \int_\Omega \chi_\epsilon\left(\frac{\rho}{r}\right) (m-2)|du|^2 dx. \end{aligned}$$

Differentiating F_ϵ and using the previous identity we see that

$$F'_\epsilon(r) = -\frac{2}{r^m} \int_\Omega \chi'_\epsilon\left(\frac{\rho}{r}\right) \rho \left|\frac{\partial u}{\partial \rho}\right|^2 dx,$$

which implies that F_ϵ is non-decreasing, and hence, passing to the limit when $\epsilon \to 0$, that F_0 is also non-decreasing. □

3.5 Weakly stationary maps with values into a sphere

Remark 3.5.7 *The quantity $E_{x,r}(u)$ satisfies another property which will play an important role: it is homogeneous, i.e. dilation invariant. This means that for every map u defined in the ball $B(a,r)$, if we denote by $T_{a,r}u$ the map defined in $B(0,1)$ by*

$$T_{a,r}u(x) = u(rx+a), \quad \forall x \in B(0,1), \qquad (3.61)$$

then

$$E_{0,1}(T_{a,r}u) = E_{a,r}(u).$$

Finally, the knowledge of $E_{x,r}(u)$ enables us to control the BMO norm of u since, by the Sobolev–Poincaré inequality, we have that for any map v

$$\frac{1}{r^m} \int_{B(a,r)} |v - v_{a,r}| dx \leq CE_{a,r}(v),$$

where $v_{a,r} := \frac{1}{|B(a,r)|} \int_{B(a,r)} v \, dx$. Thus, if the $E_{a,r}(v)$ are uniformly bounded, v will be bounded in BMO.

Proof of theorem 3.5.3 It relies on the following "discretized" version of theorem 3.5.3. □

Theorem 3.5.8 *Let $u \in H^1(\Omega, S^n)$ be a weakly stationary map. Then, there exist $\epsilon > 0$ and $\tau \in (0,1)$ such that for any $y \in \Omega$ and $r > 0$ such that $B(y,r) \subset \Omega$, if*

$$E_{y,r}(u) \leq 2^{m-2}\epsilon^2, \qquad (3.62)$$

then

$$E_{y,\tau r}(u) \leq \frac{1}{2} E_{y,r}(u). \qquad (3.63)$$

We will end by proving this result.

Stage A How theorem 3.5.3 follows from theorem 3.5.8?

The essential tool here is the family of Morrey–Campanato spaces.

Definition 3.5.9 *Let Ω be an open subset of $(\mathcal{M}, g)$, an m-dimensional manifold. For $q \in [1, +\infty)$ and $\lambda \in (0, +\infty)$, the Morrey–Campanato space $\mathcal{L}^{q,\lambda}(\Omega)$ is*

$$\mathcal{L}^{q,\lambda}(\Omega) := \{u \in L^q(\Omega) \mid [u]_{q,\lambda} < +\infty\},$$

where

$$[u]_{q,\lambda} := \sup_{\substack{x \in \Omega \\ 0 < r < \mathrm{diam}(\Omega)}} \left(\frac{1}{r^\lambda}\int_{B(x,r)} |u - u_{x,r}|^q \, d\mathrm{vol}_g\right)^{\frac{1}{q}}, \qquad (3.64)$$

and

$$u_{x,r} := \frac{1}{|B(x,r)|}\int_{B(x,r)} u \, d\mathrm{vol}_g.$$

This space is a Banach space with the norm

$$\|u\|_{\mathcal{L}^{q,\lambda}(\Omega)} = \|u\|_{L^q(\Omega)} + [u]_{q,\lambda},$$

Morrey–Campanato spaces can be used to obtain the $\mathcal{C}^{0,\alpha}$ regularity in theorem 3.5.3 thanks to the following result.

Theorem 3.5.10 *[117], [69] For each $m < \lambda \leq m+q$, the space $\mathcal{L}^{q,\lambda}(\Omega)$ is isomorphic to the space of Hölder continuous functions $\mathcal{C}^{0,\alpha}(\Omega)$, where*

$$\alpha = \frac{\lambda - m}{q}. \qquad (3.65)$$

If $m + q < \lambda$, then $\mathcal{L}^{q,\lambda}(\Omega)$ contains only the constant functions.

The missing link for completing the proof of stage A is given by:

Lemma 3.5.11 *Let $\theta, \tau \in (0,1)$, and $u \in H^1(\Omega, S^n)$. We suppose that at a point y of Ω, there exists $r > 0$ such that $B(y,r) \subset \Omega$ and $\forall \rho \in (0,r)$*

$$E_{y,\tau\rho}(u) \leq \theta E_{y,\rho}(u). \qquad (3.66)$$

Then, for all $0 < \rho < r$,

$$E_{y,\rho}(u) \leq \frac{1}{\tau^{m-2\theta}}\left(\frac{\rho}{r}\right)^{\frac{\log \theta}{\log \tau}} E_{r,y}(u), \qquad (3.67)$$

which implies, using the Sobolev–Poincaré inequality, that for all $1 \leq q \leq \frac{2m}{m-2}$,

3.5 Weakly stationary maps with values into a sphere

$$\frac{1}{\rho^m}\int_{B(y,\rho)}|u-u_{x,\rho}|^q dx \leq C\left(\frac{\rho}{r}\right)^{q\frac{\log\theta}{\log\tau}} E_{y,r}(u), \tag{3.68}$$

where C is a constant that depends only on τ.

Proof Iterating inequality (3.66) p times, we obtain that for all $p \in \mathbb{N}$,

$$E_{y,\tau^p r}(u) \leq \theta^p E_{y,r}. \tag{3.69}$$

Let $0 < \rho < r$. There exists a unique $p \in \mathbb{N}$ such that

$$\frac{\log\left(\frac{\rho}{\tau r}\right)}{\log \tau} \leq p < \frac{\log\left(\frac{\rho}{r}\right)}{\log \tau}, \tag{3.70}$$

and for this value of p we have $\tau^{p+1} r \leq \rho < \tau^p r$, and hence

$$E_{y,\rho}(u) \leq \frac{1}{\tau^{m-2}} E_{y,\tau^p r}(u) \leq \frac{\theta^p}{\tau^{m-2}} E_{y,r}(u). \tag{3.71}$$

Still using (3.70), we have

$$\frac{\theta^p}{\tau^{m-2}} \leq \frac{1}{\tau^{m-2}\theta}\left(\frac{\rho}{r}\right)^{\frac{\log\theta}{\log\tau}} \tag{3.72}$$

and we realize that (3.67) is a consequence of (3.71) and (3.72). Using the Sobolev–Poincaré inequality (see [186]) we obtain that for all $1 \leq q \leq \frac{2m}{m-2}$,

$$\left(\frac{1}{\rho^m}\int_{B(y,\rho)}|u-u_{y,\rho}|^q\right)^{\frac{1}{q}} \leq C(E_{y,\rho}(u))^{\frac{1}{2}},$$

which, together with (3.67), yields (3.68). □

Proof of theorem 3.5.3: end of stage A Assume the conclusion of theorem 3.5.8. Let $x \in \Omega$ and $\rho > 0$ be such that $B(x, 2r) \subset \Omega$ and

$$E_{x,2r}(u) \leq \epsilon^2. \tag{3.73}$$

Then an easy calculation shows that for any $y \in B(x,r)$, we have $B(y,r) \subset B(x,2r)$ and

$$E_{y,r}(u) \leq 2^{m-2} E_{x,2r}(u). \tag{3.74}$$

The monotonicity formula and inequalities (3.73) and (3.74) imply that $\forall \rho \in (0, r)$

$$E_{y,\rho}(u) \leq E_{y,r}(u) \leq 2^{m-2} \epsilon^2.$$

We now use theorem 3.5.8 to deduce that at the point y we satisfy the hypothesis of lemma 3.5.11 with $\theta = \frac{1}{2}$. Applying this lemma, and in particular inequality (3.68) which is valid for all y in $B(x, r)$, we conclude that the restriction of u to $B(x, r)$ belongs to the Morrey–Campanato space $\mathcal{L}^{q,\lambda}(B(x, r))$, where $1 \leq q \leq \frac{2m}{m-2}$ and

$$\lambda = m + \frac{q \, \log \theta}{2 \log \tau}.$$

Then, theorem 3.5.10 implies that u is in $\mathcal{C}^{0,\alpha}$ in $B(x, r)$, where $\alpha = \frac{\log \theta}{2 \log \tau}$. □

Remark 3.5.12 *The slightly artificial use of the coefficient 2^{m-2} in front of ϵ^2 in theorem 3.5.8, is what allows us to prove the Hölder continuity of u in a neighborhood of x, instead of continuity at the point x. Notice also that we should necessarily have $\theta < \tau^2$, because if not u would have to be constant.*

Stage B PROOF OF THEOREM 3.5.8
It breaks down into several steps.

Step 1 ARGUING BY CONTRADICTION
We will suppose that the conclusion of theorem 3.5.8 is false and reason by contradiction. This means that we suppose that there exists a sequence of balls $B(x_k, r_k)$ included in Ω and such that

$$E_{x_k, r_k}(u) = \lambda_k^2 \to 0 \tag{3.75}$$

and

$$E_{x_k, \tau r_k}(u) > \frac{1}{2} \lambda_k^2, \tag{3.76}$$

where τ may be arbitrarily chosen in $(0, 1)$. Let us do a scaling: for each z in $B(0, 1)$, we let

3.5 Weakly stationary maps with values into a sphere

$$v_k(z) := \frac{u(x_k + r_k z) - u_{x_k, r_k}}{\lambda_k},$$

where $u_{x_k, r_k} = \frac{1}{|B(x_k, r_k)|} \int_{B(x_k, r_k)} u(x) dx = \overline{u_k}$. We deduce from (3.75) and (3.76) that

$$E_{0,1}(v_k) = \int_{B(0,1)} |dv_k|^2 dz = 1, \tag{3.77}$$

and

$$E_{0,\tau}(v_k) = \frac{1}{\tau^{m-2}} \int_{B(0,\tau)} |dv_k|^2 dz > \frac{1}{2}. \tag{3.78}$$

Moreover, since the mean value of v_k over $B(0,1)$ is zero, it follows from (3.77) that

$$\sup_{k \in \mathbb{N}} \int_{B(0,1)} |v_k|^2 dz < +\infty. \tag{3.79}$$

Estimates (3.77) and (3.79) imply that there exists a subsequence of k — which we will still denote by k — such that

$$v_k \to v \quad \text{strongly in} \quad L^2(B(0,1), \mathbb{R}^{n+1}), \tag{3.80}$$

$$v_k \rightharpoonup v \quad \text{weakly in} \quad H^1(B(0,1), \mathbb{R}^{n+1}). \tag{3.81}$$

Step 2 THE LIMIT v OF v_k IS AN $\mathbb{R}^{n+1}$-VALUED HARMONIC MAP
In order to prove it, we take any test function $\phi \in \mathcal{C}_c^\infty(B(0,1), \mathbb{R}^{n+1})$, and let

$$\phi_k(x) = T_{-\frac{x_k}{r_k}, \frac{1}{r_k}} \phi(x) = \phi\left(\frac{x - x_k}{r_k}\right).$$

The hypothesis of u being weakly harmonic yields

$$\int_{B(x_k, r_k)} \langle du, d\phi_k \rangle \, dx = \int_{B(x_k, r_k)} |du|^2 \langle u, \phi_k \rangle dx. \tag{3.82}$$

We make the change of variable $x = r_k z + x_k$ in both these integrals, and write u in terms of v_k (where $u(x) = \lambda_k v_k\left(\frac{x - x_k}{r_k}\right) + \overline{u_k}$):

$$\int_{B(0,1)} \langle dv_k, d\phi \rangle dz = \int_{B(0,1)} \lambda_k |dv_k|^2 \langle u(x_k + r_k z), \phi \rangle dz. \qquad (3.83)$$

When we pass to the limit in (3.83), we obtain, using (3.75), that

$$\int_{B(0,1)} \langle dv, d\phi \rangle \, dz = 0, \qquad (3.84)$$

i.e. that $\Delta v = 0$. Therefore, we can use lemma 3.3.12 to obtain

$$\frac{1}{\tau^{m-2}} \int_{B(0,\tau)} |dv|^2 dz \le \tau^2,$$

since by (3.77) and (3.80), $\int_{B(0,1)} |dv|^2 dz = 1$. This is the moment to choose τ: if it is sufficiently small, we will certainly have

$$E_{0,\tau}(v) < \frac{1}{4}. \qquad (3.85)$$

The idea now is to show that (3.78) and (3.85) are in contradiction, but this cannot work unless we succeed in proving that v_k converges *strongly* to v in $H^1(B(0,1))$.

Step 3 STRONG CONVERGENCE OF v_k IN H^1
We will just prove this convergence in $H^1(B(0, \frac{1}{2}))$: it is clear that choosing $\tau \le \frac{1}{2}$, this will suffice to give a contradiction. Let $\zeta \in \mathcal{C}_c^\infty(B(0,1))$ be a function which is constant equal to 1 in $B(0, \frac{1}{2})$, and write

$$\phi_k = \zeta^2 (v_k - v) \in H_0^1(B(0,1)).$$

The idea is to take advantage of the facts that u is weakly harmonic, and v is harmonic, using ϕ_k as a test function, i.e.

$$\int_{B(0,1)} \langle dv_k, d\phi_k \rangle \, dz = \lambda_k \int_{B(0,1)} |dv_k|^2 \langle \overline{u_k} + \lambda_k v_k, \phi_k \rangle \, dz, \qquad (3.86)$$

$$\int_{B(0,1)} \langle dv, d\phi_k \rangle \, dz = 0. \qquad (3.87)$$

Subtracting (3.87) from (3.86), we obtain

3.5 Weakly stationary maps with values into a sphere

$$\int_{B(0,1)} \langle d(v_k - v), d\phi_k \rangle \, dz = \lambda_k \int_{B(0,1)} |dv_k|^2 \langle \overline{u_k} + \lambda_k v_k, \phi_k \rangle \, dz. \tag{3.88}$$

However, in writing these equations there is a slight problem due to the integral on the r.h.s. of (3.82) and (3.88). In fact, $|dv_k|^2(z)(\overline{u_k} + \lambda_k v_k)(z) = |dv_k|^2(z)u(r_k z + x_k)$ is only bounded in $L^1(B(0,1))$ and, a priori, ϕ_k is not bounded in $L^\infty(B(0,1))$ as $k \to \infty$. Thus this integral could tend to infinity.

We disregard this (crucial) difficulty for the moment, and try to see first, what it is possible to obtain from (3.88). Using the expression of ϕ_k as a function of v and v_k, and using (3.80) and (3.81) in the integral on the l.h.s. of (3.88), we see that the integral satisfies

$$L_k := \int_{B(0,1)} \zeta^2 |d(v_k - v)|^2 dz + 2 \int_{B(0,1)} \zeta \langle d(v_k - v), (v_k - v) d\zeta \rangle \, dz$$
$$\geq \int_{B(0,\frac{1}{2})} |d(v_k - v)|^2 dz + o(1),$$

and thus,

$$\int_{B(0,\frac{1}{2})} |d(v_k - v)|^2 dz \leq R_k + o(1), \tag{3.89}$$

where

$$R_k := \lambda_k \int_{B(0,1)} |dv_k|^2 \langle \overline{u_k} + \lambda_k v_k, \phi_k \rangle \, dz.$$

As in theorem 2.6.4, we will take advantage of the fact that $2\langle u, du \rangle = d(|u|^2) = 0$, and hence that

$$\langle \overline{u_k} + \lambda_k v_k, dv_k \rangle = 0.$$

This implies, introducing the $(n+1) \times (n+1)$ matrices

$$b_{k,\alpha} = (\overline{u_k} + \lambda_k v_k) \frac{{}^t \partial v_k}{\partial z^\alpha} - \frac{\partial v_k}{\partial z^\alpha} {}^t(\overline{u_k} + \lambda_k v_k),$$

that (summing over $\alpha = 1, \ldots, m$)

$$|dv_k|^2 (\overline{u_k} + \lambda_k v_k) = b_{k,\alpha} \cdot \frac{\partial v_k}{\partial z^\alpha},$$

and thus,

$$R_k = \lambda_k \int_{B(0,1)} \left\langle \zeta b_{k,\alpha} \frac{\partial v_k}{\partial z^\alpha}, \zeta(v_k - v) \right\rangle dz$$
$$= \lambda_k (A_k + B_k),$$

where

$$A_k := \int_{B(0,1)} \left\langle b_{k,\alpha} \frac{\partial(\zeta v_k)}{\partial z^\alpha}, \zeta(v_k - v) \right\rangle dz,$$
$$B_k := \int_{B(0,1)} \left\langle b_{k,\alpha} v_k \frac{\partial \zeta}{\partial z^\alpha}, \zeta(v_k - v) \right\rangle dz.$$

We need to show that A_k and B_k have a meaning. Moreover, knowing that λ_k converges to 0, and because of (3.89), it suffices to show that A_k and B_k are uniformly bounded in k to conclude that $v_k \to v$ in $H^1(B(0,\frac{1}{2}))$, and thus conclude the proof of the theorem.

We start by examining B_k, the simplest of the two. In fact, we can easily see that $b_{k,\alpha}$ is bounded in $L^2(B(0,1))$, and therefore it suffices to show that v_k is bounded in $L^4(B(0,1))$, for B_k to make sense and be bounded. We will show this in the next step.

The estimate for A_k will pass by a hairbreadth. We notice that

$$b_{k,\alpha}(z) = \frac{r_k}{\lambda_k} \left(u\, {}^t\frac{\partial u}{\partial x^\alpha} - \frac{\partial u}{\partial x^\alpha}\, {}^t u \right)(r_k z + x_k),$$

and thus, by Noether's theorem,

$$\sum_{\alpha=1}^m \frac{\partial}{\partial z^\alpha}(b_{k,\alpha}) = 0.$$

Since we also have that $b_{k,\alpha}$ is bounded in $L^2(B(0,1))$, we deduce, just as in lemma 3.2.10, that $b_{k,\alpha} \cdot \frac{\partial(\zeta v_k)}{\partial z^\alpha}$ is bounded in the Hardy space $\mathcal{H}^1(\mathbb{R}^m)$. We will also show in the following step that $\zeta(v_k - v)$ is bounded in $BMO(\mathbb{R}^m)$. Thus, using the Fefferman–Stein theorem on the duality between $\mathcal{H}^1(\mathbb{R}^m)$ and $BMO(\mathbb{R}^m)$ (see definition 3.2.8), we deduce that

$$A_k = \int_{B(0,1)} \left\langle b_{k,\alpha} \frac{\partial(\zeta v_k)}{\partial x^\alpha}, \zeta(v_k - v) \right\rangle dz$$
$$\leq C \|b_{k,\alpha} \cdot \frac{\partial(\zeta v_k)}{\partial x^\alpha}\|_{\mathcal{H}^1(\mathbb{R}^m)} \|\zeta(v_k - v)\|_{BMO(\mathbb{R}^m)}$$
$$\leq C < +\infty.$$

3.5 Weakly stationary maps with values into a sphere

Step 4 Estimates in L^4 and BMO

Lemma 3.5.13 *With the hypothesis we made on v_k, it is bounded in $L^4(B(0, \frac{7}{8}))$ and $\zeta(v_k - v)$ is bounded in $BMO(\mathbb{R}^m)$.*

Proof Choose $z_0 \in B(0, \frac{7}{8})$ and $0 < r \leq \frac{1}{8}$, and let

$$y_k = x_k + r_k z_0 \in B(x_k, \frac{7r_k}{8}).$$

Then, by lemma 3.5.11 (monotonicity formula), we have

$$\begin{aligned} E_{y_k, rr_k}(u) &\leq 8^{m-2} E_{y_k, \frac{r_k}{8}}(u) \\ &\leq 8^{m-2} E_{x_k, r_k}(u) \\ &= 8^{m-2} \lambda_k^2. \end{aligned}$$

Hence, coming back to v_k,

$$E_{z_0, r}(v_k) \leq 8^{m-2}.$$

Using the Sobolev–Poincaré inequality, this implies that

$$\frac{1}{r^m} \int_{B(z_0, r)} |v_k - (v_k)_{z_0, r}| dz \leq C < +\infty$$

(recall that by $(v_k)_{z_0, r}$ we mean the average of v_k on $B(z_0, r)$), and since v_k is bounded in $L^2(B(0,1))$ we deduce, using the John–Nirenberg inequality, that v_k is bounded in $L^p(B(0, \frac{7}{8}))$ for $1 \leq p < +\infty$, and thus in particular for $p = 4$. This proves the first statement. Concerning the BMO estimate, we use the fact that $\zeta \in C^\infty$ to deduce that

$$\sup_{B(z_0, r)} |(\zeta v_k)_{z_0, r} - \zeta(v)_{z_0, r}| \leq \frac{Cr}{r^m} \int_{B(z_0, r)} |v_k| dz.$$

Hence, for $z_0 \in B(0, \frac{3}{4})$ and $0 < r \leq \frac{7}{8}$,

$$\frac{1}{B(z_0, r)} \int_{B(z_0, r)} |\zeta v_k - (\zeta v_k)_{z_0, r}| dz$$

$$\leq \frac{1}{r^m} \int_{B(z_0,r)} |v_k - (v_k)_{z_0,r}| dz + \frac{Cr}{r^m} \int_{B(z_0,r)} |v_k| dz$$

$$\leq C + \frac{C}{r^{m-1}} \int_{B(z_0,r)} |v_k| dz$$

$$\leq C + \frac{C}{r^{m-1}} \left(\int_{B(0,\frac{7}{8})} |v_k|^m dz \right)^{\frac{1}{m}} r^{m-1}$$

$$\leq C .$$

A similar inequality is trivially valid for $z_0 \in \mathbb{R}^m \setminus B(0, \frac{3}{4})$ (since $\zeta = 0$ on $\mathbb{R}^m \setminus B(0, \frac{5}{8})$). This implies that ζv_k is bounded in $BMO(\mathbb{R}^m)$. □

Remark 3.5.14 *It is interesting to notice that the BMO estimates are a consequence of the symmetries of the domain manifold, via Noether's theorem 1.4.15 and lemma 3.5.6, whereas the Hardy space estimates come from the symmetries of the image manifold, via Noether's theorem 1.4.13. And these two facts complement each other precisely, thanks to the Fefferman–Stein theorem on duality between Hardy space and BMO.*

A variant of the above proof, which uses basically the same argument, but without arguing by contradiction, can be found in [71] or [87].

Let us add also that the ideas used in the proof of Theorems 2.6.4 and 3.5.1 have been applied to other questions concerning maps into spheres or symmetric manifolds: to the regularity of weakly p-harmonic maps by L. Mou and P. Yang [119], by P. Strzelecki [161], by H. Takeuchi [162]; to the compactness of weakly p-harmonic maps by T. Toro and C. Wang [173]; to subelliptic p-harmonic maps by P. Hajlasz and P. Strzelecki [79]; to biharmonic maps by S.-Y.A. Chang, L. Wang and P.C. Yang [31]. In particular it has been realized that the use of Hardy space can be replaced by other arguments as shown in [79] and one simple such proof has been found by S.-Y.A. Chang, L. Wang and P.C. Yang [30].

Lastly, beautiful results on wave maps into spheres have recently been found by T. Tao [163], [164] and generalized to maps into homogeneous manifolds by S. Klainerman and I. Rodnianski [100] using similar ingredients.

4
Harmonic maps without symmetry

We come back to the study of harmonic maps, but now we drop the symmetry hypothesis on the image manifold. It is then clear that most of the methods seen in chapter 3 are no longer valid. Nevertheless, it is tempting to try to adapt these results to our new situation. This is the naïve point of view we will adopt in this chapter.

The tool we will use the most, and which will replace the conservation laws in chapter 3, is an orthonormal moving frame on the image manifold: it turns out that this choice of representation, used a century ago by Gaston Darboux for the study of surfaces, and developed by Elie Cartan, is very efficient for studying harmonic maps. We remark that, as with all geometric coordinate systems, there is not only one, but infinitely many ways of defining orthonormal tangent frame fields. Instead of being an inconvenience, this abundance of choice is an advantage since one passes from one orthonormal frame field to another through the action of a gauge group.

In this way, symmetries re-enter, and Noether's theorem is not far away: by choosing a "Coulomb frame", the orthonormal frame selected satisfies an equation which may be written as a conservation law.

The use of "Coulomb frames" is fundamental for the regularity theorems for harmonic maps which will be presented in the first three sections. We will also need exotic function spaces (Hardy, BMO, Lorentz) and some results from the previous chapter. Likewise, in chapter 5, we will come back to Coulomb frames, and we will give an example of their use for studying surfaces.

In section 4.4, we will discuss the compactness of weakly harmonic maps in the weak topology. This problem, which was solved in the case where the image manifold is symmetric (see theorem 2.5.1), is still open in general, and we will not be able to solve it. Nevertheless, we will see that it may be possible to find conservation laws for generalized harmonic maps, which might enable us to pass to the limit in the equation satisfied by a sequence of weakly harmonic maps. "Producing" these conservation laws should be a hard problem since it is analogous to the isometric embedding of a Riemannian manifold. This is why I doubt that such a strategy could be an easy way to prove the weak compactness of weakly harmonic maps. Its interest is that it gives rise to many interesting open questions.

4.1 Regularity of weakly harmonic maps of surfaces

The purpose of this section is the following result:

Theorem 4.1.1 *[85] Let $\mathcal{N}$ be a $\mathcal{C}^2$ compact Riemannian manifold without boundary. Let $(\mathcal{M}, g)$ be a Riemannian surface, and let $u \in H^1(\mathcal{M}, \mathcal{N})$. Then, if u is weakly harmonic, u is $\mathcal{C}^{1,\alpha}$. Furthermore, if the embedding of $\mathcal{N}$ into $\mathbb{R}^N$ is $\mathcal{C}^{l,\alpha}$, for some $l \geq 2$, then u is also $\mathcal{C}^{l,\alpha}$.*

We start by noticing that to prove this theorem, it suffices to show that it is valid in the neighborhood of each point in $\mathcal{M}$. But, since every sufficiently small geodesic ball in $(\mathcal{M}, g)$ is conformally equivalent to the unit ball B^2 of $\mathbb{R}^2$ equipped with the canonical metric (theorem 1.1.3), and since harmonic maps are preserved under conformal transformation (proposition 1.1.2), it is enough to prove the result for maps $u \in H^1(B^2, \mathcal{N})$, where B^2 is the Euclidean unit ball.

PLAN OF THE PROOF OF THEOREM 4.1.1

We start by defining a smooth orthonormal tangent frame field over the manifold $(\mathcal{N}, h)$. If y is a point in $\mathcal{N}$, we denote by

$$\widetilde{e}(y) = (\widetilde{e_1}(y), ..., \widetilde{e_n}(y))$$

the "value" of the field in y: $\widetilde{e}(y)$ is an orthonormal basis of $T_y\mathcal{N}$.

We will write the equations satisfied by a weakly harmonic map $u \in H^1(B^2, \mathcal{N})$, by projecting them along the moving frame on $\mathcal{N}$. But first, it will be crucial to optimize the moving frame, adapting it to u. More precisely, we introduce the "gauge group"

$$H^1(B^2, SO(n)) = \{R \in H^1(B^2, M(n \times n, \mathbb{R})) / \ R \in SO(n) \text{ a.e.}\},$$

4.1 Regularity of weakly harmonic maps of surfaces

and for each $R = (R^a_b) \in H^1(B^2, SO(n))$, we consider

$$e(z) = (e_1(z), \ldots, e_n(z)) \text{ for almost all } z \in B^2$$

where

$$e_a(z) := \sum_{b=1}^{n} \tilde{e}_b(u(z)) R^b_a(z). \tag{4.1}$$

We choose an R that minimizes the functional

$$F(R) = \int_{B^2} \left(\sum_{a,b=1}^{n} \left\langle \frac{\partial e_a}{\partial x}, e_b \right\rangle^2 + \left\langle \frac{\partial e_a}{\partial y}, e_b \right\rangle^2 \right) dxdy.$$

The frame obtained satisfies, in particular, the conservation law

$$\frac{\partial}{\partial x}\left\langle \frac{\partial e_a}{\partial x}, e_b \right\rangle + \frac{\partial}{\partial y}\left\langle \frac{\partial e_a}{\partial y}, e_b \right\rangle = 0 \text{ in } B^2. \tag{4.2}$$

We call such a frame a *Coulomb frame*. An important consequence of equation (4.2) is that the coefficients $\left\langle \frac{\partial e_a}{\partial x}, e_b \right\rangle$ and $\left\langle \frac{\partial e_a}{\partial y}, e_b \right\rangle$ belong to the Lorentz space $L^{(2,1)}(B^2)$, instead of just $L^2(B^2)$. We remark that this slight improvement is due only to equation (4.2), and is true even if u is not weakly harmonic.

Next, we use the fact that u is weakly harmonic. Defining

$$\alpha^a := \left\langle \frac{\partial u}{\partial x}(z), e_a(z) \right\rangle - i \left\langle \frac{\partial u}{\partial y}(z), e_a(z) \right\rangle,$$

and

$$\omega^a_b := \frac{1}{2}\left\langle \frac{\partial e_b}{\partial x}(z), e_a(z) \right\rangle + \frac{i}{2}\left\langle \frac{\partial e_b}{\partial y}(z), e_a(z) \right\rangle$$

$$= \left\langle \frac{\partial e_b}{\partial \bar{z}}(z), e_a(z) \right\rangle$$

(where $z = x + iy$ and $\frac{\partial}{\partial \bar{z}} = \frac{1}{2}\left(\frac{\partial}{\partial x} + i\frac{\partial}{\partial y}\right)$), we see that α^a and ω^a_b satisfy

$$\frac{\partial \alpha^a}{\partial \bar{z}} = \sum_{b=1}^{n} \omega^a_b \alpha^b. \tag{4.3}$$

It is then possible to deduce from this equation that u is locally Lipschitz: it all boils down to showing that α^a is locally bounded in L^∞. A priori, we know that α^a and ω_b^a are in $L^2(B^2)$ and it seems hard to obtain such an estimate. However, thanks to the particular choice of orthonormal frame made above, we can show that ω_b^a belongs to $L^{(2,1)}(B^2)$, and this suffices to conclude.

We did not mention a certain number of technical difficulties: can we be sure that the frame field $\widetilde{e}$, where we start from, exists? Is it possible to minimize the functional F?

We start by considering these questions.

It is not hard to find examples of manifolds $\mathcal{N}$ which, for topological reasons, do not admit orthonormal frame fields defined everywhere: such is the case for the spheres S^n, except for $n = 1, 3$ or 7. Thus, sometimes it is impossible to construct the frame field $(\widetilde{e}_1, ..., \widetilde{e}_n)$ mentioned above. We will see two ways of getting around this difficulty: the first is to construct an isometric embedding J of $\mathcal{N}$ into another manifold $\widehat{\mathcal{N}}$, in such a way that if u is a weakly harmonic map taking values in $\mathcal{N}$, then $J \circ u$ is weakly harmonic with values in $\widehat{\mathcal{N}}$.

At the same time, $\widehat{\mathcal{N}}$ will be constructed in such a way that we will always be able to define a frame field over it (we will have $\widehat{\mathcal{N}}$ diffeomorphic to a torus). This result is the purpose of lemma 4.1.2 below. The advantage of this construction is that it is valid for any dimension of the domain manifold (of u), and we can thus use it to study the regularity of stationary weakly harmonic maps in dimension bigger than 2 (see theorem 4.3.1). The drawback is that we need to suppose $\mathcal{N}$ to be $\mathcal{C}^k$, with $k \geq 4$, to be able to construct the embedding J.

Once this result is proved, we will need to deform $\widetilde{e} \circ u$ into a Coulomb frame; that is the purpose of lemma 4.1.3, which is a generalization to arbitrary dimension of a result in [43].

The second way to remedy the absence of smooth frame fields over $\mathcal{N}$ is to directly construct a Coulomb frame adapted to u, by approximating u by smooth maps u_ϵ for which there are no problems. We then show that the Coulomb frames e_ϵ, associated to the u_ϵ, satisfy sufficiently good estimates to allow us to pass to the limit when ϵ goes to 0. This result is contained in lemma 4.1.6. It is valid as long as $\mathcal{N}$ is $\mathcal{C}^2$, but it is necessary for the map u to be defined over a surface. I do not know whether this result can be generalized to higher dimensions.

4.1 Regularity of weakly harmonic maps of surfaces

Lemma 4.1.2 *Let $\mathcal{N}$ be a compact n-dimensional submanifold, without boundary and $\mathcal{C}^k$ embedded in $(\mathbb{R}^N, \langle .,.\rangle)$, where $k \geq 4$. Then there exists an N-dimensional submanifold, $\widehat{\mathcal{N}}$, $\mathcal{C}^{k-1}$ isometrically embedded in $(\mathbb{R}^{\widehat{N}}, \langle .,.\rangle)$, and a $\mathcal{C}^{k-1}$ embedding, J, of $\mathcal{N}$ in $\widehat{\mathcal{N}}$, such that*

(i) $J : (\mathcal{N}, \langle .,.\rangle_{\mathbb{R}^N}) \longrightarrow (\widehat{\mathcal{N}}, \langle .,.\rangle_{\mathbb{R}^{\widehat{N}}})$ *is an isometric embedding.*
(ii) $\widehat{\mathcal{N}}$ *is diffeomorphic to the torus $T^N = (S^1)^N$.*
(iii) *For any Riemannian manifold $(\mathcal{M}, g)$ of dimension $m \geq 1$, for any open set Ω of $\mathcal{M}$ and for any map $u \in H^1(\Omega, \mathcal{N})$, if u is weakly harmonic, then so is $J \circ u$.*

Proof We start by introducing the objects and the notation. Since $\mathcal{N}$ is compact, there exists a cube C^N in $\mathbb{R}^N$ which contains $\mathcal{N}$. By identifying opposite faces of C^N, we obtain an N-dimensional torus T^N. Consider the Euclidean metric $\langle .,.\rangle$ on T^N. Indifferently, we consider $\mathcal{N}$ as a submanifold of C^N, T^N or $\mathbb{R}^N$. Let

$$V_{2\delta}\mathcal{N} = \{z \in C^N \mid d(z, \mathcal{N}) < 2\delta\}$$

be a tubular neighborhood of $\mathcal{N}$ in C^N. Choosing $\delta \in \mathbb{R}_+^*$ sufficiently small, we may define, using an argument based on the implicit function theorem and a partition of unity, a (not necessarily orthogonal) projection

$$P : V_{2\delta}\mathcal{N} \longrightarrow \mathcal{N},$$

which will be $\mathcal{C}^k$, if $\mathcal{N}$ is $\mathcal{C}^k$ (see exercises 1.7 to 1.11, end of section 1.4). For $y \in \mathcal{N}$, we define

$$D_y = \{z \in V_{2\delta}\mathcal{N} \mid P(z) = y\}.$$

D_y is topologically a $(N-n)$-dimensional disk, cutting $\mathcal{N}$ transversally at y. We may then define, for any $z \in V_{2\delta}\mathcal{N}$, the orthogonal projection

$$V_z : T_z C^N \simeq \mathbb{R}^N \longrightarrow T_z D_{P(z)}$$

which, to each $\xi \in T_z C^N$, associates its "vertical" component $V_z(\xi)$ in the fibration $P : V_{2\delta}\mathcal{N} \longrightarrow \mathcal{N}$.

We remark that V_z will then have a $\mathcal{C}^{k-1}$ dependence on z.

We also consider the differential of P at z,

$$dP_z : T_z C^N \simeq \mathbb{R}^N \longrightarrow T_{P(z)}\mathcal{N}$$

which is of class $\mathcal{C}^{k-1}$.

We define the $\mathcal{C}^{k-1}$ metric h_0 on $V_{2\delta}\mathcal{N}$ by, $\forall z \in V_{2\delta}\mathcal{N}$, $\forall \xi, \eta \in T_z C^N$,

$$h_0(z)(\xi, \eta) = \langle V_z(\xi), V_z(\eta) \rangle_{\mathbb{R}^N} + \langle dP_z(\xi), dP_z(\eta) \rangle_{\mathbb{R}^N}.$$

Then, choosing a function $\chi \in \mathcal{C}_c^\infty(V_{2\delta}\mathcal{N})$ which is equal to 1 over $V_\delta\mathcal{N}$, we may construct a metric h on T^N (or $\mathbb{R}^N$), which coincides with h_0 on $V_\delta\mathcal{N}$ and with the Euclidean metric outside $V_{2\delta}\mathcal{N}$, by letting

$$h = \chi h_0 + (1-\chi)\langle \cdot, \cdot \rangle_{\mathbb{R}^N}.$$

Using the Nash–Moser theorem (see [123], [102], [77]), since $k \geq 4$ and h is $\mathcal{C}^{k-1}$, we may isometrically embed the manifold (T^N, h) in the Euclidean space $(\mathbb{R}^{\widehat{N}}, \langle \cdot, \cdot \rangle)$. Let J be this embedding – it will be $\mathcal{C}^{k-1}$. We define $\widehat{\mathcal{N}} = J(T^N)$. Again, we consider the tubular neighborhood

$$\widehat{V}_{\widehat{\delta}}\widehat{\mathcal{N}} = \{z \in \mathbb{R}^{\widehat{N}} \mid d(z, \widehat{\mathcal{N}}) < \widehat{\delta}\},$$

and we suppose $\widehat{\delta}$ to be sufficiently small for us to be able to construct a $\mathcal{C}^{k-1}$ projection $\widehat{P} : \widehat{V}_{\widehat{\delta}}\widehat{\mathcal{N}} \longrightarrow \widehat{\mathcal{N}}$. We also let

$$\widehat{V}_{\widehat{\delta}}\mathcal{N} = \{z \in \mathbb{R}^{\widehat{N}} \mid d(z, \mathcal{N}) < \widehat{\delta}\}.$$

We suppose, as well, that $\widehat{\delta}$ is sufficiently small to have

$$\widehat{V}_{\widehat{\delta}}\mathcal{N} \cap \widehat{\mathcal{N}} \subset J(V_\delta\mathcal{N})$$

so that, since J is an embedding, we can define the inverse diffeomorphism of J

$$K : \widehat{V}_{\widehat{\delta}}\mathcal{N} \cap \widehat{\mathcal{N}} \longrightarrow V_\delta\mathcal{N}$$

such that $J \circ K$ is the identity over $\widehat{V}_{\widehat{\delta}}\mathcal{N} \cap \widehat{\mathcal{N}}$.

Finally, for $u \in H^1(\Omega, \mathcal{N})$, we define

$$E(u) = \frac{1}{2}\int_\Omega g^{\alpha\beta}(x) \left\langle \frac{\partial u}{\partial x^\alpha}, \frac{\partial u}{\partial x^\beta} \right\rangle_{\mathbb{R}^N} d\mathrm{vol}_g$$

4.1 Regularity of weakly harmonic maps of surfaces

(where $H^1(\Omega, \mathcal{N}) = \{u \in H^1(\Omega, \mathbb{R}^N) \mid u(x) \in \mathcal{N} \text{ a.e.}\}$) and if $v \in H^1(\Omega, \widehat{\mathcal{N}})$,

$$\widehat{E}(v) = \frac{1}{2} \int_\Omega g^{\alpha\beta}(x) \left\langle \frac{\partial v}{\partial x^\alpha}, \frac{\partial v}{\partial x^\beta} \right\rangle_{\mathbb{R}^{\widehat{N}}} d\mathrm{vol}_g .$$

Once we are this far, it just remains to check (iii) to prove the lemma. Let $u \in H^1(\Omega, \mathcal{N})$ be a weakly harmonic map. Define

$$v = J \circ u \in H^1(\Omega, \widehat{\mathcal{N}}) .$$

Let $\phi \in H^1_0(\Omega, \mathbb{R}^{\widehat{N}}) \cap L^\infty(\Omega, \mathbb{R}^{\widehat{N}})$ be a test function and estimate

$$\widehat{E}(\widehat{P}(v + t\phi))$$

to order 1 in t.

We remark that K is an isometry from $(\widehat{V_\delta \mathcal{N}} \cap \widehat{\mathcal{N}}, \langle ., . \rangle_{\mathbb{R}^{\widehat{N}}})$ to $(V_\delta \mathcal{N}, h) \subset (\mathbb{R}^N, h)$. Since $\widehat{P}(v + t\phi) \in \widehat{\mathcal{N}}$ a.e., we can consider

$$w_t = K \circ \widehat{P}(v + t\phi)$$

and deduce from the fact that K is an isometry

$$\widehat{E}(\widehat{P}(v + t\phi)) = \frac{1}{2} \int_\Omega g^{\alpha\beta}(x) h_{w_t}\left(\frac{\partial w_t}{\partial x^\alpha}, \frac{\partial w_t}{\partial x^\beta} \right) d\mathrm{vol}_g . \quad (4.4)$$

Developing w_t we have

$$\begin{aligned} w_t &= K \circ \widehat{P}(v) + t\psi_t \\ &= u + t\psi_t , \end{aligned}$$

where

$$\psi_t = \int_0^1 \frac{\partial K \circ \widehat{P}}{\partial y^j}(v + st\phi) \phi^j \, ds . \quad (4.5)$$

Since $K \circ \widehat{P}$ is bounded in $\mathcal{C}^{k-1}$ and $k \geq 4$, ψ_t is bounded in H^1. Moreover, for sufficiently small t,

$$h_{w_t}\left(\frac{\partial w_t}{\partial x^\alpha}, \frac{\partial w_t}{\partial x^\beta}\right) = \left\langle dP_{w_t}\left(\frac{\partial w_t}{\partial x^\alpha}\right), dP_{w_t}\left(\frac{\partial w_t}{\partial x^\beta}\right)\right\rangle$$
$$+ \left\langle V_{w_t}\left(\frac{\partial w_t}{\partial x^\alpha}\right), V_{w_t}\left(\frac{\partial w_t}{\partial x^\beta}\right)\right\rangle \quad (4.6)$$

with

$$dP_{w_t}\left(\frac{\partial w_t}{\partial x^\alpha}\right) = \frac{\partial}{\partial x^\alpha}(P(w_t)). \quad (4.7)$$

On the other hand,

$$V_{w_t} = V_u + t\delta V_t,$$

where

$$\delta V_t = \int_0^1 \frac{\partial V_{u+st\psi_t}}{\partial z^i} \psi_t^i \, ds$$

is bounded in $L^\infty(\Omega)$. And since $V_u\left(\frac{\partial u}{\partial x^\alpha}\right) = 0$, we have

$$V_{w_t}\left(\frac{\partial w_t}{\partial x^\alpha}\right) = (V_u + t\delta V_t)\left(\frac{\partial u}{\partial x^\alpha} + t\frac{\partial \psi_t}{\partial x^\alpha}\right)$$
$$= t\left[\delta V_t\left(\frac{\partial u}{\partial x^\alpha}\right) + V_u\left(\frac{\partial \psi_t}{\partial x^\alpha}\right) + t\delta V_t\left(\frac{\partial \psi_t}{\partial x^\alpha}\right)\right].$$

Thus,

$$\int_\Omega g^{\alpha\beta}(x)\left\langle V_{w_t}\left(\frac{\partial u}{\partial x^\alpha}\right), V_{w_t}\left(\frac{\partial w_t}{\partial x^\beta}\right)\right\rangle d\mathrm{vol}_g = O(t^2). \quad (4.8)$$

Therefore, we deduce from (4.6), (4.7) and (4.8) that

$$\widehat{E}(\widehat{P}(u+t\phi)) = E(P(u+t\psi_t)) + O(t^2). \quad (4.9)$$

But it is easy to see that when t goes to 0,

$$\psi_t \to \frac{\partial K \circ \widehat{P}}{\partial y^j}(v)\phi^j, \text{ in } H^1(\Omega, \mathbb{R}^N),$$

and we can then deduce, from the fact that u is weakly harmonic, that

4.1 Regularity of weakly harmonic maps of surfaces

$$\lim_{t\to 0} \frac{\widehat{E}(\widehat{P}(v+t\phi)) - \widehat{E}(v)}{t} = 0.$$

□

Thanks to the result we have just proved, it will be possible to reduce the study of a weakly harmonic map to the case where the image manifold is diffeomorphic to a torus (as long as the image manifold we start with is at least of class C^4). The interest will then be that we will have available a C^1 orthonormal frame field over the image manifold, $\widetilde{e}(y) = (\widetilde{e_1}(y), \ldots, \widetilde{e_n}(y))$. We denote by $\mathcal{F}$ the fiber bundle of orthonormal frames over $\mathcal{N}$. Its pull-back by u, which we will denote by $u^*\mathcal{F}$, is the bundle over Ω whose fiber over a point $x \in \Omega$ is the set of orthonormal frames of $T_{u(x)}\mathcal{N}$. It is not a smooth bundle since the fiber will depend on x in an H^1 fashion. From $\widetilde{e}$, which we may interpret as a section of $\mathcal{F}$, we construct a section $\widetilde{e} \circ u$ of $u^*\mathcal{F}$. In fact, it consists of the map $x \longmapsto (\widetilde{e_1}(u(x)), \ldots, \widetilde{e_n}(u(x)))$. We will then need to construct, using $\widetilde{e} \circ u$, a Coulomb frame over Ω, i.e. a finite energy harmonic section of the fiber bundle $u^*\mathcal{F}$. This is the purpose of the following result.

Lemma 4.1.3 *Let $\mathcal{N}$ be an n-dimensional compact manifold without boundary, C^2 embedded in $\mathbb{R}^N$. Let $(\mathcal{M}, g)$ be a Riemannian manifold and Ω an open subset of $(\mathcal{M}, g)$. Let $u \in H^1(\Omega, \mathcal{N})$ and $\overline{e} = (\overline{e_1}, \ldots, \overline{e_n})$ be any finite energy section of $u^*\mathcal{F}$. Then, there exists a Coulomb frame over Ω associated to u, $e = (e_1, \ldots, e_n)$, i.e. a finite energy section of $u^*\mathcal{F}$ such that*

$$\begin{cases} \dfrac{\partial}{\partial x^\alpha}\left(g^{\alpha\beta}(x)\sqrt{\det g(x)}\left\langle \dfrac{\partial e_a}{\partial x^\beta}, e_b\right\rangle\right) = 0 & \text{in } \Omega \\ n^\alpha(x)\left\langle \dfrac{\partial e_a}{\partial x^\alpha}, e_b\right\rangle = 0 & \text{on } \partial\Omega, \end{cases} \quad (4.10)$$

where n is the normal vector on $\partial\Omega$. Moreover, we have the estimates

$$\|{}^t ede\|^2_{L^2(\Omega)} \leq \|{}^t\overline{e}d\overline{e}\|^2_{L^2(\Omega)}, \quad (4.11)$$

and

$$\|de\|^2_{L^2(\Omega)} \leq \|{}^t ede\|^2_{L^2(\Omega)} + C\|du\|^2_{L^2(\Omega)}, \quad (4.12)$$

where C is a constant that depends only on $\mathcal{N}$. In case $\overline{e} = \widetilde{e} \circ u$, where $\widetilde{e}$ is a C^1 section of $\mathcal{F}$, we deduce from (4.11) and (4.12) that

$$\|de\|^2_{L^2(\Omega)} \leq C\|du\|^2_{L^2(\Omega)}. \tag{4.13}$$

Remark 4.1.4 *We can view e in two different ways. An intrinsic point of view consists of seeing e as a harmonic section of $u^*\mathcal{F}$. But we can also consider e as a map in $H^1(\Omega,(\mathbb{R}^N)^n)$, which to nearly all $x \in \Omega$ associates a family of n orthonormal vectors in $\mathbb{R}^N$ constituting a basis for $T_{u(x)}\mathcal{N}$. In what follows, we will in fact represent e as the $n \times N$ matrix of the components in the canonical basis of $\mathbb{R}^N$ of the n vectors $e_1,\ldots,e_n$, i.e.*

$$e = (e_1,\ldots,e_n) = \begin{pmatrix} e_1^1 & \cdots & e_n^1 \\ \vdots & & \vdots \\ e_1^N & \cdots & e_n^N \end{pmatrix}.$$

Thus, in inequalities (4.11) and (4.12), ${}^t e d e$ should be interpreted as a matrix product; in this case it will be the $n \times n$ matrix whose elements are the 1-forms $\langle de_a, e_b\rangle$. Likewise, the action of the gauge group $H^1(\Omega, SO(n))$ on $u^*\mathcal{F}$ will be represented by the matrix product

$$e \longmapsto eR$$

for any $R \in H^1(\Omega, SO(n))$ (here $e_a \longmapsto \sum_{b=1}^n e_b R_a^b$).

Proof of lemma 4.1.3 For any map $R \in H^1(\Omega, SO(n))$, we consider the moving frame e defined by

$$e_a(x) = \overline{e_b}(x) R_a^b(x),$$

and we consider the quantity

$$F(R) = \frac{1}{2}\int_\Omega |{}^t e d e|^2 d\mathrm{vol}_g,$$

where

$$|{}^t e d e|^2 = g^{\alpha\beta}(x) \sum_{a,b=1}^n \left\langle \frac{\partial e_a}{\partial x^\alpha}, e_b\right\rangle \left\langle \frac{\partial e_a}{\partial x^\beta}, e_b\right\rangle.$$

4.1 Regularity of weakly harmonic maps of surfaces

Step 1 EXISTENCE OF A MINIMUM FOR F
Let $(R_k)_{k\in\mathbb{N}}$ be a sequence in $H^1(\Omega, SO(n))$, minimizing F. Define

$$e_k(x) = \bar{e}(x)R_k(x).$$

We start by showing that (R_k) is bounded in $H^1(\Omega, SO(n))$. We can easily see that

$$\begin{aligned}|{}^t e_k de_k|^2 &= |{}^t \bar{e} d\bar{e}|^2 + 2\langle {}^t \bar{e} d\bar{e}, dR_k\, {}^t R_k\rangle + |dR_k\, {}^t R_k|^2\\ &\geq (|{}^t \bar{e} d\bar{e}| - |dR_k|)^2.\end{aligned} \quad (4.14)$$

But $|{}^t \bar{e} d\bar{e}|$ is bounded in $L^2(\Omega)$, and since (R_k) minimizes F, $|{}^t e_k de_k|$ is also bounded in $L^2(\Omega)$. Thus, we deduce using (4.14) that $|dR_k|$ is bounded in $L^2(\Omega)$. Therefore, we may extract a subsequence of k (which, to simplify, we still call k) such that there exists $R \in H^1(\Omega, SO(n))$, satisfying

$$\begin{aligned}dR_k &\rightharpoonup dR \text{ weakly in } L^2(\Omega, M(n\times n, \mathbb{R}))\\ R_k &\to R \text{ in } L^2(\Omega, SO(n))\end{aligned} \quad (4.15)$$

$$R_k \to R \text{ a.e.} \quad (4.16)$$

Using the dominated convergence theorem, we deduce from (4.16) that $|{}^t\bar{e}d\bar{e}R_k - {}^t\bar{e}d\bar{e}R|^2$ goes to 0 in $L^1(\Omega)$, and thus that

$${}^t\bar{e}d\bar{e}R_k \to {}^t\bar{e}d\bar{e}R \quad \text{in } L^2(\Omega, M(n\times n, \mathbb{R})). \quad (4.17)$$

Since $\langle {}^t\bar{e}d\bar{e}, dR_k\, {}^t R_k\rangle = \langle {}^t\bar{e}d\bar{e}R_k, dR_k\rangle$, the convergences (4.15) and (4.17) imply that

$$\lim_{k\to+\infty} \int_\Omega \langle {}^t\bar{e}d\bar{e}, dR_k\, {}^t R_k\rangle d\text{vol}_g = \int_\Omega \langle {}^t\bar{e}d\bar{e}, dR\, {}^t R\rangle d\text{vol}_g. \quad (4.18)$$

Likewise, we deduce from (4.15) that

$$\lim_{k\to+\infty} \int_\Omega |dR_k\, {}^t R_k|^2 d\text{vol}_g \geq \int_\Omega |dR\, {}^t R|^2 d\text{vol}_g. \quad (4.19)$$

Therefore, we may conclude, putting together (4.14), (4.18) and (4.19), that

$$\lim_{k \to +\infty} F(R_k) \geq F(R). \tag{4.20}$$

This implies that R minimizes F. We denote by $e = (e_1, \ldots, e_n)$ the frame associated to R.

Step 2 EULER–LAGRANGE EQUATION
Given a test function $\phi \in \mathcal{C}^\infty(\Omega, so(n))$, and for ϵ close to zero, let

$$\begin{aligned} R_\epsilon &= R(x)\exp(\epsilon\phi(x)) \\ &= R(x)(\mathbb{1} + \epsilon\phi(x) + o(\epsilon)), \end{aligned}$$

and

$$\begin{aligned} e_\epsilon(x) &= \overline{e}(x)R_\epsilon(x) \\ &= e(x)(\mathbb{1} + \epsilon\phi(x) + o(\epsilon)). \end{aligned}$$

Since R minimizes F, we necessarily have that

$$F(R_\epsilon) = F(R) + o(\epsilon).$$

A calculation yields

$$\,^t e_\epsilon de_\epsilon = \,^t e d e + \epsilon(d\phi - \phi \,^t e d e) + o(\epsilon).$$

Thus,

$$\int_\Omega \langle \,^t e d e, d\phi - \phi \,^t e d e \rangle d\mathrm{vol}_g = 0. \tag{4.21}$$

We now notice that, because ϕ takes values in $so(n)$, the quantity $\langle \,^t e d e, \phi \,^t e d e \rangle$ always vanishes. This simplifies (4.21), and, integrating by parts, we obtain

$$\int_{\partial\Omega} \left\langle \phi, \,^t e \frac{\partial e}{\partial n} \right\rangle d\sigma - \int_\Omega \langle \phi, \mathrm{div}_g(\,^t e d e)\rangle d\mathrm{vol}_g = 0, \tag{4.22}$$

where

$$\mathrm{div}_g(\,^t e d e) = \frac{1}{\sqrt{\det g}} \frac{\partial}{\partial x^\alpha}\left(g^{\alpha\beta}(x)\sqrt{\det g}\left(\,^t e \frac{\partial e}{\partial x^\beta}\right)\right).$$

It is now obvious that (4.22) implies (4.10).

4.1 Regularity of weakly harmonic maps of surfaces

Step 3 ESTIMATES (4.11) AND (4.12)
Estimate (4.11) is a consequence of the construction of e, obtained by minimization. We will deduce (4.12).

For any $y \in \mathcal{N}$, we denote by $P(y)$ the orthogonal projection of $\mathbb{R}^N$ onto $T_y\mathcal{N}$, and $P^\perp(y) = 1\!\!1_N - P(y)$. It follows from the hypothesis on $\mathcal{N}$ that P and $P^\perp$ are $\mathcal{C}^1$ maps on $\mathcal{N}$. Using the identity

$$de_i = P(u)(de_i) + P^\perp(u)(de_i)$$

we deduce

$$\|de\|^2_{L^2(\Omega)} = \sum_{a=1}^n \|P(u)(de_a)\|^2_{L^2(\Omega)} + \sum_{a=1}^n \|P^\perp(u)(de_a)\|^2_{L^2(\Omega)}$$

$$= \|{}^tede\|^2_{L^2(\Omega)} + \sum_{a=1}^n \|P^\perp(u)(de_a)\|^2_{L^2(\Omega)}. \qquad (4.23)$$

Since $d(P^\perp(u)e_a) = 0$, we obtain

$$\sum_{a=1}^n \|P^\perp(u)(de_a)\|^2_{L^2(\Omega)} = \sum_{a=1}^n \|d(P^\perp(u))(e_a)\|^2_{L^2(\Omega)}$$

$$\leq \|P\|^2_{W^{1,\infty}} \|du\|^2_{L^2(\Omega)}, \qquad (4.24)$$

and (4.23) and (4.24) together imply that

$$\|de\|^2_{L^2(\Omega)} \leq \|{}^tede\|^2_{L^2(\Omega)} + \|P\|^2_{W^{1,\infty}} \|du\|^2_{L^2(\Omega)}$$

which is (4.12). □

Remark 4.1.5 *We see that the above proof of equation (4.10) reproduces that of Noether's theorem (theorem 1.3.1) with the symmetry group $SO(n)$. This explains why (4.10) is a conservation law.*

Lemma 4.1.6 *Let $\mathcal{N}$ be a compact submanifold without boundary, $\mathcal{C}^2$ embedded in $\mathbb{R}^N$. Let B^2 be the unit ball in $\mathbb{R}^2$ and u a map in $H^1(B^2, \mathcal{N})$. There exists a constant γ_0 depending only on $\mathcal{N}$ such that if*

$$\|du\|^2_{L^2(B^2)} \leq \gamma_0, \qquad (4.25)$$

then we can construct a finite energy Coulomb orthonormal frame for u (i.e. a harmonic section of u^*F).

Proof This result is proved by essentially following the proof of lemma 5.1.4, of the next chapter. We leave it to the reader as an exercise. The plan is as follows:

Step 1 We start by working with a map $u \in C^2(B^2, \mathcal{N}) \cap H^1(B^2, \mathcal{N})$ satisfying (4.25). Since u is smooth and B^2 is contractible, it is possible to construct a section $\bar{e}$ of $u^*\mathcal{F}$ (where $\mathcal{F}$ is the bundle of orthonormal tangent frames on $\mathcal{N}$) belonging to $C^1(B^2) \cap H^1(B^2)$.

We will work with $\bar{e}$ as a map in $H^1(B^2, M(n \times N, \mathbb{R}))$.

Step 2 We let $e(z) = \bar{e}(z)R(z)$, where $R \in H^1(B^2, SO(n))$, and we minimize $F(R) = \frac{1}{2}\|{}^t e d e\|^2_{L^2(B^2)}$ as in lemma 4.1.3. In this way we obtain a Coulomb frame which we still denote by $e(z) = (e_1(z), \ldots, e_n(z))$. Thanks to equation (4.10), there exist maps A_a^b in $H^1(B^2)$ such that

$$\left(\frac{\partial A_a^b}{\partial x}, \frac{\partial A_a^b}{\partial y}\right) = \left(\left\langle\frac{\partial e_a}{\partial y}, e_b\right\rangle, -\left\langle\frac{\partial e_a}{\partial x}, e_b\right\rangle\right). \quad (4.26)$$

A direct calculation then yields

$$-\Delta A_a^b = \sum_{k=1}^{N}\{e_a^k, e_b^k\} \text{ in } B^2 \quad (4.27)$$

and the boundary condition in (4.10) implies that we may choose

$$A_a^b = 0 \text{ on } \partial B^2. \quad (4.28)$$

Using theorem 3.1.2, we deduce from (4.27) and (4.28) that there exists a universal constant C_1 such that

$$\|dA\|^2_{L^2(B^2)} := \sum_{i,j=1}^{n} \|dA_i^j\|^2_{L^2(B^2)} \leq C_1 \|de\|^4_{L^2(B^2)}. \quad (4.29)$$

Step 3 We put together inequality (4.12) of lemma 4.1.3, i.e.

$$\|de\|^2_{L^2(B^2)} \leq \|dA\|^2_{L^2(B^2)} + C_2\|du\|^2_{L^2(B^2)}$$

with inequality (4.29). This yields

$$C_1 t^2 - t + C_2\|du\|^2_{L^2(B^2)} \geq 0, \quad (4.30)$$

4.1 Regularity of weakly harmonic maps of surfaces

where $t = \|de\|^2_{L^2(B^2)}$. We remark that if

$$\|du\|^2_{L^2(B^2)} < \gamma_0 := (4C_1C_2)^{-1},$$

then the polynomial in (4.30) would take negative values in a neighborhood of $t = (2C_1)^{-1}$, and hence $\|de\|^2_{L^2(B^2)}$ cannot approach this value.

Step 4 We go over the previous steps once again, but this time over balls of variable size B_r^2, $0 < r < 1$. We show that $\|de_r\|^2_{L^2(B_r)}$ is a continuous function of r, and that if (4.25) is true, then $\|de_r\|^2_{L^2(B_r)}$ cannot approach $t = (2C_1)^{-1}$. By a continuity argument we deduce that

$$\|de\|^2_{L^2(B^2)} \leq (2C_1)^{-1}.$$

Step 5 We use a result of [145]: the set $\mathcal{C}^2(B^2, \mathcal{N})$ is dense in $H^1(B^2, \mathcal{N})$, and applying the previous steps to a sequence u_k in $\mathcal{C}^2(B^2, \mathcal{N})$ approaching u in $H^1(B^2, \mathcal{N})$, we construct a sequence of bounded energy harmonic sections $u_k^*\mathcal{F}$. Passing to the limit we obtain our result. □

The following result will specify what is the gain of regularity obtained by using a Coulomb frame in dimension 2. Here we will encounter once more compensation phenomena for Jacobian determinants, as well as Hardy and Lorentz spaces.

Lemma 4.1.7 *Let $\mathcal{N}$ be a compact Riemannian manifold without boundary, C^2 embedded in $\mathbb{R}^N$. Let $u \in H^1(B^2, \mathcal{N})$ and e be a Coulomb frame, i.e. a section of $u^*\mathcal{F}$ belonging to $H^1(B^2)$, and satisfying (4.10). Then, all the coefficients of ${}^tede$ belong to $L^{(2,1)}(B^2)$. Moreover, there exists a constant C such that*

$$\|{}^tede\|^2_{L^{(2,1)}(B^2)} \leq C\|de\|^4_{L^2(B^2)}. \tag{4.31}$$

Furthermore, in the case where the frame e is obtained by minimization, in the gauge orbit of a frame $\bar{e} = \widetilde{e} \circ u$ (see lemma 4.1.3), we have the estimate

$$\|{}^tede\|^2_{L^{(2,1)}(B^2)} \leq C\|du\|^4_{L^2(B^2)}. \tag{4.32}$$

Proof Inequality (4.32) is an immediate consequence of (4.31) and of inequality (4.13) in lemma 4.1.3.

In order to prove (4.31) we introduce the quantities $A_a^b \in H^1(B^2)$ defined by (4.26). Using equations (4.27) and (4.28) satisfied by A_a^b, and applying theorem 3.4.1, we obtain that $A_a^b \in L^{(2,1)}(B^2)$ and

$$\|dA_a^b\|^2_{L^{(2,1)}(B^2)} \leq C \|de_a\|^2_{L^2(B^2)} \|de_b\|^2_{L^2(B^2)} \qquad (4.33)$$

which, because of (4.26), immediately yields (4.31). □

Proof of theorem 4.1.1 Let $u \in H^1(B^2, \mathcal{N})$ be a weakly harmonic map. We will apply lemma 4.1.6, and for this we need to assume that

$$\|du\|^2_{L^2(B^2)} \leq \gamma_0.$$

If such is not the case, it suffices to place ourselves in a smaller ball, so that this hypothesis is satisfied. Thus, we obtain a finite energy Coulomb frame e for u (a section of $u^*\mathcal{F}$). Moreover, by lemma 4.1.7 we know that ${}^t\!ede$ is in $L^{(2,1)}(B^2)$. Let

$$\alpha^a = \left\langle \frac{\partial u}{\partial x}, e_a \right\rangle - i \left\langle \frac{\partial u}{\partial y}, e_b \right\rangle \quad \text{for } a = 1, ..., n$$

$$\omega_b^a = \left\langle \frac{\partial e_b}{\partial \bar{z}}, e_a \right\rangle \quad \text{for } a, b = 1, ..., n.$$

Writing

$$du = \sum_{a=1}^n \langle du, e_a \rangle e_a \qquad (4.34)$$

and using the relations

$$d^2 u = 0, \qquad (4.35)$$

$$d(\star du) = d\left(\frac{\partial u}{\partial x}dy - \frac{\partial u}{\partial y}dx\right) = \Delta u \, dx \wedge dy \perp T_u \mathcal{N}, \qquad (4.36)$$

we obtain that

4.1 Regularity of weakly harmonic maps of surfaces

$$\frac{\partial \alpha^a}{\partial \bar z} = \sum_{b=1}^{n} \omega_b^a \alpha^b, \qquad (4.37)$$

where the real part of (4.37) corresponds to (4.36) and the imaginary part to (4.35). Putting

$$\alpha = \begin{pmatrix} \alpha^1 \\ \vdots \\ \alpha^n \end{pmatrix} \in \mathbb{C}^n,$$

$$\omega = \begin{pmatrix} 0 & \omega_2^1 & \cdots & \omega_n^1 \\ \omega_1^2 & 0 & \cdots & \omega_n^2 \\ \vdots & \vdots & & \vdots \\ \omega_1^n & \omega_2^n & \cdots & 0 \end{pmatrix} \in SO(n) \otimes \mathbb{C},$$

we can also write (4.37) as

$$\frac{\partial \alpha}{\partial \bar z} = \omega \alpha. \qquad (4.38)$$

The important fact to retain is that $\alpha \in L^2(B^2, \mathbb{C}^n)$, and that, because of lemma 4.1.7 (and $\omega = \frac{1}{2}\left({}^t e \frac{\partial e}{\partial x} + i\, {}^t e \frac{\partial e}{\partial y}\right)$), $\omega \in L^{(2,1)}(B^2, so(n) \otimes \mathbb{C})$. Using this information we will study equation (4.38).

EXISTENCE OF SOLUTIONS OF (4.38) IN L^∞_{loc}

For any $z_0 \in B^2$, we choose a ball centered at z_0 and contained in B^2, $B^2_{z_0}$, such that

$$\|\omega_{|B^2_{z_0}}\|_{L^{(2,1)}} < \frac{1}{3\sqrt{2\pi}}.$$

We will construct $\beta \in L^\infty(B^2_{z_0}, GL(n, \mathbb{C}))$, a solution of

$$\frac{\partial \beta}{\partial \bar z} = \omega \beta \text{ in } B^2_{z_0}. \qquad (4.39)$$

We temporarily denote by ω the function on $\mathbb{R}^2 = \mathbb{C}$ that coincides with ω in $B^2_{z_0}$, and takes the value 0 outside $B^2_{z_0}$. We always have $\|\omega\|_{L^{(2,1)}} < \frac{1}{2\sqrt{2\pi}}$.

We now define the linear operator T on $L^\infty(\mathbb{C}, M(n, \mathbb{C}))$ by

$$T(\beta)(z) = \int_V \frac{\omega(v)\beta(v)}{\pi(z-v)} dv^1 dv^2$$

(we remark that T is the composition of two operators, the first one being the convolution product by $\frac{1}{\pi z}$, the kernel of $\frac{\partial}{\partial \bar{z}}$, and the second one being the product on the left by ω).

Here we use the fact that $L^{(2,1)}(\mathbb{R}^2)$ and $L^{(2,\infty)}(\mathbb{R}^2)$ are in duality (see [158]), and that $\frac{1}{\pi z}$ is in $L^{(2,\infty)}(\mathbb{R}^2)$. This implies that T maps $L^\infty(\mathbb{C}, M(n,\mathbb{C}))$ continuously to itself, with the bound

$$\|T\|_{L^\infty} \leq \|\frac{1}{\pi z}\|_{L^{(2,\infty)}} \|\omega\|_{L^{(2,1)}} < \frac{1}{3}. \tag{4.40}$$

Applying T to both sides of equation (4.39), we see that any solution β of (4.39) should be a solution of

$$\beta - T(\beta) = H, \tag{4.41}$$

where $H : \mathbb{C} \longrightarrow M(n,\mathbb{C})$ is holomorphic. Using estimate (4.40), we deduce, using a fixed point argument, that if H is in $L^\infty(\mathbb{C})$ (and in particular if it is constant), equation (4.41) has a unique solution. We choose $H = \mathbb{1}$, so that

$$\|\beta - \mathbb{1}\|_{L^\infty} \leq \frac{2}{3},$$

and β takes values in the set of regular matrices, $GL(n, \mathbb{C})$. Furthermore, it is clear that β is also a solution of (4.39).

Conclusion We use over $B^2_{z_0}$ the solution β of (4.39) which we have just constructed. We deduce from (4.38) that

$$\frac{\partial}{\partial \bar{z}}(\beta^{-1}\alpha) = \beta^{-1}(\omega - \omega)\alpha = 0.$$

Hence $\beta^{-1}\alpha$ is holomorphic in $B^2_{z_0}$, and, in particular, is locally bounded. This yields that $\alpha \in L^\infty_{loc}(B^2, \mathbb{C}^n)$, i.e. that u is locally Lipschitz. Using equation (1.65), we deduce that $\Delta u \in L^\infty_{loc}$, and hence, by the Sobolev embedding theorems (see [19]), this implies that u is $\mathcal{C}^{1,\alpha}$. All higher regularity follows from theorem 1.5.1. $\square$

4.1 Regularity of weakly harmonic maps of surfaces

A VARIANT OF THE PROOF

It is possible to obtain this result using, instead of complex analysis, estimates in Morrey–Campanato spaces (see section 3.5). These techniques are more complicated but have the advantage of being more flexible when we are interested in generalizing theorem 4.1.1 (as we will see in section 4.2). The method proposed here is a simplification of the result in [38].

Proof of theorem 4.1.1 (variant) The beginning is similar to the previous proof: we construct a Coulomb frame $e = (e_1, ..., e_n)$. Then, we know by lemma 4.1.7 that ${}^t e d e$ is in $L^{(2,1)}(B^2)$, and that $\|{}^t e d e\|_{L^{(2,1)}(B^2)}$ is controlled by $\|du\|_{L^2(B^2)}$. Up to working on an even smaller ball, we can always suppose that

$$\|{}^t e d e\|_{L^{(2,1)}(B^2)} \leq \frac{\varepsilon_0}{4}, \qquad (4.42)$$

where $\varepsilon_0 > 0$ is a constant which will be chosen later.

Step 1 REWRITING
For each $z_0 \in B^2$ and $r > 0$ such that $B^2(z_0, r) \subset B^2$, we decompose each 1-form $\langle du, e_i \rangle$ over $B_r := B^2(z_0, r)$ according to

$$\begin{cases} \left\langle \dfrac{\partial u}{\partial x}, e_a \right\rangle = \dfrac{\partial w^a}{\partial x} - \dfrac{\partial v^a}{\partial y} \\ \left\langle \dfrac{\partial u}{\partial y}, e_a \right\rangle = \dfrac{\partial w^a}{\partial y} + \dfrac{\partial v^a}{\partial x} \end{cases} \text{in } B_r, \qquad (4.43)$$

where $v^a = \widetilde{\langle u, e_a \rangle}$, i.e.

$$\begin{cases} -\Delta v^a = \{u, e_a\} & \text{in } B_r \\ v^a = 0 & \text{on } \partial B_r. \end{cases} \qquad (4.44)$$

We deduce, by a simple calculation using (4.43), that $\|du\|^2_{L^2(B_r)} = \|dv\|^2_{L^2(B_r)} + \|dw\|^2_{L^2(B_r)}$.

Having in mind (4.44), we might feel like estimating dv^a in $L^{(2,1)}(B_r)$, using the results of chapter 3. We will proceed otherwise, rewriting (4.44) as

$$-\Delta v^a = \sum_{b=a}^n \left(\left\langle \frac{\partial u}{\partial x}, e_b \right\rangle \left\langle e_b, \frac{\partial e_a}{\partial y} \right\rangle - \left\langle \frac{\partial u}{\partial y}, e_b \right\rangle \left\langle e_b, \frac{\partial e_a}{\partial x} \right\rangle \right). \quad (4.45)$$

Likewise, using the fact that u is weakly harmonic and (4.43), we have

$$-\Delta w^a = -\sum_{b=a}^n \left(\left\langle \frac{\partial u}{\partial x}, e_b \right\rangle \left\langle e_b, \frac{\partial e_a}{\partial x} \right\rangle + \left\langle \frac{\partial u}{\partial y}, e_b \right\rangle \left\langle e_b, \frac{\partial e_a}{\partial y} \right\rangle \right). \quad (4.46)$$

For convenience, we condense equations (4.45) and (4.46), using (4.43), as

$$-\Delta U = A dU \quad \text{in} \quad B_r \quad (4.47)$$

where U and $A = (A^x, A^y)$ are defined by

$$U = \begin{pmatrix} v^1 \\ \vdots \\ v^n \\ w^1 \\ \vdots \\ w^n \end{pmatrix} \in \mathbb{R}^{2n},$$

$$A^x = \begin{pmatrix} -{}^t e \frac{\partial e}{\partial x} & -{}^t e \frac{\partial e}{\partial y} \\ {}^t e \frac{\partial e}{\partial y} & {}^t e \frac{\partial e}{\partial x} \end{pmatrix}, \quad A^y = \begin{pmatrix} {}^t e \frac{\partial e}{\partial y} & {}^t e \frac{\partial e}{\partial x} \\ -{}^t e \frac{\partial e}{\partial x} & {}^t e \frac{\partial e}{\partial y} \end{pmatrix}.$$

Pause IDEA OF THE PROOF

We will look for Morrey–Campanato-type estimates for U in a well-chosen norm. The information we will use will be the fact that U is in $L^2(B_r)$ and A has its coefficients in $L^{(2,1)}(B_r)$, with the estimate

$$\|A\|_{L^{(2,1)}(B_r)} \leq \varepsilon_0, \quad (4.48)$$

thanks to (4.42). The idea is to treat equation (4.47) as a perturbation of $-\Delta U = 0$. In order to do so, we decompose U as

$$U = U_0 + U_1,$$

where

4.1 Regularity of weakly harmonic maps of surfaces

$$\begin{cases} -\Delta U_0 = 0 & \text{in } B_r \\ U_0 = U & \text{on } \partial B_r, \end{cases} \quad (4.49)$$

and

$$\begin{cases} -\Delta U_1 = AdU & \text{in } B_r \\ U_1 = 0 & \text{on } \partial B_r. \end{cases} \quad (4.50)$$

A second idea will be to work with the $L^{(2,\infty)}$ norm. The reason for this is that because the r.h.s. of (4.47) belongs to L^1, we will (unfortunately) not have an estimate of dU in L^2 (see theorem 3.3.6).

Step 2 ESTIMATE FOR U_0
Choose $\alpha \in (0,1]$ and apply lemma 3.3.13 to U_0, with $p = 2$ and $q = +\infty$. We obtain that for each constant γ such that $0 < \gamma < \frac{2}{p}$, there exists a constant $C(\gamma)$, such that

$$\|dU_0\|_{L^{(2,\infty)}(B_{\alpha r})} \leq C(\gamma)\alpha^\gamma \, \| \, dU \, \|_{L^{(2,\infty)}(B_r)} . \quad (4.51)$$

It will suffice in what follows to choose, for instance, $\gamma = \frac{1}{2}$.

Step 3 ESTIMATE FOR U_1
We will use the fact that the r.h.s. of equation (4.50) is in $L^1(B_r)$, together with theorem 3.3.6, to estimate dU_1 in $L^{(2,\infty)}$:

$$\|dU_1\|_{L^{(2,\infty)}(B_r)} \leq C\|AdU\|_{L^1(B_r)}. \quad (4.52)$$

But instead of using the inequality $\|AdU\|_{L^1} \leq \|A\|_{L^2}\|dU\|_{L^2}$, we will take advantage of the fact that $L^{(2,1)}$ and $L^{(2,\infty)}$ are in duality:

$$\|AdU\|_{L^1(B_r)} \leq \|A\|_{L^{(2,1)}(B_r)}\|dU\|_{L^{(2,\infty)}(B_r)}. \quad (4.53)$$

Then, it follows from (4.52), (4.53) and (4.48) that

$$\|dU_1\|_{L^{(2,\infty)}(B_r)} \leq C\varepsilon_0 \|dU\|_{L^{(2,\infty)}(B_r)}. \quad (4.54)$$

Step 4 CONCLUSION
We put together estimates (4.51) and (4.54). They imply that

$$\|dU\|_{L^{(2,\infty)}(B_{\alpha r})} \leq (C\varepsilon_0 + C(\gamma)\alpha^\gamma)\|dU\|_{L^{(2,\infty)}(B_r)}.$$

It remains only to choose ε_0 and α sufficiently small for $\theta := C\varepsilon_0 + C(\gamma)\alpha^\gamma$ to be strictly smaller than 1. By an argument analogous to that of lemma 3.5.11, we deduce that there is a constant C such that

$$\|dU\|_{L^{(2,\infty)}(B_r)} \leq Cr^{\frac{\log \theta}{\log \alpha}}. \qquad (4.55)$$

Then, using the estimate

$$\|dU\|_{L^p(B_r)} \leq Cr^{\frac{2-p}{p}} \|dU\|_{L^{(2,\infty)}(B_r)}$$

valid for all $1 \leq p < 2$ (see [95]), and the Sobolev–Poincaré inequality

$$\left(\frac{1}{r^2} \int_{B_r} |U - U_{z_0,r}|^q \right)^{\frac{1}{q}} \leq Cr^{1-\frac{2}{p}} \|dU\|_{L^p}$$

(for $1 \leq q < \frac{p}{2-p}$), we deduce from (4.55) that U belongs (locally) to the Morrey–Campanato space $\mathcal{L}^{q,2+q\frac{\log \theta}{\log \alpha}}$. Hence, by theorem 3.5.10, U is locally in $C^{0,\frac{\log \theta}{\log \alpha}}$. We remark that this reasoning is the same as that used to conclude the proof of theorem 3.5.3. Using theorem 1.5.1, we deduce our result. □

Remark 4.1.8 *Theorem 4.1.1 was extended to the case of weakly harmonic maps on a surface $\mathcal{M}$ with boundary by Jie Qing ([132]).*

Remark 4.1.9 (a heuristic justification for the use of Coulomb frames) *We can give two reasons why the use of Coulomb frames is well adapted to the study of weakly harmonic maps. First, we may notice that the essential difficulty of the problem is connected to the image manifold being curved. The Riemannian curvature is precisely what measures the lack of flatness. But, when we use Elie Cartan's description of a Riemannian manifold in terms of moving frames, the curvature is represented by a 2-form with coefficients in the Lie algebra so(n). Hence, if we put ourselves in the place of a map u from B^2 to the manifold (which may require a certain effort), the curvature effects are felt through the pull-back by u of this curvature form, which is essentially a combination of order 2 minors of the Jacobian matrix. But then, Wente's theorem tells us that this type of non-linearity is not very bad. A second reason is the difficulty in taking advantage of the Euler equation $\Delta u \perp T_u \mathcal{N}$. Recall that it is an equation in the sense of distributions which needs to be tested with functions ϕ in $H_0^1(\Omega, \mathbb{R}^N) \cap L^\infty(\Omega, \mathbb{R}^N)$. But we may* as well suppose ϕ to be smooth. Moreover, if we see things on a very small scale, we can finally think of ϕ as being a constant: then u can have a violent behavior, without our being able to measure it using ϕ. We are in the situation of the "Young man intrigued by the theft of the non-euclidean fly" (a drawing by Max Ernst): the mean displacement of the fly in one direction is nearly null but, nevertheless, it moves a lot. It is thus necessary to be able to follow the map u in all its movements, even if it tries to escape. We have seen two procedures to achieve this: when $\mathcal{N}$ is symmetric, the function ϕ that should be used is given by composing a Killing vector field on $\mathcal{N}$ with u; in the general case, the test function is constructed using a Coulomb frame.

4.2 Generalizations in dimension 2

The result of the previous section has been extended to other 2-dimensional variational problems for which the Lagrangian is a quadratic function of the gradient. One possibility is, for instance, to consider that the metric over the image manifold depends on both the domain and the image variables, i.e. that we look at maps from B^2 (to simplify) to a manifold $\mathcal{N}$, which are critical points of

$$F(u) = \int_\Omega g^{\alpha\beta}(z, u(z)) \left(\frac{\partial u}{\partial z^\alpha}, \frac{\partial u}{\partial z^\beta} \right) dxdy,$$

where $g^{\alpha\beta}(z, u(z))$ is a positive definite symmetric bilinear form on $T_{u(z)}\mathcal{N}$. We can think of such maps as (weakly harmonic) sections of a fiber bundle over B^2 or any other surface. This problem was studied by Gilles Carbou [28], who proved the regularity of the critical points of F in the case where $g^{\alpha\beta}(z, u(z)) = g(z, u(z))\delta^{\alpha\beta}$. Another example of a problem is to study the maps u in $H^1(B^2, \mathcal{N})$ which are critical points of the functional

$$\begin{aligned}E_\omega(u) &= \int_{B^2} \left(\frac{|du|^2}{2} + \omega(u)\left(\frac{\partial u}{\partial x}, \frac{\partial u}{\partial y}\right) \right) dxdy \\ &= \int_{B^2} \frac{|du|^2}{2} dxdy + \int_{B^2} u^*\omega, \end{aligned} \quad (4.56)$$

where ω is a 2-form on $\mathcal{N}$. We will suppose that $\mathcal{N}$ is a $\mathcal{C}^2$ submanifold of $\mathbb{R}^N$, and that we extended ω to a $\mathcal{C}^2$ 2-form on $\mathbb{R}^N$. A calculation of the first variation of E_ω yields, for any function $\phi \in H^1_0(B^2, \mathbb{R}^N) \cap L^\infty(B^2, \mathbb{R}^N)$ such that $\langle \phi, u \rangle = 0$ a.e.,

$$\delta E_\omega(u)(\phi) = \int_\Omega \left(\langle d\phi, du \rangle + d\omega \left(\phi, \frac{\partial u}{\partial x}, \frac{\partial u}{\partial y} \right) \right) dxdy. \qquad (4.57)$$

Consequently, the critical points of E_ω are distributional (weak) solutions of

$$-\Delta u + F \perp T_u \mathcal{N}, \qquad (4.58)$$

where F is the map from B^2 to $\mathbb{R}^N$ such that $F(z) \in T_{u(z)}\mathcal{N}$ a.e. and

$$\forall V \in T_u \mathcal{N}, \quad \langle V, F(z) \rangle = d\omega(u) \left(V, \frac{\partial u}{\partial x}, \frac{\partial u}{\partial y} \right). \qquad (4.59)$$

We remark that in the case where $\mathcal{N}$ is 3-dimensional, $d\omega$ may be written as $d\omega = 2H d\text{vol}$, where $d\text{vol}$ is the volume form on $\mathcal{N}$ and H is a Lipschitz function on $\mathcal{N}$. We have already seen this type of equation in chapter 2, for simple cases ($\mathcal{N} = \mathbb{R}^n$): this equation is connected to prescribed mean curvature surfaces because if u is a conformal immersion satisfying (4.58), then the image of u will have a mean curvature equal to $H(u)$ (see [78]).

We will prove the following result, due to Philippe Choné, which generalizes a previous result due to Fabrice Bethuel [9].

Theorem 4.2.1 *[38] Suppose that ω is a $\mathcal{C}^2$ 2-form on $\mathbb{R}^N$, and that $\mathcal{N}$ is a $\mathcal{C}^2$ compact submanifold of $\mathbb{R}^N$ with no boundary. Then, any map $u \in H^1(B^2, \mathcal{N})$, which is a weak solution of solution (4.58), is smooth.*

Proof We start in the same way as in the proof of theorem 4.1.1. Using lemma 4.1.6 we construct a Coulomb frame e over B^2 (up to replacing B^2 by a smaller ball in order to have $\|du\|^2_{L^2(B^2)} \leq \gamma_0$). Then, using lemma 4.1.7, we deduce that ${}^t\!ede$ is bounded in $L^{(2,1)}$. Up to changing once more the size of the ball, we will suppose, once and for all, that

$$\| {}^t\!ede \|_{L^{(2,1)}(B_2)}, \|de\|_{L^2(B^2)}, \|du\|_{L^2(B^2)} \leq \varepsilon_0 \qquad (4.60)$$

where ε_0 will be chosen later.

Step 1 Using Cartesian coordinates $y^1, \ldots, y^N$ on $\mathbb{R}^N$, we may write

$$\omega = \sum_{1\leq i\leq j\leq N} \omega_{ij}(y)dy^i \wedge dy^j,$$

and

$$d\omega = \sum_{1\leq i<j<k\leq N} (d\omega)_{ijk}(y)dy^i \wedge dy^j \wedge dy^k.$$

We introduce on B^2 the functions

$$B_{ij}^a(z) = \sum_{k=1}^{N} (d\omega)_{ijk}(u(z))e_a^k(z), \tag{4.61}$$

for $a = 1, ..., n$ and $i, j = 1, ..., N$, so that

$$\langle F, e_a \rangle = \sum_{1\leq i<j\leq N} B_{ij}^a \{u^i, u^j\} \text{ for } a = 1, \ldots, n.$$

We remark that since the $(d\omega)_{ijk}$ are Lipschitz functions, and because of (4.60), we have

$$\|dB_{ij}^a\|_{L^2(B^2)} \leq C\varepsilon_0. \tag{4.62}$$

Having introduced the notation B_{ij}^a we can rewrite equation (4.58) as

$$-\langle \Delta u, e_a \rangle + \sum_{1\leq i<j\leq N} B_{ij}^a \{u^i, u^j\} = 0, \; \forall a = 1, \ldots, n. \tag{4.63}$$

Step 2 Let $z_0 \in B^2$ and $r > 0$ be such that $B_r := B^2(z_0, r)$ is included in B^2. We decompose the forms $\langle du, e_a \rangle$ over B_r as

$$\begin{cases} \left\langle \dfrac{\partial u}{\partial x}, e_a \right\rangle = \dfrac{\partial w^a}{\partial x} - \dfrac{\partial v^a}{\partial y} \\ \left\langle \dfrac{\partial u}{\partial y}, e_a \right\rangle = \dfrac{\partial w^a}{\partial y} - \dfrac{\partial v^a}{\partial x}, \end{cases} \tag{4.64}$$

where

$$v^a = \sum_{k=1}^{N} \widetilde{u^k e_a^k}_{B_r} \tag{4.65}$$

and

$$-\Delta w^a = -\sum_{b=1}^{n} \left(\left\langle \frac{\partial u}{\partial x}, e_b \right\rangle \left\langle e_b, \frac{\partial e_a}{\partial x} \right\rangle + \left\langle \frac{\partial u}{\partial y}, e_b \right\rangle \left\langle e_b, \frac{\partial e_a}{\partial y} \right\rangle \right)$$
$$- \sum_{1 \leq i < j \leq N} B_{ij}^a \{u^i, u^j\} \quad \text{in } B_r. \tag{4.66}$$

We also decompose, for $1 \leq a \leq n$ and $1 \leq j \leq N$,

$$\sum_{i=1}^{N} B_{ij}^a du^i = ds_j^a + \star dt_j^a \tag{4.67}$$

where

$$t_j^a = \sum_{i=1}^{N} \widetilde{B_{ij}^a u^i}_{B_r}. \tag{4.68}$$

We deduce from (4.67) that

$$\sum_{1 \leq i < j \leq N} B_{ij}^a \{u^i, u^j\} = \frac{1}{2} \sum_{j=i}^{N} \{s_j^a, u^j\} - \frac{1}{2} \sum_{j=i}^{N} \langle dt_j^a, du^j \rangle. \tag{4.69}$$

We remark that, by the definition of t_j^a, hypothesis (4.60) implies

$$\|dt_j^a\|_{L^{(2,1)}(B_r)} \leq C\varepsilon_0^2, \tag{4.70}$$

which implies, again using (4.60) and (4.67),

$$\|ds_j^a\|_{L^2(B_r)} \leq C\varepsilon_0. \tag{4.71}$$

We may now condense the system of equations (4.65), (4.66) and (4.69) in the form

$$-\Delta U = \langle AdU \rangle - \frac{1}{2}\{S, u\} + \frac{1}{2}\langle dT \cdot du \rangle \tag{4.72}$$

where

$$U = \begin{pmatrix} v \\ w \end{pmatrix}, \quad A = (A^x, A^y),$$

$$A^x = \begin{pmatrix} -{}^t\!e\dfrac{\partial e}{\partial x} & -{}^t\!e\dfrac{\partial e}{\partial y} \\ {}^t\!e\dfrac{\partial e}{\partial y} & {}^t\!e\dfrac{\partial e}{\partial x} \end{pmatrix} \; ; \quad A^y = \begin{pmatrix} {}^t\!e\dfrac{\partial e}{\partial y} & {}^t\!e\dfrac{\partial e}{\partial x} \\ -{}^t\!e\dfrac{\partial e}{\partial x} & {}^t\!e\dfrac{\partial e}{\partial y} \end{pmatrix},$$

$$S = \begin{pmatrix} & 0_{n\times N} & \\ s_1^1 & \cdots & s_N^1 \\ \vdots & & \vdots \\ s_1^n & \cdots & s_N^n \end{pmatrix} \; ; \quad T = \begin{pmatrix} & 0_{n\times N} & \\ t_1^1 & \cdots & t_N^1 \\ \vdots & & \vdots \\ t_1^n & \cdots & t_N^n \end{pmatrix},$$

and $\{S, u\} = \dfrac{\partial S}{\partial x}\dfrac{\partial u}{\partial y} - \dfrac{\partial S}{\partial y}\dfrac{\partial u}{\partial x}$.

We remark that, thanks to (4.64), the norms of dU and du are all equivalent, i.e. $\forall p \in (1, +\infty), \forall q \in [1, +\infty], \exists m, M > 0$

$$m\|dU\|_{L^{(p,q)}} \le \|du\|_{L^{(p,q)}} \le M\|dU\|_{L^{(p,q)}}.$$

Step 3 We decompose U as $U = U_0 + U_1 + U_2$, where

$$\begin{cases} -\Delta U_0 &= 0 \quad \text{in } B_r \\ U_0 &= U \quad \text{on } \partial B_r, \end{cases}$$

$$\begin{cases} -\Delta U_1 &= \langle AdU \rangle + \tfrac{1}{2}\langle dT, du \rangle \quad \text{in } B_r \\ U_1 &= 0 \quad \text{on } \partial B_r, \end{cases}$$

$$\begin{cases} -\Delta U_2 &= -\tfrac{1}{2}\{S, u\} \quad \text{in } B_r \\ U_2 &= 0 \quad \text{on } \partial B_r. \end{cases}$$

We now estimate U_0, U_1 and U_2 in $L^{(2,\infty)}$.

(i) Using lemma 3.3.13 with U_0, $p = 2$, $q = +\infty$ and $\gamma \in (0,1)$, we obtain

$$\|dU_0\|_{L^{(2,\infty)}(B_{\alpha r})} \le C(\gamma)\alpha^\gamma \|dU\|_{L^{(2,\infty)}(B_r)}, \tag{4.73}$$

for all $\alpha \in (0,1)$.

(ii) Using (4.60) and the duality between $L^{(2,1)}$ and $L^{(2,\infty)}$, we have

$$\begin{aligned} \|\langle A, dU\rangle\|_{L^1(B_r)} &\le \|A\|_{L^{(2,1)}(B_r)} \|dU\|_{L^{(2,\infty)}(B_r)} \\ &\le C\varepsilon_0 \|dU\|_{L^{(2,\infty)}(B_r)}, \end{aligned}$$

and by (4.70)

$$\|\langle dT, du\rangle\|_{L^1(B_r)} \leq \|dT\|_{L^{(2,1)}(B_r)}\|du\|_{L^{(2,\infty)}(B_r)}$$
$$\leq C\varepsilon_0\|du\|_{L^{(2,\infty)}(B_r)}.$$

These two inequalities yield, using theorem 3.3.6,

$$\|dU_1\|_{L^{(2,\infty)}(B_r)} \leq C\varepsilon_0\|dU\|_{L^{(2,\infty)}(B_r)}. \tag{4.74}$$

(iii) Finally, to deduce from the equation on U_2 an estimate for dU_2, we use theorem 3.4.5 to obtain

$$\|dU_2\|_{L^{(2,\infty)}(B_r)} \leq \|dU_2\|_{L^2(B_r)}$$
$$\leq C\|dS\|_{L^2(B_r)}\|du\|_{L^{(2,\infty)}(B_r)}$$
$$\leq C\varepsilon_0\|du\|_{L^{(2,\infty)}(B_r)}. \tag{4.75}$$

Step 4 We put together inequalities (4.73), (4.74) and (4.75) to obtain that for every $\theta \in (0,1)$, we may choose ε_0 and α sufficiently small in order to have

$$\|dU\|_{L^{(2,\infty)}(B_{\alpha r})} \leq \theta\|dU\|_{L^{(2,\infty)}(B_r)}.$$

Then we can finish the proof as in the variant of the proof of theorem 4.1.1. □

This result is a step towards the proof of the following conjecture due to Stefan Hildebrandt: in dimension 2, the maps in H^1 which are critical points of a functional which is invariant under conformal transformation, are smooth. We recall that every functional which is invariant under conformal transformation on a domain Ω in $\mathbb{R}^2$ is of the form

$$\mathcal{L}(u) = \int_\Omega L(u, du) dx dy,$$

where it is reasonable to suppose that L is a $\mathcal{C}^1$ function and satisfies

$$\sum_i \frac{\partial u^i}{\partial x} \frac{\partial L}{\partial \frac{\partial u^i}{\partial x}} + \frac{\partial u^i}{\partial y} \frac{\partial L}{\partial \frac{\partial u^i}{\partial y}} = 2L$$

$$\sum_i \frac{\partial u^i}{\partial x} \frac{\partial L}{\partial \frac{\partial u^i}{\partial y}} - \frac{\partial u^i}{\partial y} \frac{\partial L}{\partial \frac{\partial u^i}{\partial x}} = 0.$$

(See [88] for more details on this class of variational problems.)

4.3 Regularity results in arbitrary dimension

In this section we present a result due to Fabrice Bethuel which generalizes the theorem of L.C. Evans (3.5.1) to the case of stationary harmonic maps with values in arbitrary manifolds. A part of the procedure is essentially the same, but the lack of symmetry makes the problem more complex and we need to use Coulomb frames like those introduced for theorem 4.1.1.

Theorem 4.3.1 *[10] Let Ω be an open domain in an m-dimensional Riemannian manifold $(\mathcal{M}, g)$, and let $\mathcal{N}$ be a compact Riemannian submanifold, isometrically embedded in a $\mathcal{C}^5$ way into $(\mathbb{R}^N, \langle .,. \rangle)$. We assume g to be $\mathcal{C}^{0,\alpha}$. Let $u \in H^1(\Omega, \mathcal{N})$ be a weakly stationary map. Then, there exists a closed subset $\mathcal{S}$ of Ω, whose $(m-2)$-dimensional Hausdorff measure is zero, and such that u is $\mathcal{C}^{1,\alpha}$ in $\Omega \backslash \mathcal{S}$.*

Remark 4.3.2 *In the case where it is possible to construct an orthonormal tangent frame field on $\mathcal{N}$, it suffices to suppose $\mathcal{N}$ to be $\mathcal{C}^2$ for the result to be valid. Moreover, depending on the regularity of $(\mathcal{M}, g)$ and of $\mathcal{N}$, the regularity of u outside $\mathcal{S}$ may be improved (see theorem 1.5.1).*

For simplicity, we will just prove this result for the case where Ω is an open subset of $\mathbb{R}^m$. Part of the proof of this result is identical to that of theorem 3.5.1: it consists of the arguments used for deducing theorem 3.5.1 from theorem 3.5.3. We will not repeat this part and will concentrate on proving the following (analogous to theorem 3.5.3).

Theorem 4.3.3 *With the same hypotheses as in theorem 4.3.1, there exists $\epsilon > 0$ such that if for $x \in \Omega$ and $R > 0$,*

$$E_{x,4R}(u) \leq \epsilon^2, \tag{4.76}$$

then u is Hölder continuous in a neighborhood of x.

Recall that if $B(x,r) \subset \Omega$, we write

$$E_{x,r} = \frac{1}{r^{m-2}} \int_{B(x,r)} |du|^2 dx.$$

As in theorem 3.5.1, the hypothesis of u being Noether harmonic is used only in order to have the monotonicity formula

$$E_{x,r_1}(u) \leq E_{x,r_2}(u), \ \forall r_1 < r_2, \tag{4.77}$$

proved in lemma 3.5.4.

We will use the following quantity: for $x_0 \in \Omega$, $r > 0$ such that $B(x_0, r) \subset \Omega$,

$$M(x_0, r) = \sup \left\{ \frac{1}{\rho^{m-1}} \int_{B(x,\rho)} |du| \, dy \mid B(x, \rho) \subset B(x_0, r) \right\}.$$

Notice that, by the Poincaré–Sobolev inequality, $M(x_0, r)$ bounds the BMO norm of u in $B(x_0, r)$. But the quantity $M(x_0, r)$ is itself controlled by $E_{x_0, 2r}(u)$, since by the Cauchy–Schwarz inequality,

$$\frac{1}{\rho^{m-1}} \int_{B(x,\rho)} |du| \, dy \leq \left(\frac{C}{\rho^{m-2}} \int_{B(x,\rho)} |du|^2 dy \right)^{1/2}$$
$$= C \left(E_{x,\rho}(u) \right)^{1/2}.$$

And, because of (4.77), and the fact that $B(x_0, 2r) \subset \Omega$,

$$E_{x,\rho}(u) \leq E_{x,\rho+r}(u) \leq 2^{m-2} E_{x_0, 2r}(u),$$

and thus, by the definition of $M(x_0, r)$,

$$M(x_0, r) \leq C \left(E_{x_0, 2r}(u) \right)^{1/2} < +\infty. \tag{4.78}$$

Theorem 4.3.3 follows from:

Theorem 4.3.4 *Under the hypotheses of theorem 4.3.1, there exist $\epsilon > 0$, $\theta \in (0, 1)$ and $\tau < \frac{1}{2}$ such that if $B(y, 2r) \subset \Omega$ and*

$$E_{y, 2r}(u) \leq 2^{2-m} \epsilon^2, \tag{4.79}$$

then

$$M(y, \tau r) \leq \theta M(y, r). \tag{4.80}$$

HOW DOES THEOREM 4.3.3 FOLLOW FROM THEOREM 4.3.4?

Let $x \in \Omega$, and choose $R > 0$ such that $B(x, 4R) \subset \Omega$ and

$$E_{x, 4R}(u) \leq \epsilon^2.$$

Then an elementary calculation yields that for all $y \in B(x, 2R)$, we have that $B(y, 2R) \subset \Omega$ and
$$E_{y,2R}(u) \leq 2^{m-2} E_{x,4R}(u) \leq 2^{m-2}\epsilon^2.$$
Using the monotonicity formula (4.77), we deduce that $\forall r \in (0, R)$,
$$E_{y,2r}(u) \leq E_{y,2R}(u) \leq 2^{m-2}\epsilon^2. \tag{4.81}$$

Applying theorem 4.3.4 we deduce from (4.81) that $\exists \tau \in (0, 1/2)$, $\exists \theta \in (0,1)$, $\forall r \in (0, R)$,
$$M(y, \tau r) \leq \theta M(y, r).$$
By a reasoning identical to that of lemma 3.5.11, this implies that $\forall r \in (0, R)$,
$$M(y, r) \leq \frac{1}{\theta} \left(\frac{r}{R}\right)^{\frac{\log \theta}{\log \tau}} M(y, R). \tag{4.82}$$

In particular,
$$\frac{1}{r^{m-1}} \int_{B(y,r)} |du|\, dz \leq C \left(\frac{r}{R}\right)^{\frac{\log \theta}{\log \tau}} M(y, R).$$

Using the Poincaré–Sobolev inequality, we deduce the following inequality, valid for all $y \in B(x, 2R)$, $r \in (0, R)$, and $1 \leq q \leq \frac{m}{m-1}$:
$$\left(\frac{1}{r^m} \int_{B(y,r)} |u - u_{y,r}|^q\right)^{\frac{1}{q}} \leq C r^{\frac{\log \theta}{\log \tau}}.$$

This means that the restriction $u_{|B(x,2R)}$ belongs to the Morrey–Campanato space $\mathcal{L}^{q,\lambda}(B(x, 2R))$, where $\lambda = m + q \frac{\log \theta}{\log \tau}$ and $1 \leq q \leq \frac{m}{m-1}$. By theorem 3.5.10, we deduce that u is $C^{0, \frac{\log \theta}{\log \tau}}$ in $B(x, 2R)$. We conclude using theorem 1.5.1.

HODGE DECOMPOSITION OF DIFFERENTIAL FORMS

Recall that for weakly harmonic maps on surfaces, in the variant of the proof of theorem 4.2.1, we used the Hodge decomposition
$$\begin{cases} \left\langle \dfrac{\partial u}{\partial y}, e_a \right\rangle = \dfrac{\partial w^a}{\partial x} - \dfrac{\partial v^a}{\partial y} \\ \left\langle \dfrac{\partial u}{\partial x}, e_a \right\rangle = \dfrac{\partial w^a}{\partial y} + \dfrac{\partial v^a}{\partial x}. \end{cases} \tag{4.83}$$

Note that Hodge decomposition arguments can also be used for proving theorem 2.6.4 without using Wente's lemma or Hardy spaces. This was actually the original method used in [83] for proving theorem 2.6.4, and in [84] where theorem 2.6.4 was generalized to weakly harmonic maps into homogeneous manifolds.

In the proof of theorem 4.3.4, we will be led to consider the generalization of this decomposition to arbitrary dimension. We start by noticing that, using the language of differential forms, (4.83) may be written as

$$\langle du, e_a \rangle = dw^a + \star dv^a,$$

where $\star dv^a = \star(\frac{\partial v^a}{\partial x}dx + \frac{\partial v^a}{\partial y}dy) = -\frac{\partial v^a}{\partial y}dx + \frac{\partial v^a}{\partial x}dy$. We will work with this formalism. A general definition of the Hodge $\star$ operator is the following.

Definition 4.3.5 *For $0 \leq k \leq m$, we denote by $\Lambda^k \mathbb{R}^m$ the vector space of skew-symmetric multilinear forms on $\mathbb{R}^m$. The Hodge $\star$ operator is the unique linear operator from $\Lambda^k \mathbb{R}^m$ to $\Lambda^{m-k} \mathbb{R}^m$ such that*

(i) $\star 1 = dx^1 \wedge ... \wedge dx^m$
(ii) $\forall \alpha, \beta \in \Lambda^k \mathbb{R}^m, \ \alpha \wedge (\star \beta) = \beta \wedge (\star \alpha) = \langle \alpha, \beta \rangle dx^1 \wedge ... \wedge dx^m.$

We also use the following notation: if ϕ belongs to $L^2(\mathbb{R}^m, \Lambda^k \mathbb{R}^m)$,

$$\delta \phi = (-1)^{mk+m+1} \star d \star \phi$$

(δ is the adjoint of d for the scalar product in $L^2(\mathbb{R}^m, \Lambda^k \mathbb{R}^m)$). A generalized version of (4.83) is:

Lemma 4.3.6 *Let $\phi \in L^2(\mathbb{R}^m, \Lambda^1 \mathbb{R}^m)$ be a differential 1-form with coefficients in $L^2(\mathbb{R}^m)$. Then, there exists a unique function $w \in L^{2^\star}(\mathbb{R}^m)$ and a unique 2-form $v \in L^{2^\star}(\mathbb{R}^m, \Lambda^2 \mathbb{R}^m)$ (where $2^\star := \frac{2m}{m-2}$) such that*

$$\phi = dw + \delta v \qquad (4.84)$$

$$dv = 0. \qquad (4.85)$$

Furthermore, if we write $v = \sum_{1 \leq \alpha < \beta \leq m} v_{\alpha\beta} dx^\alpha \wedge dx^\beta$, then dw and $dv_{\alpha\beta}$ belong to $L^2(\mathbb{R}^m)$ and we have the estimate

$$||(w, v_{\alpha\beta})||_{L^{2^\star}}^2 + ||d(w, v_{\alpha\beta})||_{L^2}^2 \leq C ||\phi||_{L^2}^2. \qquad (4.86)$$

4.3 Regularity results in arbitrary dimension

Remark 4.3.7 *We will define*

$$L_1^2(\mathbb{R}^m) := \{f \in L^{2^*} \mid df \in L^2(\mathbb{R}^m)\}.$$

This space is the closure of $\mathcal{C}_c^\infty(\mathbb{R}^m)$ for the seminorm $||df||_{L^2}$ (Gagliardo–Nirenberg–Sobolev inequality, see [19]). The restriction of a function in this space to any open bounded subset of $\mathbb{R}^m$ belongs to H^1.

Proof of lemma 4.3.6 Recall that for any $B \in L^2(\mathbb{R}^m, \Lambda^k \mathbb{R}^m)$, a differential k-form, if

$$B = \sum_{1 \leq \alpha_1 < \ldots < \alpha_k \leq m} B_{\alpha_1 \ldots \alpha_k} dx^{\alpha_1} \wedge \ldots \wedge dx^{\alpha_k},$$

then

$$(d\delta + \delta d)B = - \sum_{1 \leq \alpha_1 < \ldots < \alpha_k \leq m} \Delta B_{\alpha_1 \ldots \alpha_k} dx^{\alpha_1} \wedge \ldots \wedge dx^{\alpha_k},$$

which justifies the following notation which we will use below:

$$-\Delta_k := (\delta d + d\delta)_{|L^2(\mathbb{R}^m, \Lambda^k \mathbb{R}^m)}.$$

We look at relation (4.84): $\phi = dw + \delta v$. Applying δ we obtain

$$\Delta w = \sum_{\alpha=1}^m \frac{\partial \phi_\alpha}{\partial x^\alpha} \quad \text{in } \mathbb{R}^m. \tag{4.87}$$

If we impose the condition $w \in L^{2^*}(\mathbb{R}^m)$, this implies the uniqueness of w. The existence of w may be proved by minimizing in $L_1^2(\mathbb{R}^m)$ the functional

$$\int_{\mathbb{R}^m} |df - \phi|^2 dx$$

(exercise!) We also have the inequality

$$||dw||^2_{L^2(\mathbb{R}^m)} \leq ||\phi||^2_{L^2(\mathbb{R}^m)},$$

and using the Gagliardo–Nirenberg–Sobolev inequality, we deduce the upper bound for $||w||_{L^{2^*}(\mathbb{R}^m)}$ given by (4.86). Likewise, applying the operator d to equation (4.84), and using (4.85), we obtain

$$-\Delta_2 v = d\phi \quad \text{in } \mathbb{R}^m, \tag{4.88}$$

or equivalently,

$$-\Delta v_{\alpha\beta} = \frac{\partial \phi_\beta}{\partial x^\alpha} - \frac{\partial \phi_\alpha}{\partial x^\beta}.$$

The existence, uniqueness and estimates for v in $L_1^2(\mathbb{R}^m, \Lambda^2 \mathbb{R}^m)$ are obtained as for w.

To conclude, let us check that we answered the question correctly: is relation (4.84) satisfied? To see this, it suffices to notice that the 1-form $\phi - dw - \delta v$ is harmonic and has coefficients in $L_1^2(\mathbb{R}^m)$, and thus is identically zero. $\square$

Proof of theorem 4.3.4

STRATEGY OF THE PROOF

Choose a point y in Ω, and a real $r > 0$, such that $B(y, 2r) \subset \Omega$ and

$$E_{y,2r}(u) \leq 2^{2-m} \epsilon^2, \tag{4.89}$$

where ϵ will be chosen later. We start by deducing from (4.89) and the monotonicity formula, that for any ball $B(z, \rho)$ contained in $B(y, r)$,

$$E_{z,\rho}(u) \leq 2^{m-2} E_{y,2r}(u) \leq \epsilon^2. \tag{4.90}$$

The aim is, for a well-chosen real $\tau \in (0, \frac{1}{2})$, to estimate $M(y, \tau r)$ using $M(y, r)$. Attending to the definition of $M(y, \tau r)$, this involves estimating

$$\frac{1}{(\tau \rho)^{m-1}} \int_{B(z, \tau\rho)} |du| dx$$

using $M(y, r)$, for each z and ρ such that $B(z, \tau\rho) \subset B(y, \tau r)$. We notice that then $B(z, \rho) \subset B(y, r)$ and hence (by definition)

$$M(z, \rho) \leq M(y, r). \tag{4.91}$$

Thus, it suffices to prove the intermediate estimate

$$\frac{1}{(\tau\rho)^{m-1}} \int_{B(z,\tau\rho)} |du| dx \leq \theta M(z, \rho). \tag{4.92}$$

We will then be able to conclude as follows: (4.91) and (4.92) imply that

4.3 Regularity results in arbitrary dimension

$$\frac{1}{(\tau\rho)^{m-1}}\int_{B(z,\tau\rho)}|du|dx \leq \theta M(y,r), \quad \forall B(z,\tau\rho) \subset B(y,r).$$

Taking the upper bound of the l.h.s. of this inequality, we obtain inequality (4.80):

$$M(y,\tau r) \leq \theta M(y,r).$$

We will now work over $B(z,\rho)$, in order to prove (4.92).

Step 1 REWRITING THE EQUATION
Using lemmas 4.1.2 and 4.1.3, we construct a Coulomb frame $e = (e_1,...,e_m)$, a section of the bundle $u^*\mathcal{F}$ over $B(z,\rho)$. We then know that

$$||de||^2_{L^2(B(z,\rho))} \leq C||du||^2_{L^2(B(z,\rho))}. \tag{4.93}$$

We use lemma 4.3.6 with the 1-forms

$$\phi^a = \langle d[\zeta(u - u_{z,\rho})], e_a \rangle,$$

where $\zeta \in \mathcal{C}^\infty_c(\mathbb{R}^m)$ is a function with support contained in $B(z, \frac{5\rho}{8})$, taking the value 1 on $B(z, \frac{3\rho}{4})$, and whose gradient is bounded by $C\rho^{-1}$, and $u_{z,\rho}$ is the mean value of u over $B(z,\rho)$. This yields

$$\langle d[\zeta(u - u_{z,\rho})], e_a \rangle = dw^a + \delta v^a. \tag{4.94}$$

We remark that in $\mathbb{R}^m$,

$$-\Delta w^a = -\sum_{\alpha=1}^m \frac{\partial \phi^a_\alpha}{\partial x^\alpha},$$

which implies that in $B(z, \frac{\rho}{2})$ (using the fact that u is weakly harmonic)

$$\begin{aligned}-\Delta w^a &= -\sum_{\alpha=1}^m \frac{\partial}{\partial x^\alpha}\left\langle \frac{\partial u}{\partial x^\alpha}, e_a \right\rangle \\ &= -\sum_{\alpha=1}^m \left\langle \frac{\partial u}{\partial x^\alpha}, \frac{\partial e_a}{\partial x^\alpha} \right\rangle\end{aligned}$$

$$= -\sum_{\alpha=1}^{m}\sum_{b=1}^{n}\left\langle\frac{\partial u}{\partial x^\alpha},\left\langle\frac{\partial e_a}{\partial x^\alpha},e_b\right\rangle,e_b\right\rangle. \tag{4.95}$$

We now use the fact that e is a Coulomb frame. This means that $\delta\langle de_a,e_b\rangle = 0$, for all a and b, and hence there exist 2-forms $A_a^b \in H^1(B(z,\rho),\Lambda^2\mathbb{R}^m)$, such that

$$\langle de_a,e_b\rangle = -\delta A_a^b. \tag{4.96}$$

Moreover, we may choose A_a^b in such a way that

$$\|A_a^b\|_{H^1(B(z,\rho))}^2 \leq C\|de\|_{L^2(B(z,\rho))}^2 \leq C\|du\|_{L^2(B(z,\rho))}^2. \tag{4.97}$$

Putting $A_a^b = \sum_{1\leq\alpha<\beta\leq m} A_{a\alpha\beta}^b dx^\alpha \wedge dx^\beta$, relation (4.96) can be written as

$$\langle de_a,e_b\rangle = \sum_{\beta=1}^{m}\left(\sum_{\alpha=1}^{m}\frac{\partial A_{a\alpha\beta}^b}{\partial x^\alpha}\right) dx^\beta.$$

Substituting this expression into (4.95), we obtain that in $B(z,\frac{\rho}{2})$

$$-\Delta w^a = -\sum_{\alpha=1}^{m}\sum_{\beta=1}^{m}\sum_{b=1}^{n}\left\langle\frac{\partial A_{a\beta\alpha}^b}{\partial x^\beta}\frac{\partial u}{\partial x^\alpha},e_b\right\rangle$$

$$= \sum_{1\leq\alpha<\beta\leq m}\sum_{b=1}^{n}\langle\{u,A_{a\alpha\beta}^b\}_{\alpha\beta},e_b\rangle, \tag{4.98}$$

where $\{f,g\}_{\alpha\beta} := \frac{\partial f}{\partial x^\alpha}\frac{\partial g}{\partial x^\beta} - \frac{\partial f}{\partial x^\beta}\frac{\partial g}{\partial x^\alpha}$.

Likewise, we may obtain an elliptic equation for v^a: applying the operator d to (4.94) (and using $dv^a = 0$), we see that in $\mathbb{R}^m$

$$-\sum_{1\leq\alpha<\beta\leq m}\Delta v_{\alpha\beta}^a dx^\alpha \wedge dx^\beta = -\Delta_2 v^a = d\langle e_a, d(\zeta(u-u_{z,\rho}))\rangle$$

$$= \sum_{1\leq\alpha<\beta\leq m}\{e_a,\zeta(u-u_{z,\rho})\}_{\alpha\beta}dx^\alpha \wedge dx^\beta. \tag{4.99}$$

Finally, we decompose w^a in $B(z,\frac{\rho}{2})$ as follows:

$$w^a = w_0^a + w_1^a \text{ in } B(z,\frac{\rho}{2}) \tag{4.100}$$

and

$$\Delta w_0^a = 0 \text{ in } B(z,\frac{\rho}{2}),$$

and
$$w_1^a = 0 \text{ on } \partial B(z, \frac{\rho}{2}).$$

And, using (4.94), we have in $B(z, \frac{\rho}{2})$

$$|du| \le C \sum_{a=1}^{n} \left(|dw_0^a| + |dw_1^a| + \sum_{1 \le \alpha < \beta \le m} |dv_{\alpha\beta}| \right). \qquad (4.101)$$

Step 2 ESTIMATE FOR $dv_{\alpha\beta}^a$
We remark that here it is possible to estimate $E_{z,\rho}(v_{\alpha\beta}^a)$ using $M(z,\rho)$, which is stronger than what we need. In order to achieve this, it suffices to multiply equation (4.99) by $v_{\alpha\beta}^a$, and to integrate over $\mathbb{R}^m$. However, a difficulty shows up (already found in theorem 3.5.8) when trying to give a meaning to

$$\int_{\mathbb{R}^m} \{e_a, \zeta(u - u_{z,\rho})\}_{\alpha\beta} v_{\alpha\beta}^a dx.$$

It is relevant to recall that if $f, g, h \in L_1^2(\mathbb{R}^m)$, then $\int_{\mathbb{R}^m} \{f, g\}_{\alpha\beta} h dx$ generally does not have a meaning, unless $m = 2$. On the other hand, if h belongs to $BMO(\mathbb{R}^m)$ (instead of $L^{\frac{2m}{m-2}}(\mathbb{R}^m)$), the Fefferman–Stein theorem enables us to interpret this integral as a duality product between Hardy and BMO. As a matter of fact, we can do better:

Lemma 4.3.8 *If $f, g, h \in L_1^2(\mathbb{R}^m)$, and if one of these functions, for instance h, belongs to $BMO(\mathbb{R}^m)$, then we can give a meaning to the quantities*

$$\int_{\mathbb{R}^m} \{f, g\}_{\alpha\beta} h dx = \int_{\mathbb{R}^m} \{h, f\}_{\alpha\beta} g dx = \int_{\mathbb{R}^m} \{g, h\}_{\alpha\beta} f dx, \qquad (4.102)$$

and, furthermore, we have the estimate

$$\left| \int_{\mathbb{R}^m} \{f, g\}_{\alpha\beta} h dx \right| \le C \|df\|_{L^2} \|dg\|_{L^2} \|h\|_{BMO}. \qquad (4.103)$$

Proof Inequality (4.103) is essentially a consequence of theorem 3.2.3 and the Fefferman–Stein $\mathcal{H}^1(\mathbb{R}^m) - BMO(\mathbb{R}^m)$ duality theorem (we remark that there is a direct proof of (4.103) without passing through Hardy space, due to Sagun Chanillo [32]). The proof of (4.102) is straightforward in case f, g, h belong to $\mathcal{C}_c^\infty(\mathbb{R}^m)$. Using this identity in the smooth case and estimate (4.103), we may define the integrals in

(4.102) using precisely this relation as definition, and using the density of $C_c^\infty(\mathbb{R}^m)$ in $L_1^2(\mathbb{R}^m)$. □

BACK TO STEP 2

After these precautions, we deduce from (4.99) that

$$\begin{aligned}
\int_{\mathbb{R}^m} |dv_{\alpha\beta}^a|^2 dx &= \int_{\mathbb{R}^m} \{e_a, \zeta(u - u_{z,\rho})\}_{\alpha\beta} v_{\alpha\beta}^a dx \\
&\leq C \|dv_{\alpha\beta}^a\|_{L^2(\mathbb{R}^m)} \|de_a\|_{L^2(B(z,\rho))} \|\zeta(u - u_{z,\rho})\|_{BMO}.
\end{aligned} \quad (4.104)$$

By estimate (4.86) in lemma 4.3.6,

$$\begin{aligned}
\|dv_{\alpha\beta}^a\|_{L^2(\mathbb{R}^m)}^2 &\leq C \|d(\zeta(u - u_{z,\rho}))\|_{L^2(\mathbb{R}^m)}^2 \\
&\leq C \|du\|_{L^2(B(z,\rho))}^2.
\end{aligned}$$

Likewise, (4.93) yields an analogous estimate for $\|de_a\|_{L^2(B(z,\rho))}^2$. Using (4.90), we deduce that

$$\|dv_{\alpha\beta}^a\|_{L^2(\mathbb{R}^m)} \|de_a\|_{L^2(B(z,\rho))} \leq C\rho^{m-2}\epsilon^2. \quad (4.105)$$

It remains to estimate $\zeta(u - u_{z,\rho})$ in BMO. We leave it to the reader to show that

$$\|\zeta(u - u_{z,\rho})\|_{BMO} \leq CM(z,\rho). \quad (4.106)$$

The proof is essentially the same as step 4 of the proof of theorem 3.5.8 (i.e. lemma 3.5.13).

We deduce from (4.104), (4.105) and (4.106) that

$$\int_{\mathbb{R}^m} |dv_{\alpha\beta}^a|^2 dx \leq C\rho^{m-2}\epsilon^2 M(z,\rho),$$

which implies, by the Cauchy–Schwarz inequality,

$$\frac{1}{\rho^{m-1}} \int_{B(z,\rho)} |dv_{\alpha\beta}^a| dx \leq C\epsilon M(z,\rho). \quad (4.107)$$

Step 3 ESTIMATE FOR dw_1^a

We deduce from (4.95) and (4.100) that w_1^a is a solution of

$$\begin{cases} -\Delta w_1^a = \sum_{1 \leq \alpha < \beta \leq m} \sum_{b=1}^n \langle \{u, A_{a\alpha\beta}^b\}_{\alpha\beta}, e_b \rangle & \text{in } B(z, \frac{\rho}{2}) \\ w_1^a = 0 & \text{on } \partial B(z, \frac{\rho}{2}). \end{cases} \quad (4.108)$$

The situation is more delicate than before. In fact, if we multiply (4.108) by ψ and integrate by parts over $B(z, \frac{\rho}{2})$, we obtain

$$\int_{B(z,\frac{\rho}{2})} \langle dw_1^a, d\psi \rangle = \sum_{1 \leq \alpha < \beta \leq m} \sum_{b=1}^n \int_{B(z,\frac{\rho}{2})} \langle \{u, A_{\alpha\beta}^b\}_{\alpha\beta}, \psi e_b \rangle dx. \quad (4.109)$$

It would be pleasant to choose $\psi = w_1^a$, but unfortunately this is not possible. In fact, we cannot show that the product $w_1^a e_b$ belongs to H^1 (unless for instance w_1^a is bounded in L^∞).

To get around this difficulty, we will take as ψ a function which is more regular than w_1^a: it will be the solution of

$$\begin{cases} \Delta \psi = \text{div}\left(\frac{\nabla w_1^a}{|\nabla w_1^a|}\right) & \text{in } B(z, \frac{\rho}{2}) \\ \psi = 0 & \text{on } \mathbb{R}^m \setminus B(z, \frac{\rho}{2}). \end{cases} \quad (4.110)$$

Since $\frac{\nabla w_1^a}{|\nabla w_1^a|}$ is uniformly bounded in L^∞, classic elliptic estimates (see [69], [155]) imply that ψ belongs to all $W^{1,q}$ spaces, for $1 < q < +\infty$, and hence in particular $\psi \in L^\infty(\mathbb{R}^m) \cap H^1(\mathbb{R}^m)$, with the estimates

$$\|\psi\|_{L^\infty} \leq C\rho, \quad (4.111)$$

$$\|d\psi\|_{L^2} \leq C\rho^{\frac{m}{2}}. \quad (4.112)$$

A consequence of this is that, since $e_b \in H^1(B(z,\rho)) \cap L^\infty(B(z,\rho))$, the product ψe_b belongs to $H^1(\mathbb{R}^m)$ (we extend it by 0 outside $B(z, \frac{\rho}{2})$), and

$$\|d(\psi e_b)\|_{L^2(\mathbb{R}^m)} \leq \|\psi\|_{L^\infty}\|de_b\|_{L^2(B(z,\rho))} + \|d\psi\|_{L^2}.$$

Using (4.90) and (4.93), we obtain

$$\|de\|_{L^2(B(z,\rho))}^2 \leq C\|du\|_{L^2(B(z,\rho))}^2 \leq C\epsilon^2 \rho^{m-2}.$$

Hence we conclude, using (4.111) and (4.112), that

$$\|d(\psi e_b)\|_{L^2(\mathbb{R}^m)}^2 \leq C\rho^m. \quad (4.113)$$

We may now come back to (4.109): we replace u by $(u - u_{z,\rho})\zeta$ (which is bounded in BMO) in the integral on the l.h.s. and we use lemma 4.3.8. This yields

$$\int_{B(z,\frac{\rho}{2})} \langle dw_1^a, d\psi \rangle = \sum_{1 \leq \alpha < \beta \leq m} \sum_{b=1}^n \int_{\mathbb{R}^m} \{\zeta(u - u_{z,\rho}), \zeta A_{a\alpha\beta}^b\}_{\alpha\beta} \psi e_b dx. \tag{4.114}$$

As in the previous step, we deduce that

$$\begin{aligned}\int_{B(z,\frac{\rho}{2})} \langle dw_1^a, d\psi \rangle &\leq C||\zeta(u - u_{z,\rho})||_{BMO}||d(\zeta A_{a\alpha\beta}^b)||_{L^2}||d(\psi e_b)||_{L^2} \\ &\leq C\epsilon \rho^{m-1} M(z,\rho),\end{aligned}$$

using (4.90), (4.97), (4.106) and (4.113).

On the other hand, we show, by integrating by parts and using (4.110), that the integral on the l.h.s. of (4.114) equals $\int_{B(z,\frac{\rho}{2})} |dw_1^a|$. Thus, we conclude that

$$\frac{1}{\rho^{m-1}} \int_{B(z,\frac{\rho}{2})} |dw_1^a| \leq C\epsilon M(z,\rho). \tag{4.115}$$

Step 4 ESTIMATE FOR dw_0^a
Since we have almost everywhere

$$|dw_0^a| \leq |dw_1^a| + |dw^a| \leq |dw_1^a| + C(|du| + |dv_{\alpha\beta}^a|),$$

we deduce from (4.107) and (4.115) that

$$\frac{1}{\rho^{m-1}} \int_{B(z,\frac{\rho}{2})} |dw_0^a| \leq CM(z,\rho).$$

Using the fact that w_0^a is harmonic and lemma 3.3.13, we see that $\forall \tau \in (0, \frac{1}{2})$,

$$\frac{1}{(\tau\rho)^{m-1}} \int_{B(z,\tau\rho)} |dw_0^a| \leq C\tau M(z,\rho). \tag{4.116}$$

Step 5 CONCLUSION
From (4.107), (4.115) and (4.116), it follows that

$$\frac{1}{(\tau\rho)^{m-1}} \int_{B(z,\tau\rho)} |du| \leq C(\epsilon + \tau) M(z,\rho),$$

and choosing ϵ and τ sufficiently small, we obtain (4.92). □

4.4 Conservation laws for harmonic maps without symmetry

We come back to an analysis problem already treated in chapter 2: that of weak compactness of weakly harmonic maps, i.e. that of determining whether the limit in the weak topology of $H^1(\mathcal{M},\mathcal{N})$ of a sequence of weakly harmonic maps is still weakly harmonic. In view of the positive result obtained in chapter 2 (theorem 2.5.1) for S^n-valued maps, we have the right to be optimistic, and to think, for instance, that the use of moving frames might solve the problem. Unfortunately, this problem still remains open for the moment.

In the case where the image manifold is arbitrary, the only thing we can do is to "pass to the limit" for sequences of solutions defined on surfaces. In fact, when the domain is 2-dimensional, we know from theorem 4.1.1 that every weak solution is smooth. Going over the proof of this result, we see that we can obtain estimates in norms that are stronger than H^1 (for instance H^2) for sequences of weak solutions, wherever the energy density does not concentrate excessively (see, for instance, [132] to see how such estimates can be obtained). On the other hand, we can show that the energy density does not concentrate in the domain Ω, except maybe over a subset Σ which is a finite union of points. Thus, we can deduce that there is strong convergence outside Σ, and hence that the limit is harmonic outside Σ. Using a result in [141], we conclude that the limit is harmonic everywhere (see also [11]).

Recently, a result was obtained for the wave equation on a surface, with values in a manifold, by A. Freire, S. Müller and M. Struwe. The proof relies on the use of Coulomb frames, the duality between Hardy and BMO spaces as in the regularity results of L.C. Evans and F. Bethuel, and a detailed analysis of the concentration locus for the energy density.

Theorem 4.4.1 *[61] Let $\mathcal{N}$ be a $\mathcal{C}^5$ submanifold, embedded in $(\mathbb{R}^N, \langle .,. \rangle)$. We also consider the Minkowski space $\mathbb{R} \times \mathbb{R}^2 = \{(t,x,y) \in \mathbb{R}^3\}$, over which we define the d'Alembertian*

$$\Box := \frac{\partial^2}{\partial t^2} - \frac{\partial^2}{\partial x^2} - \frac{\partial^2}{\partial y^2}.$$

Let u_n be a sequence of maps in $L^2_{loc}(\mathbb{R} \times \mathbb{R}^2, \mathcal{N})$, such that $\frac{\partial u_n}{\partial t}$, $\frac{\partial u_n}{\partial x}$, $\frac{\partial u_n}{\partial y}$ belong to $L^\infty(\mathbb{R}, L^2(\mathbb{R}^2, \mathbb{R}^N))$. We suppose that for any n, u_n is a

solution of
$$\Box u_n \perp T_{u_n}\mathcal{N} \quad \text{in } \mathcal{D}'(\mathbb{R}\times\mathbb{R}^2). \tag{4.117}$$

We denote the total energy by
$$E(u(t)) = \frac{1}{2}\int_{\mathbb{R}^2}\left(\left|\frac{\partial u}{\partial t}(t)\right|^2 + \left|\frac{\partial u}{\partial x}(t)\right|^2 + \left|\frac{\partial u}{\partial y}(t)\right|^2\right)dxdy,$$

and moreover, we suppose that the energy is uniformly bounded in time, i.e.
$$\exists E_0 \text{ such that } \forall t \in \mathbb{R}, \ \forall n \in \mathbb{N}, \quad E(u_n(t)) \leq E_0. \tag{4.118}$$

Finally, we suppose that
$$\begin{aligned} u_n &\to u & &\text{in } L^2_{loc}(\mathbb{R}\times\mathbb{R}^2, \mathcal{N}) \\ du_n &\rightharpoonup du & &\text{weak } \star \text{ in } L^\infty(\mathbb{R}, L^2(\mathbb{R}^2, \mathbb{R}^N)). \end{aligned} \tag{4.119}$$

Then u is still a weak solution of the wave equation
$$\Box u \perp T_u\mathcal{N} \quad \text{in } \mathcal{D}'(\mathbb{R}\times\mathbb{R}^2). \tag{4.120}$$

At this time, we do not know more.

Not being able to answer these questions, I will present a possible strategy to tackle them. This is the purpose of section 4.4.1 below. The idea, motivated by what we saw in chapter 2, is to try to construct conservation laws for harmonic maps, even when the image manifold is not symmetric. This will naturally lead us to consider a class of new (open) problems, which will be presented in section 4.4.2. These questions are a generalization of the isometric embedding problem for Riemannian manifolds, to know how to embed vector-bundle-valued covariant closed differentials. In the special case of isometric embedding of surfaces, we will propose in section 4.4.3 a variational approach. Finally, the aim of section 4.4.4 is to suggest the existence of a particular structure for harmonic maps of surfaces, which remains totally undiscovered.

4.4.1 Conservation laws

Let $\mathcal{N}$ be a compact manifold without boundary, over which there is a smooth orthonormal frame field $\widehat{e} = (\widehat{e}_1, ..., \widehat{e}_n)$. We may always assume that this is the case thanks to lemma 4.1.2. We also suppose that $\mathcal{N}$ is isometrically embedded in $(\mathbb{R}^N, \langle.,.\rangle)$, and that for each point $y \in \mathcal{N}$

4.4 Conservation laws for harmonic maps without symmetry

we can complete the family $\hat{e}$ to make an orthonormal basis of $\mathbb{R}^N$, $(\hat{e}_1, ..., \hat{e}_N)$, where $(\hat{e}_{n+1}, ..., \hat{e}_N)$ is an orthonormal basis of the normal space to $T_y\mathcal{N}$.

Let Ω be an open subset of $(\mathcal{M}, g)$ (or, to simplify, of $\mathbb{R}^m$), u be an H^1 map from Ω to $\mathcal{N}$, and define

$$e_a(x) := \hat{e}_a \circ u(x), \quad 1 \leq a \leq N, \tag{4.121}$$

$$\alpha^a := \langle du, e_a \rangle, \quad 1 \leq a \leq n, \tag{4.122}$$

$$\omega_b^a := \langle de_b, e_a \rangle, \quad 1 \leq a, b \leq N. \tag{4.123}$$

These quantities satisfy the structure equations

$$d\alpha^a + \omega_b^a \wedge \alpha^b = 0, \quad \forall 1 \leq a \leq n, \tag{4.124}$$

$$\omega_b^a \wedge \alpha^b = 0, \quad \forall n+1 \leq a \leq N, \tag{4.125}$$

$$d\omega_b^a + \omega_c^a \wedge \omega_b^c = 0, \quad \forall 1 \leq a, b \leq N \tag{4.126}$$

(we agree that $\alpha^b = 0$ if $n+1 \leq b \leq N$) and

$$du = e_a \alpha^a, \tag{4.127}$$

where we sum over repeated indices.

If moreover u is weakly harmonic, we have the additional equations

$$0 = \langle d(\star du), e_b \rangle = d(\star \alpha^b) + \omega_a^b \wedge (\star \alpha^a), \quad \forall 1 \leq b \leq n. \tag{4.128}$$

This equation is analogous to the structure equation (4.124). A natural problem is to try to see (4.128) as part of a system of compatibility conditions to "integrate" the α^b, just as (4.124), (4.125) and (4.126) (Cartan equations) constitute the necessary (and sufficient) condition for (4.127): (4.126) implies the existence of a solution e of $de_a = e_b \omega_a^b$, and then (4.124) and (4.125) together imply that $d(e_a \alpha^a) = 0$.

We should thus find $\mathcal{N}' \in \mathbb{N}$, and construct a map $e^\star$ from Ω to the set of orthonormal frames in $\mathbb{R}^{N'}$ (which will be identified with $SO(N')$), where $N' > n$, such that

$$d(e_b^\star(\star\alpha^b)) = 0. \tag{4.129}$$

This equation is equivalent to

$$
\begin{aligned}
0 &= \sum_{b=1}^{n}\left[e_b^\star d(\star\alpha^b) + \sum_{a=1}^{N'} e_a^\star \omega_b^{\star a} \wedge (\star\alpha^b)\right] \\
&= \sum_{a=1}^{n} e_a^\star \left[d(\star\alpha^a) + \sum_{b=1}^{n} \omega_b^{\star a} \wedge (\star\alpha^b)\right] + \sum_{a=n+1}^{N'} e_a^\star \sum_{b=1}^{n} \omega_b^{\star a} \wedge (\star\alpha^b),
\end{aligned}
$$

where

$$\omega_b^{\star a} = \langle de_b^\star, e_a^\star \rangle.$$

Hence, (4.129) is also equivalent to the system

$$d(\star\alpha^a) + \sum_{b=1}^{n} \omega_b^{\star a} \wedge (\star\alpha^a) = 0, \quad \forall 1 \le a \le n, \tag{4.130}$$

$$\omega_b^{\star a} \wedge (\star\alpha^b) = 0, \quad \forall n+1 \le a \le N. \tag{4.131}$$

Using (4.128), we see that (4.130) is satisfied if, for instance, $\omega_b^{\star a} = \omega_b^a$, for $1 \le a, b \le n$.

Problem 4.4.2 *Given a weakly harmonic map $u \in H^1(\Omega, \mathcal{N})$, and $e = \widehat{e} \circ u$, is it possible to find a map $e^\star \in H^1(\Omega, SO(N'))$ such that if we write*

$$\omega_b^{\star,a} := \langle de_b^\star, e_a^\star \rangle,$$

the following conditions are satisfied:

$$||de^\star||_{H^1(\Omega)} \le C||du||_{H^1(\Omega)}, \tag{4.132}$$

$$\omega_b^{\star,a} = \omega_b^a, \quad \text{if } 1 \le a, b \le n, \tag{4.133}$$

$$\omega_b^{\star,a} \wedge (\star\alpha^b) = 0 \quad \text{if } n+1 \le a \le N'\,? \tag{4.134}$$

4.4 Conservation laws for harmonic maps without symmetry

Remark 4.4.3 *This problem is probably hard: apparently the difficulties are similar to those encountered when constructing an isometric embedding of a Riemannian manifold, with in addition very weak regularity hypotheses on the metric. It is not certain that it will always be possible to make such a construction, with estimate (4.132). We will come back to this point below.*

Remark 4.4.4 *We may always dream that we know how to solve this problem. Let us then show how we can use it to deduce a weak-compactness result for weakly harmonic maps.*

Suppose Ω to be bounded. Let u_k be a sequence of weakly harmonic maps in $H^1(\Omega, \mathcal{N})$ such that

$$u_k \rightharpoonup u \quad \text{weakly in} \quad H^1(\Omega, \mathcal{N}). \tag{4.135}$$

Let $e_k = (e_{k,1}, ..., e_{k,N}) \in H^1(\Omega, SO(N))$ with $e_{k,a} = \widehat{e}_a \circ u_k$ and $\alpha_k^a = \langle du_k, e_{k,a} \rangle$. We have

$$||de_k||_{L^2(\Omega)} \leq ||du_k||_{L^2(\Omega)} \leq C. \tag{4.136}$$

Now suppose that there exists a sequence $e_k^\star \in H^1(\Omega, SO(N'))$ with

$$||de_k^\star||_{L^2(\Omega)} \leq ||du_k||_{L^2(\Omega)} \leq C, \tag{4.137}$$

$$\langle de_{k,a}^\star, e_{k,b}^\star \rangle = \langle de_{k,a}, e_{k,b} \rangle, \quad \forall 1 \leq a, b \leq n, \tag{4.138}$$

$$\sum_{b=1}^{n} \langle de_{k,a}^\star, e_{k,b}^\star \rangle \wedge (\star \alpha_k^b) = 0, \quad \forall n+1 \leq a \leq N', \tag{4.139}$$

i.e., that it is a solution of problem 4.4.2.

Using the compact embedding of $H^1(\Omega)$ in $L^2(\Omega)$ and the Rellich–Kondrakov theorem (see [19]), we deduce from (4.136) and (4.137) that, up to extracting a subsequence k,

$$e_{k,a}, e_{k,a}^\star \rightharpoonup e_a, e_a^\star \quad \text{weakly in} \quad H^1(\Omega), \tag{4.140}$$

$$e_{k,a}, e_{k,a}^\star \to e_a, e_a^\star \quad \text{in} \quad L^2(\Omega). \tag{4.141}$$

Using also the fact that $e_{k,a}$ and $e^\star_{k,a}$ are bounded in L^∞, we deduce the following. (4.135) and (4.141) imply that

$$\alpha^a_k \rightharpoonup \alpha^a \quad \text{weakly in} \quad L^2(\Omega),$$

and also

$$\star\alpha^a_k \rightharpoonup \star\alpha^a \quad \text{weakly in} \quad L^2(\Omega).$$

Moreover, using (4.141), we have

$$\star\alpha^a_k e^\star_{k,a} \rightharpoonup \star\alpha^a e^\star_a \quad \text{weakly in} \quad L^2(\Omega). \tag{4.142}$$

On the other hand, using (4.138), (4.139) and the fact that u is weakly harmonic, we deduce that

$$d(\star\alpha^a_k e^\star_{k,a}) = 0 \quad \text{in} \quad \mathcal{D}'(\Omega). \tag{4.143}$$

Hence (4.142) and (4.143) yield

$$d(\star\alpha^a e^\star_a) = 0 \quad \text{in} \quad \mathcal{D}'(\Omega). \tag{4.144}$$

Finally, we can pass to the limit in (4.138), using (4.140) and (4.141), to obtain

$$\langle de^\star_a, e^\star_b \rangle = \langle de_a, e_b \rangle, \quad \forall 1 \leq a, b \leq n. \tag{4.145}$$

We just have to develop (4.144) to deduce, using (4.145), that

$$d(\star\alpha^a) + \omega^a_b \wedge (\star\alpha^b) = 0, \quad \forall 1 \leq a \leq n,$$

which means that u is weakly harmonic (see (4.128)).

Example 4.4.5 In case $\mathcal{N} = S^{n-1}$, problem 4.4.2 has an obvious solution given by $N' = \frac{n(n-1)}{2}$, $\mathbb{R}^{N'} \simeq so(n)$ and

$$e^\star_j = e_j \times e_0$$

(see problem 2.4.5).

We may formulate a little better the statement of problem 4.4.2, not to simplify it, but to understand better its geometrical content. In doing so, we will come across a series of problems which, in my opinion, have

4.4.2 Isometric embedding of vector-bundle-valued differential forms

Let $\mathcal{M}$ be an m-dimensional manifold, and $\mathcal{F}$ be an n-dimensional vector bundle over $\mathcal{M}$ (at each point $x \in \mathcal{M}$, the fiber, $\mathcal{F}_x$, is a vector space isomorphic to $\mathbb{R}^n$). We denote by $P : \mathcal{F} \longrightarrow \mathcal{M}$ the projection associated to this bundle. We are given a metric h on $\mathcal{F}$, i.e. a scalar product h_x on each $\mathcal{F}_x$ (h_x and $\mathcal{F}_x$ depend smoothly on x). Moreover, we choose a connection ∇ on $\mathcal{F}$, which is compatible with the metric h. (For an introduction to fiber bundles see, for instance, [101] or [47]).

To be more explicit, we use local coordinates. Without loss of generality, we will thus restrict our attention to an open subset Ω of $\mathcal{M}$, on which is defined a chart taking its values in an open subset of $\mathbb{R}^m$. If Ω has a trivial topology (which we will assume), we can find a trivialization of $\mathcal{F}$ over Ω, in the sense that if $\mathcal{F}_\Omega := P^{-1}(\Omega)$, then there exists a diffeomorphism Φ_Ω between $\mathcal{F}_\Omega$ and $\Omega \times \mathbb{R}^n$, such that the restriction of Φ_Ω to each fiber $\mathcal{F}_x$ is a vector space isometry from $(\mathcal{F}_x, h_x)$ to $(\mathbb{R}^n, \langle .,. \rangle)$.

Every section X of $\mathcal{F}$ (i.e. a map X from $\mathcal{M}$ to $\mathcal{F}$ such that $P \circ X = Id_\mathcal{M}$) is represented in this trivialization by a function $\widehat{X}$ from Ω to $\mathbb{R}^n$ such that

$$\forall x \in \Omega, \quad \Phi_\Omega(X(x)) = (x, \widehat{X}(x)).$$

For instance, we consider the n sections of $\mathcal{F}_\Omega$ $E_1, ..., E_n$, whose expressions in this trivialization are

$$\widehat{E}_1 = \begin{pmatrix} 1 \\ 0 \\ \vdots \\ 0 \end{pmatrix}, ..., \widehat{E}_n = \begin{pmatrix} 0 \\ \vdots \\ 0 \\ 1 \end{pmatrix}.$$

It is clear that for all $x \in \Omega$, $E(x) = (E_1(x), ..., E_n(x))$ constitutes an orthonormal basis of $\mathcal{F}_x$.

The covariant derivative or connection ∇ is known once we have specified the 1-forms ω_a^b such that

$$\forall \xi \in T_x\mathcal{M}, \quad \nabla_\xi E_a = E_b \omega_a^b(\xi). \tag{4.146}$$

These 1-forms, called connection 1-forms, satisfy

$$\omega_b^a + \omega_a^b = 0, \quad \forall 1 \leq a,b \leq n.$$

Thus, for any section $X = \sum_{a=1}^n X^a E_a$ of $\mathcal{F}_\Omega$,

$$\nabla_\xi X = E_a dX^a(\xi) + E_b \omega_a^b(\xi) X^a, \qquad (4.147)$$

or equivalently,

$$(\nabla_\xi X)^a = dX^a(\xi) + \omega_b^a(\xi) X^b. \qquad (4.148)$$

We also consider for each integer p between 0 and m the bundle $\Lambda^p \mathcal{M} \otimes \mathcal{F}$ of skew-symmetric p-forms on $\mathcal{M}$, with coefficients in $\mathcal{F}$. A point in this bundle corresponds to specifying a point x in $\mathcal{M}$ and a skew-symmetric p-linear form of $T_x \mathcal{M}$ taking values in the vector space $\mathcal{F}_x$. A section of this bundle is a differential p-form with coefficients in $\mathcal{F}$. We may thus define the covariant differential

$$d_\nabla : \Lambda^p \otimes \mathcal{F} \longrightarrow \Lambda^{p+1} \otimes \mathcal{F},$$

by letting

$$d_\nabla E_a = E_b \omega_a^b, \qquad (4.149)$$

$$\forall X : \mathcal{M} \longrightarrow \mathcal{F}, \ \forall \beta \in \Lambda^p \mathcal{M}, \quad d_\nabla(X\beta) = (d_\nabla X) \wedge \beta + X d\beta. \quad (4.150)$$

In the trivialization over Ω we know, for instance, that for any section $X : \Omega \longrightarrow \mathcal{F}_\Omega$,

$$(d_\nabla X) = E_a dX^a + E_b \omega_a^b X^a = E_a(dX^a + \omega_b^a X^b), \qquad (4.151)$$

$$d_\nabla \circ d_\nabla X = E_a \Omega_b^a X^b, \qquad (4.152)$$

where $\Omega_b^a = d\omega_b^a + \omega_c^a \wedge \omega_b^c$ is the curvature form.

We are interested in the $\mathcal{F}$-valued *closed covariant p-forms*. We start with the case $p = 1$. A 1-form $\phi : \mathcal{M} \longrightarrow \Lambda^1 \otimes \mathcal{F}$ is said to be closed covariant if

$$d_\nabla \phi = 0. \qquad (4.153)$$

Writing $\phi = E_a \phi^a$, where $\phi^a \in \Lambda^1 \mathcal{M}$, equation (4.153) implies, using (4.146),

4.4 Conservation laws for harmonic maps without symmetry

$$0 = E_a d\phi^a + d_\nabla E_a \wedge \phi^a = E_a(d\phi^a + \omega_b^a \wedge d\phi^b). \tag{4.154}$$

Therefore, we obtain the system

$$d\phi^a + \omega_b^a \wedge \phi^b = 0. \tag{4.155}$$

We remark that this system of equation is similar to that satisfied by the Maurer–Cartan forms on a Riemannian manifold (see equation (4.124)). This resemblance will be made more precise below.

Example 4.4.6 Let $(\mathcal{N}, h)$ be an n-dimensional Riemannian manifold, and $u : \mathcal{M} \longrightarrow \mathcal{N}$ be a differentiable map. Then the bundle

$$\mathcal{F} = u^*T\mathcal{N} = \{(x, V) \mid x \in \mathcal{M}, \ V \in T_{u(x)}\mathcal{N}\}$$

(pull-back of $T\mathcal{N}$ by u) is equipped with the metric u^*h and the connection $\nabla_\xi V = \nabla_{u(\xi)}^h V$ where ∇^h is the Levi-Civita connection on $(\mathcal{N}, h)$.

We can then see $\phi = du$ as an $\mathcal{F}$-valued 1-form. Moreover, ϕ is a solution of (4.153). Using an orthonormal moving frame $e = (e_1, ..., e_n)$ on $\mathcal{N}$, and defining $\omega_b^a = u^*\langle de_b, e_a \rangle$ and $\alpha^a = \langle du, e_a \rangle$, we remark that these quantities are solutions of equation (4.154).

A natural question is to look for a converse to this example.

Problem 4.4.7 Let $\mathcal{M}$ be an m-dimensional manifold, $\mathcal{F}$ be an n-dimensional vector bundle with metric h, and ∇ be a covariant derivative on $\mathcal{F}$ compatible with the metric. We suppose that ϕ is a closed covariant $\mathcal{F}$-valued differentiable p-form on $\mathcal{M}$, and hence

$$d_\nabla \phi = 0. \tag{4.156}$$

Find $N \in \mathbb{N}$ and an embedding T of $\mathcal{F}$ into $\mathcal{M} \times \mathbb{R}^N$ given by $T(x, X) = (x, T_x(X))$, where T_x is a linear map from $\mathcal{F}_x$ to $\mathbb{R}^N$, such that

(i) T is isometric, i.e. for every $x \in \mathcal{M}$, T_x is an isometry.
(ii) If $T(\phi)$ is the image of ϕ by T, then

$$dT(\phi) = 0. \tag{4.157}$$

It is not superfluous to write a local version, using local coordinates, of the above. Let $(E_1, ..., E_n)$ be an orthonormal family of sections of $\mathcal{F}$, then

$$e_1 = T_x(E_1), \ldots, \ e_n = T_x(E_n)$$

are an orthonormal family in $\mathbb{R}^N$, for all x. We complete it to obtain an orthonormal basis $(e_1, ..., e_N)$ (depending on x) of $\mathbb{R}^N$. Then if

$$\phi = E_a \phi^a$$

equation (4.157) means that

$$d(e_a \phi^a) = 0. \tag{4.158}$$

We write $\widehat{\omega}_b^a = \langle de_b, e_a \rangle$ and develop (4.158). This yields

$$\sum_{a=1}^{n} e_a(d\phi^a + \widehat{\omega}_b^a \wedge \phi^b) + \sum_{a=n+1}^{N} e_a(\widehat{\omega}_b^a \wedge \phi^b) = 0,$$

condition which, given (4.156), is satisfied if, for instance,

$$\omega_b^a = \widehat{\omega}_b^a, \quad \forall 1 \leq a, b \leq n, \tag{4.159}$$

$$\widehat{\omega}_b^a \wedge \phi^b = 0, \quad \forall n+1 \leq a \leq N. \tag{4.160}$$

Thus, it is sufficient (and in generic cases also necessary) to solve the following problem.

Problem 4.4.8 *With the hypotheses of problem 4.4.7, find $N \in \mathbb{N}$, and $e : \mathcal{M} \longrightarrow SO(N)$ (the set of positively oriented orthonormal frames of $\mathbb{R}^N$), such that*

$$\langle de_a, e_b \rangle = \omega_a^b, \quad \forall 1 \leq a, b \leq n, \tag{4.161}$$

$$\langle de_b, e_a \rangle \wedge \phi^b = 0, \quad \forall n+1 \leq a \leq N. \tag{4.162}$$

Remark 4.4.9 *We invite the reader to check that problem 4.4.2, related to the construction of a conservation law for a harmonic map is – up to regularity questions – a particular case of problem 4.4.8, for $p = m - 1$.*

Example 4.4.10 CASE $m = n$ AND $p = 1$
Suppose that ϕ is a sufficiently regular ($\mathcal{C}^3$ is sufficient) section with rank m everywhere. This means that for any point x of $\mathcal{M}$, ϕ is an isomorphism from $T_x\mathcal{M}$ to $\mathcal{F}_x$. We can then construct a Riemannian metric g on $\mathcal{M}$ by letting

$$g = (\phi)^2 = \sum_{a=1}^{m} \phi^a \otimes \phi^a, \quad \text{where} \quad \phi = E_a \phi^a.$$

This metric is of class $\mathcal{C}^3$. Using the Nash–Moser theorem we see that there exists an isometric embedding, u, of $(\mathcal{M}, g)$ into the Euclidean

4.4 Conservation laws for harmonic maps without symmetry

space $(\mathbb{R}^N, \langle.,.\rangle)$. For each $x \in \mathcal{M}$, we consider the basis of $T_x\mathcal{M}$, $f = (f_1, ..., f_m)$, dual to the basis $(\phi^1, ..., \phi^m)$ of $T_x^*\mathcal{M}$, i.e.

$$\phi^a(f_b) = \delta_b^a. \tag{4.163}$$

Let

$$e_a := du(f_a). \tag{4.164}$$

Clearly, f is an orthonormal basis of $(T_x\mathcal{M}, g)$, and hence $(e_1, ..., e_n)$ is an orthonormal family in $\mathbb{R}^N$. Moreover, we deduce from (4.163) and (4.164) that

$$du = e_a \phi^a,$$

and hence that $d(e_a \phi^a) = 0$. Thus, we have found a map T satisfying (4.157).

4.4.3 A variational formulation for the case $m = n = 2$ and $p = 1$

From example 4.4.10 above, we know that in case $m = n = 2$ and $p = 1$, problem 4.4.7 consists essentially in isometrically embedding a Riemannian surface. We also remark that the solution of this problem enables us to produce conservation laws for harmonic maps of surfaces. Below, Ω will denote an open subset of a surface $\mathcal{M}$, $\mathcal{F}$ a plane-bundle over $\mathcal{M}$, and we consider an $\mathcal{F}$-valued 1-form, such that

$$\phi = E_1 \phi^1 + E_2 \phi^2 \text{ in } \Omega$$

is a solution of $d_\nabla \phi = 0$, i.e., over Ω,

$$\begin{cases} d\phi^1 + \omega_2^1 \wedge \phi^2 = 0 \\ d\phi^2 + \omega_1^2 \wedge \phi^1 = 0. \end{cases} \tag{4.165}$$

We will study problem 4.4.7 supposing that $N = 3$. We consider the set $\mathcal{E}$ of maps from Ω to the set of positively oriented orthonormal frames of $\mathbb{R}^3$ — which we identify with $SO(3)$. Over $\mathcal{E}$ we define the functional

$$F(e) = \int_\Omega \phi^1 \wedge \langle de_3, e_2 \rangle + \langle de_3, e_1 \rangle \wedge \phi^2.$$

Since I do not have a clear idea on this subject, I will not specify which topology should be used over $\mathcal{E}$, and what follows is essentially formal (notice that the H^1 topology might be suitable).

Lemma 4.4.11 *Let $e \in \mathcal{E}$ be a critical point of F. Then, we have*

$$\begin{cases} d\phi^1 + \langle de_2, e_1 \rangle \wedge \phi^2 & = 0 \\ d\phi^2 + \langle de_1, e_2 \rangle \wedge \phi^1 & = 0 \\ \phi^1 \wedge \langle de_3, e_1 \rangle + \phi^2 \wedge \langle de_3, e_2 \rangle & = 0. \end{cases} \quad (4.166)$$

These equations formally imply that e is a solution of problem 4.4.7, i.e.

$$d(e_1\phi_1 + e_2\phi_2) = 0.$$

Proof It consists of calculating the first variation of F. Let $A = (A_b^a) \in C_c^\infty(\Omega, so(3))$, and calculate the effect on $F(e)$ of a variation

$$e \longmapsto e_t = e.\exp(tA) = e.(Id + tA) + o(t).$$

This means that

$$e_{t,a} = e_a + te_b.A_a^b + o(t). \quad (4.167)$$

Write $\widehat{\omega}_b^a = \langle de_b, e_a \rangle$ and $\widehat{\omega}_{t,b}^a = \langle de_{t,b}, e_{t,a} \rangle$. It follows from (4.167) that

$$\widehat{\omega}_{t,b}^a = \widehat{\omega}_b^a + t(dA_b^a + \widehat{\omega}_c^a A_b^c - A_c^a \widehat{\omega}_b^c) + o(t).$$

Substituting this expression into $F(e_t)$, we obtain

$$F(e_t) = F(e) + t \int_\Omega d(A_3^1 \phi^2 - A_3^2 \phi^1)$$

$$+ t \int_\Omega A_2^1(\phi^1 \wedge \widehat{\omega}_3^1 + \phi^2 \wedge \widehat{\omega}_3^2)$$

$$+ A_1^3(d\phi^2 + \widehat{\omega}_1^2 \wedge \phi^1) + A_3^2(d\phi^1 + \widehat{\omega}_2^1 \wedge \phi^2) + o(t).$$

Thus,

$$\delta E(e)(A) = \int_\Omega A_1^3(d\phi^2 + \widehat{\omega}_1^2 \wedge \phi^1) + A_3^2(d\phi^1 + \widehat{\omega}_2^1 \wedge \phi^2)$$

$$+ A_2^1(\phi^1 \wedge \widehat{\omega}_3^1 + \phi^2 \wedge \widehat{\omega}_3^2). \quad (4.168)$$

And we see that e is a critical point of F if and only if it satisfies equations (4.166). □

4.4 Conservation laws for harmonic maps without symmetry

This variational formulation does not seem very convenient since we might expect critical points of F to be neither minimizing nor maximizing. A possible use for this variational structure is to try to use linking or to construct a Floer theory (see [59]). A first step in this direction consists of calculating the flow equation for F.

To do so, we will suppose ϕ to be non-degenerate, i.e. for all $x \in \mathcal{M}$, $\phi_x : T_x\mathcal{M} \longrightarrow \mathcal{F}_x$ is an isomorphism. We define on Ω the metric $g = (\phi)^2 = \phi^1 \otimes \phi^1 + \phi^2 \otimes \phi^2$, as in example 4.4.10. We also consider the basis $f = (f_1, f_2)$ of $T_x\mathcal{M}$, dual to (ϕ^1, ϕ^2), which is orthonormal for g.

A solution of the flow equation for F will be a map defined on (a subset of) $\{(x,t) \in \Omega \times \mathbb{R}\}$, taking its values in $SO(3)$, such that for any t

$$\delta F(e)(A) = \int_\Omega \left\langle A, \widehat{\omega}(\frac{\partial}{\partial t}) \right\rangle \phi^1 \wedge \phi^2, \qquad (4.169)$$

where $\langle A, \widehat{\omega} \rangle = \sum_{1 \leq a < b \leq 3} A_b^a \widehat{\omega}_b^a$ and $\widehat{\omega}_b^a = \langle de_b, e_a \rangle$.

Lemma 4.4.12 e is a solution of (4.169) if and only if

$$\begin{cases} \widehat{\omega}_2^1(\frac{\partial}{\partial t}) &= -\widehat{\omega}_3^2(f_1) - \widehat{\omega}_1^3(f_2) \\ \widehat{\omega}_1^3(\frac{\partial}{\partial t}) &= \widehat{\omega}_2^1(f_2) - \omega_2^1(f_2) \\ \widehat{\omega}_3^2(\frac{\partial}{\partial t}) &= \widehat{\omega}_2^1(f_1) - \omega_2^1(f_1). \end{cases} \qquad (4.170)$$

Proof We leave the proof (using (4.169) and (4.168)) to the reader. □

We can also formulate equations (4.170) by introducing a map from Ω to the $Spin(3)$ group, the universal covering of $SO(3)$ (which is also the set of norm 1 quaternions). In order to do so, we will suppose Ω to be simply connected.

We start by remarking that $\widehat{\omega} = e^{-1}de$ satisfies

$$d\widehat{\omega} + \widehat{\omega} \wedge \widehat{\omega} = 0,$$

which we may rewrite as

$$d\theta + \theta \wedge \theta = 0, \qquad (4.171)$$

where this time

$$\theta = \frac{1}{2} \begin{pmatrix} i\widehat{\omega}_2^1 & \widehat{\omega}_3^2 - i\widehat{\omega}_1^3 \\ -\widehat{\omega}_3^2 - i\widehat{\omega}_1^3 & -i\widehat{\omega}_2^1 \end{pmatrix} \in su(2). \qquad (4.172)$$

Equation (4.171) implies that there is a map $u : \Omega \times \mathbb{R} \longrightarrow SU(2)$ such that
$$du = u\,\theta. \tag{4.173}$$

We introduce the matrices
$$I = \frac{1}{2}\begin{pmatrix} 0 & -1 \\ 1 & 0 \end{pmatrix},\ J = \frac{1}{2}\begin{pmatrix} 0 & i \\ i & 0 \end{pmatrix},\ K = \frac{1}{2}\begin{pmatrix} -i & 0 \\ 0 & i \end{pmatrix},$$
so that $\theta = -\widehat{\omega}_3^2 I - \widehat{\omega}_1^3 J - \widehat{\omega}_2^1 K$. We check by a direct calculation that (4.170) is equivalent to the following equation of "$SU(2)$-valued Dirac" type,

$$[u^{-1}du(f_1), J] - [u^{-1}du(f_2), I] + u^{-1}\frac{\partial u}{\partial t} = \omega_2^1(f_1)I + \omega_2^1(f_2)J. \tag{4.174}$$

4.4.4 Hidden symmetries for harmonic maps on surfaces?

We come back to harmonic maps on surfaces. We saw in chapter 2 that when the image manifold $\mathcal{N}$ is symmetric, the set of symmetries which act on harmonic maps on a surface is very big: apart from the trivial symmetries (like the action of $SO(n)$ on S^{n-1}), there are also the dressing action of the loop group of the manifold, and the action of the circle. The quasi-algebraic description of the action of these groups is an exceptional fact, which is characteristic of completely integrable Hamiltonian systems. It is obvious that all this collapses if the symmetry of $\mathcal{N}$ is broken. A naive question is: what happens to these symmetries? We will see that something remains from the action of the circle. For the moment, I find it hard to identify what this residual symmetry is, and to specify which is the right geometrical setting. We will limit ourselves to examining two points of view, clearly incomplete, about this circle action. I hope to convey to the reader a feeling for this phenomenon, not being able to really explain it.

Remark 4.4.13 *In example 4.4.6 above, we saw that to every map u from a manifold $\mathcal{M}$ to a Riemannian manifold $(\mathcal{N}, h)$, we can associate the bundle $\mathcal{F} = u^*T\mathcal{N}$ over $\mathcal{M}$, which is naturally equipped with a metric and a connection. Then, $\phi := du$ is an $\mathcal{F}$-valued closed covariant 1-form. Assume, in addition, that $\mathcal{M}$ is equipped with a metric g and that u is a harmonic map. Then, we know that $\psi := \star du$ (where $\star$ is the Hodge*

4.4 Conservation laws for harmonic maps without symmetry 219

operator on $(\mathcal{M}, g)$) is an $\mathcal{F}$-valued closed covariant $(m-1)$ differential form: this is exactly what equations (4.124) ($\Leftrightarrow d_\nabla \phi = 0$) and (4.128) ($\Leftrightarrow d_\nabla \psi = 0$) tell us. In the case where $\mathcal{M}$ is 2-dimensional, it is possible to use a conformal complex coordinate $z = x + iy$ over $\mathcal{M}$, in which equations (4.124) and (4.128) may be condensed into a single equation:

$$d\alpha_\lambda^a + \omega_b^a \wedge \alpha_\lambda^b = 0, \quad \forall \lambda \in \mathbb{C}^\star, \tag{4.175}$$

where

$$\alpha_\lambda^a = \lambda^{-1} \alpha^a \left(\frac{\partial}{\partial z}\right) dz + \lambda \alpha^a \left(\frac{\partial}{\partial \bar{z}}\right) d\bar{z}.$$

This form recalls the formulation given for harmonic maps between a surface and a sphere in section 3.2. In particular, denoting by $\phi_\lambda = \cos \lambda \phi + \sin \lambda \psi$ the $\mathcal{F}$-valued 1-form whose coordinate representation is α_λ^a, equation (4.175) is simply

$$d_\nabla \phi_\lambda = 0.$$

Noticing that $\phi_1 = \phi$ and $\phi_i = \psi$, we thus interpret ϕ_λ as an S^1-parametrized deformation of ϕ, in a family of closed covariant 1-forms.

Remark 4.4.14 HARMONIC DIFFEOMORPHISMS
If u is a diffeomorphism from an open subset Ω of a surface, to an open subset of $(\mathcal{N}, h)$, then it is possible to identify the image of u with Ω equipped with the pull-back metric of h by u, $u^*h = \alpha^1 \otimes \alpha^1 + \alpha^2 \otimes \alpha^2$. Hence, u is represented by the identity map from (Ω, g) to $(\Omega, u^*h) \simeq (\mathcal{N}, h)$. For $\lambda \in S^1$, we let

$$h'_\lambda = \alpha_\lambda^1 \otimes \alpha_\lambda^1 + \alpha_\lambda^2 \otimes \alpha_\lambda^2,$$

and $\mathcal{N}_\lambda = (\Omega, h'_\lambda)$. Then, we define the map u_λ from (Ω, g) to $\mathcal{N}_\lambda$ as being the one which is represented by the identity map from (Ω, g) to (Ω, h'_λ).

Since u is a diffeomorphism, h'_λ is a metric for all $\lambda \in S^1$, and we associate to u a family, parametrized by S^1, of harmonic diffeomorphisms taking values in a manifold $\mathcal{N}_\lambda$, which depends on λ. We check (exercise) that if the curvature of $\mathcal{N}$ is constant, then the curvature of $\mathcal{N}_\lambda$

will be equal to this same constant, and hence $\mathcal{N}_\lambda$ will be locally isometric to $\mathcal{N}$ (actually, the curvature of the manifold is pointwise invariant under this action). In this way we recover the results that were known for sphere-valued maps.

5
Surfaces with mean curvature in L^2

We present here another example of the geometric application of compensation phenomena. We will again use Wente's inequality and the representation of equations in moving frames. We will see that orthonormal Coulomb frames on a surface are linked to conformal coordinates. This link will be clearly stated in the last section.

The starting point here is the following result, obtained in 1991 by Tatiana Toro, in her thesis.

Theorem 5.0.15 *[171] Let $\Omega \subset \mathbb{R}^2$ be a domain with Lipschitz boundary, and u a map in $H^2(\Omega, \mathbb{R})$. Let $\Gamma_u = \{(z, u(z)) \in \mathbb{R}^3 \mid z \in \overline{\Omega}\}$ be the graph of u.*

Then Γ_u is homeomorphic to a subdomain Ω' of $\mathbb{R}^2$ through a bilipschitz homeomorphism. More precisely, there exists a homeomorphism $\Phi : \Omega' \longrightarrow \Gamma_u$, and $L > 0$ such that

$$\begin{cases} |\Phi(z) - \Phi(z')| & \leq \quad L|z - z'|, \quad \forall x, y \in \Omega' \\ |\Phi^{-1}(Z) - \Phi^{-1}(Z')| & \leq \quad L|Z - Z'|, \quad \forall Z, Z' \in \Gamma_u \end{cases} \quad (5.1)$$

and furthermore,

$$L \leq (1 + C\|u\|_{H^2(\Omega)}^2)^{\frac{1}{2}}. \quad (5.2)$$

Moreover, on Ω', the inverse image by Φ of the metric induced on Γ_u, $g = \Phi^*\langle .,.\rangle$, is comparable to the standard metric on Ω', in the sense that

$$\sup_{z \in \Omega'} |g_{ij}(z) - \delta_{ij}| \leq C\|u\|_{H^2(\Omega)}^2. \tag{5.3}$$

This result is rather surprising. In fact, a map $u \in H^2(\Omega)$ is generally Hölder continuous, but might not be Lipschitz. A counterexample is the map $u \in H^2(B(0, \frac{1}{2}), \mathbb{R})$ defined by

$$u(x, y) = x \log |\log \sqrt{x^2 + y^2}|. \tag{5.4}$$

This map is not Lipschitz, since its gradient is not bounded in a neighborhood of 0. In fact, close to 0 we have

$$\begin{aligned} \frac{\partial u}{\partial x} &= \log |\log r| + O(|\log r|^{-1}), \\ \frac{\partial u}{\partial y} &= O(|\log r|^{-1}), \end{aligned} \tag{5.5}$$

where $r = \sqrt{x^2 + y^2}$. We could conclude naively that the graph of u is not Lipschitz. But this is false, as is stated in the previous theorem (we may show, in this particular example, that the graph of u is a C^1 surface). Other examples given in [171] show that theorem 5.0.15 is optimal, in the sense that there exist functions in $H^2(B(0, \frac{1}{2}), \mathbb{R})$ whose graph is a Lipschitz surface, but not C^1.

This theorem follows from the more general result shown in [171].

Theorem 5.0.16 *There exists $\epsilon > 0$, such that every surface S in $\mathbb{R}^n$ satisfying the following hypotheses, is Lipschitz:*

(i) *There exists a sequence of smooth surfaces S_k in $\mathbb{R}^n$, which converges in measure in $\mathbb{R}^n$ to S.*
(ii) *$\forall x \in \mathbb{R}^n$, there exists a ball $B(x, \rho)$ of $\mathbb{R}^n$, and $\beta > 0$ such that*

$$\mathcal{H}^2(S_k \cap B(x, \rho)) \leq \beta, \tag{5.6}$$

i.e. the 2-dimensional Hausdorff measure ($=$ the area) of $S_k \cap B(x, \rho)$ is uniformly bounded, and

$$\int_{S_k \cap B(x, \rho)} |A_k|^2 d\mathcal{H}^2 \leq \epsilon^2, \tag{5.7}$$

where A_k is the second fundamental form of the embedding of $\mathcal{S}_k$ in $\mathbb{R}^n$.

In reality, a precise statement of this result requires working in the varifold setting, but we will not follow this path. The proof of this result is very delicate, and relies on the construction of a sequence of piecewise affine bilipschitz maps, which converge to a bilipschitz homeomorphism on $\mathcal{S}$.

After this theorem, other proofs and more general results were obtained. T. Toro, in the spirit of the Reifenberg theorem [135], showed that if a locally compact subset of $\mathbb{R}^n$, of Hausdorff dimension m, is "locally well approximated" by pieces of m-dimensional affine spaces, then this subset is a Lipschitz submanifold [172]. It was also noticed by Stephen Semmes that this result implies theorem 5.0.16.

Remark 5.0.17 *An important motivation for the study of surfaces satisfying (5.7) is the Willmore surface problem. A brief description of this subject will be given in section 5.3.*

In another direction, Stefan Müller and Vladimir Šverák have shown that all conformal parametrizations of surfaces with square-integrable second fundamental form are locally bilipschitz homeomorphisms [121]. Their proof reveals the presence of compensation phenomena, and uses Hardy space. In sections 5.1 and 5.2, we present variants of their results. Their approach is based on the following observation.

Let $X : \Omega \subset \mathbb{R}^2 \longrightarrow \mathbb{R}^3$ be a smooth conformal immersion. We write the first fundamental form of X in the form

$$g = e^{2f}[(dx)^2 + (dy)^2]. \tag{5.8}$$

Then f is a solution of the equation

$$-\Delta f = K e^{2f}, \tag{5.9}$$

where K is the Gauss curvature of the surface. But the quantity Ke^{2f} is precisely equal to the inverse image by $u \circ X$ of the volume form on S^2 (where $u : X(\Omega) \longrightarrow S^2$ is the Gauss map). The volume form on S^2 is a closed (but not exact) 2-form, and hence its inverse image by $u \circ X$ is locally expressed in terms of Jacobian determinants of maps in H^1.

By the results of chapter 3, we deduce that f is continuous and bounded, and hence that X is a bilipschitz immersion, either by Wente's inequality, or via Hardy spaces.

The approach we adopt here uses an orthonormal moving frame (e_1, e_2) on the surface $X(\Omega)$, preferably in the "Coulomb gauge"

$$d \star \langle de_2, e_1 \rangle = 0. \tag{5.10}$$

An interesting point here is that $(e_1, e_2) := \frac{1}{|dX|}\left(\frac{\partial X}{\partial x}, \frac{\partial X}{\partial y}\right)$ is such a frame, as long as X is a conformal immersion, and in this case we even have

$$\star \langle de_2, e_1 \rangle = df, \tag{5.11}$$

where f is defined by (5.8). Thus, the "Coulomb" orthonormal moving frames on a surface are closely related to the conformal coordinates on this surface.

We will study this point in more detail in section 5.4.

5.1 Local results

We start by defining for all $\gamma > 0$ the set

$$\mathcal{C}_\gamma(B) := \{X \in C^\infty(\overline{B}, \mathbb{R}^n) \mid X \text{ is a conformal immersion,} \atop \text{and } \int_B |A|^2 d\sigma \leq \gamma\} \tag{5.12}$$

where B is the unit ball of $\mathbb{R}^2 \simeq \mathbb{C}$, $d\sigma$ represents the area element over the surface $\mathcal{S} := X(\overline{B})$, and A is the second fundamental form of the immersion of $\mathcal{S}$ in $\mathbb{R}^n$. We also define $\overline{\mathcal{C}_\gamma(B)}$ as being the closure of $\mathcal{C}_\gamma(B)$ for the weak topology of $H^1(B, \mathbb{R}^n)$. Then, we have

Theorem 5.1.1 *For all $\gamma \leq \frac{8\pi}{3}$, every map $X \in \overline{\mathcal{C}_\gamma(B)}$ is either*

(i) *a constant map; or*

(ii) *a bilipschitz conformal immersion. In this case, for any compact subset K of B, there is a constant $C > 0$, such that*

$$\forall z \in K, \quad \frac{1}{C} \leq |dX(z)| \leq C, \tag{5.13}$$

and the inverse image of the Euclidean metric in $\mathbb{R}^n$ by X is a continuous metric.

Remark 5.1.2 *We recall the definition of the second fundamental form A of a surface: A is the symmetric bilinear form over $\mathcal{S}$, taking values in the normal bundle of $\mathcal{S}$ in $\mathbb{R}^n$, such that if $e := (e_1, e_2, e_3, ..., e_n)$ is an orthonormal moving frame defined locally over $\mathcal{S}$, in such a way that $b := (e_1, e_2)$ is an orthonormal basis of the tangent plane $T_m\mathcal{S}$, then*

$$\forall (V, W) \in (T_m\mathcal{S})^2, \quad A_{(m)}(V, W) = \sum_{a=3}^{n} \langle V, D_W e_a \rangle e_a$$
$$= \sum_{b=1}^{2} \sum_{a=3}^{n} \langle V, e_b \rangle \langle de_a, e_b \rangle (W) e_a. \quad (5.14)$$

Thus we let

$$|A|^2 := \sum_{b=1}^{2} \sum_{a=3}^{n} |\langle de_a, e_b \rangle|^2. \quad (5.15)$$

An equivalent definition for $|A|^2$ consists of introducing the Gauss map, defined over $\mathcal{S}$, which assigns to each point m of $\mathcal{S}$ the (oriented) tangent plane to $\mathcal{S}$ at m. The image space is $Gr_2(\mathbb{R}^n)$, the grassmannian of 2-dimensional oriented subspaces of $\mathbb{R}^n$. We denote this map by

$$u : \mathcal{S} \longrightarrow Gr_2(\mathbb{R}^n). \quad (5.16)$$

It is then convenient to identify $Gr_2(\mathbb{R}^n)$ with $SO(n)/(SO(2) \times SO(n-2))$ as follows. Choose a positively oriented reference orthonormal basis $\mathbf{E} := (\mathbf{E}_1, \mathbf{E}_2, ..., \mathbf{E}_n)$ in $\mathbb{R}^n$, and for each $g \in SO(n)$, denote by $e = \mathbf{E}.g$ the frame obtained from $\mathbf{E}$ using the rotation g. Introduce the following equivalence relation on $SO(n)$,

$$g\mathcal{R}g' \iff (g(\mathbf{E}_1), g(\mathbf{E}_2)) \text{ and } (g'(\mathbf{E}_1), g'(\mathbf{E}_2)) \text{ are two bases of the same plane of } \mathbb{R}^n, \text{ with the same orientation.} \quad (5.17)$$

Then, it is easy to see that $SO(2) \times SO(n-2)$ is the equivalence class of $\mathbb{1}$, and to identify $Gr_2(\mathbb{R}^n)$ with $SO(n)/\mathcal{R}$. In this way we can define a homogeneous metric on $Gr_2(\mathbb{R}^n)$, by letting

$$|du|^2 = \sum_{b=1}^{2} \sum_{a=3}^{n} |\langle de_a, e_b \rangle|^2, \quad (5.18)$$

where $e = (e_1, e_2, ..., e_n) = \mathbf{E}.g$ is such that (e_1, e_2) is a positively oriented basis of u. Comparing (5.15) and (5.18), we see that

$$|A|^2 = |du|^2. \tag{5.19}$$

The interest of this definition is that, since X is a conformal immersion and the Dirichlet integral is invariant under conformal transformations, we have

$$\int_S |A|^2 d\sigma = \int_S |du|^2 d\sigma = \int_B |d(u \circ X)|^2 dxdy. \tag{5.20}$$

Remark 5.1.3 *It is natural to wonder what happens to theorem 5.1.1 when we no longer assume that $\gamma \leq \frac{8\pi}{3}$. I do not know if the result is valid for larger γ, but I think that it should be possible to replace the condition $\gamma \in [0, \frac{8\pi}{3}]$ by $\gamma \in [0, 8\pi)$.*

A partial global result will be given in section 5.2 (theorem 5.2.4).

Before proving theorem 5.1.1, we will need the following lemma.

Lemma 5.1.4 *For every map $u \in H^1(B, Gr_2(\mathbb{R}^n))$, such that*

$$\|du\|_{L^2(B)}^2 \leq \gamma := \frac{8\pi}{3}, \tag{5.21}$$

there exists a map $b := (e_1, e_2)$ in $H^1(B, \mathbb{R}^n \times \mathbb{R}^n)$, such that for almost every $z \in B$, $b(z)$ is a positively oriented basis of $u(z)$. Furthermore, the energy of b is controlled by

$$\|db\|_{L^2(B)}^2 \leq a(\|du\|_{L^2(B)}^2), \tag{5.22}$$

where

$$
\begin{array}{rcl}
a : [0, \frac{8\pi}{3}] & \longrightarrow & \mathbb{R} \\
t & \longmapsto & \frac{16\pi}{3}\left[1 - \sqrt{1 - \frac{3t}{8\pi}}\right].
\end{array}
\tag{5.23}
$$

Moreover, b satisfies the equation

$$d(\star \langle e_1, de_2 \rangle) = 0. \tag{5.24}$$

Remark 5.1.5 *Below, we will prefer to consider b as a section of a fiber bundle. Let $\mathcal{F} := \{(u, b) \in Gr_2(\mathbb{R}^n) \times \mathbb{R}^n \times \mathbb{R}^n \mid b := (e_1, e_2) \text{ is a positively oriented orthonormal basis of } u\}$.*

$\mathcal{F}$ is a fiber bundle over $Gr_2(\mathbb{R}^n)$, with fiber S^1. The projection defining this fibration is just $(u, b) \longmapsto u$.

There exists a canonical connection on this bundle. For each $u \in Gr_2(\mathbb{R}^n)$, denote by P_u the orthogonal projection from $\mathbb{R}^n$ onto u. Then, every local section $b : Gr_2(\mathbb{R}^n) \longrightarrow \mathcal{F}$ may be seen as an $\mathbb{R}^n \times \mathbb{R}^n$-valued map. We then let

$$\nabla(e_1, e_2) = (P_u \circ de_1, P \circ de_2). \tag{5.25}$$

Surely, this connection has non-vanishing curvature — if not, locally it would be possible to define parallel sections of $\mathcal{F}$.

The pull-back of $\mathcal{F}$ by u is the bundle $u^*\mathcal{F} = \{(z, b) \in B \times \mathbb{R}^n \times \mathbb{R}^n \mid b \text{ is an orthonormal basis of } u(z)\}$. The lemma above essentially says that if the energy of u is sufficiently small, we are able to construct sections of $u^*\mathcal{F}$, whose energy may be controlled.

Remark 5.1.6 *Again, we may wonder if lemma 5.1.4 remains true without the assumption $\|du\|_{L^2(B)}^2 < \frac{8\pi}{3}$. We will prove such a result in section 5.2, but then we will lose control of the energy of the frame constructed (theorem 5.2.1).*

Proof of lemma 5.1.4
Step 1 CONSTRUCTION OF A FRAME IN THE SMOOTH CASE
We will consider in this step, as well as in steps 2, 3, 4, 5, that $u \in H^1(B, Gr_2(\mathbb{R}^n))$ is $\mathcal{C}^\infty$ on $\overline{B}$. Then the bundle $u^*\mathcal{F}$ is smooth, and since the base manifold, B_x, is contractible, we can trivialize it. This means that there exists a section $\widetilde{b} = (\widetilde{e_1}, \widetilde{e_2})$ of $u^*\mathcal{F}$.

Step 2 CHOICE OF A COULOMB FRAME
In spite of being smooth, the frame $\widetilde{b}$ may have an arbitrarily large energy. To avoid this, we will consider, for each $\theta \in H^1(B, \mathbb{R})$, the frame $b = (e_1, e_2)$ obtained by letting

$$b := \widetilde{b} \begin{pmatrix} \cos \theta & -\sin \theta \\ \sin \theta & \cos \theta \end{pmatrix} \tag{5.26}$$

and look for the Coulomb frames in the gauge orbit of $H^1(B, \mathbb{R})$. Actually, we need to consider a family of such problems: for each $r \in (0,1]$, we will minimize over $H^1(B, \mathbb{R})$ the functional

$$\begin{aligned} F_r(\theta) &= \frac{1}{2} \int_{B_r} (|\nabla e_1|^2 + |\nabla e_2|^2) dx dy \\ &= \int_{B_r} |\omega_2^1|^2 dx dy, \end{aligned} \quad (5.27)$$

where $\omega_2^1 = \langle de_2, e_1 \rangle$.

As in the proof of theorem 4.1.1, we may show that for each r, the minimum of F_r is attained, and that the frame, $b_r := (e_{r,1}, e_{r,2})$, minimizing F_r, satisfies

$$d(\star \omega_2^1) = 0 \quad \text{in} \quad B_r, \quad (5.28)$$

$$\omega_2^1 \left(\frac{\partial}{\partial n} \right) = 0 \quad \text{on} \quad \partial B_r. \quad (5.29)$$

Using equation (5.28), we deduce that there exists $f_r \in H^1(B_r, \mathbb{R})$ such that

$$df_r = \star \omega_2^1 \quad \text{in} \quad B_r. \quad (5.30)$$

Then, it follows from (5.29) that f_r is constant on ∂B_r, and hence it is possible to choose f_r such that

$$f_r = 0 \quad \text{on} \quad \partial B_r. \quad (5.31)$$

A direct calculation using (5.30) yields

$$\begin{aligned} -\Delta f_r &= \left\langle \frac{\partial e_{r,1}}{\partial x}, \frac{\partial e_{r,2}}{\partial y} \right\rangle - \left\langle \frac{\partial e_{r,1}}{\partial y}, \frac{\partial e_{r,2}}{\partial x} \right\rangle \\ &= \{e_{r,1} \cdot e_{r,2}\}. \end{aligned} \quad (5.32)$$

In particular, (5.31) and (5.32) show that we may write $f_r = f_{r,1} + \cdots + f_{r,n}$ where the $f_{r,j}$ are solutions of

$$\begin{cases} -\Delta f_{r,j} = \{e_{r,1}^j, e_{r,2}^j\} & \text{in } B_r \\ f_{r,j} = 0 & \text{on } \partial B_r. \end{cases} \quad (5.33)$$

Thus we deduce, using theorem 3.1.7, that

$$\|df_{r,j}\|^2_{L^2(B_r)} \leq \frac{3}{16\pi} \|de^j_{r,1}\|^2_{L^2(B_r)} \|de^j_{r,2}\|^2_{L^2(B_r)} \tag{5.34}$$

which implies, summing the inequalities over j, that

$$2\|df_r\|_{L^2(B_r)} \leq \sqrt{\frac{3}{16\pi}} (\|de_{r,1}\|^2_{L^2(B_r)} + \|de_{r,2}\|^2_{L^2(B_r)}). \tag{5.35}$$

Step 3 A PRIORI ESTIMATE FOR THE ENERGY OF f_r
We complete the basis $b_r = (e_{r,1}, e_{r,2})$ to obtain a moving frame of $\mathbb{R}^n$, $e_r = (e_{r,1}, e_{r,2}, e_{r,3}, \ldots, e_{r,n})$, and write $\omega^i_{r,j} = \langle de_{r,j}, e_{r,i} \rangle$.

From the identities

$$de_{r,1} = \sum_{k=1}^{n} \omega^k_{r,1} e_k \tag{5.36}$$

and

$$de_{r,2} = \sum_{k=2}^{n} \omega^k_{r,2} e_k \tag{5.37}$$

we deduce that

$$\begin{aligned}|de_{r,1}|^2 + |de_{r,2}|^2 &= 2|\omega^1_2|^2 + \sum_{k=1}^{2}\sum_{j=3}^{n} |\omega^j_{r,k}|^2 \\ &= 2|\omega^1_2|^2 + |du|^2 \end{aligned} \tag{5.38}$$

and hence that

$$\|de_{r,1}\|^2_{L^2(B_r)} + \|de_{r,2}\|^2_{L^2(B_r)} = 2\|df_r\|^2_{L^2(B_r)} + \|du\|^2_{L^2(B_r)}. \tag{5.39}$$

Eliminating $\|db_r\|^2_{L^2(B_r)}$ between (5.35) and (5.39), we obtain

$$2\|df_r\|_{L^2(B_r)} \leq \sqrt{\frac{3}{16\pi}} (2\|df_r\|^2_{L^2(B_r)} + \|du\|^2_{L^2(B_r)}). \tag{5.40}$$

Hence, letting $t = \|df_r\|_{L^2(B_r)}$, we notice that t is a solution of

$$P(t) = 2t^2 - 8\sqrt{\frac{\pi}{3}} t + \|du\|^2_{L^2(B_r)} \geq 0. \tag{5.41}$$

The reduced discriminant of P is $\Delta' = \frac{16\pi}{3} - 2\|du\|^2_{L^2(B_r)}$. It follows that

- if $\|du\|^2_{L^2(B_r)} \geq \frac{8\pi}{3}$, P is always non-negative, and (5.41) gives no information.
- if $\|du\|^2_{L^2(B_r)} < \frac{8\pi}{3}$, the polynomial P takes negative values between its two roots $\alpha < \beta$, and it follows from (5.41) that

$$\|df_r\|_{L^2(B_r)} = t \in [0, \alpha] \cup [\beta, +\infty), \tag{5.42}$$

where $\alpha = \sqrt{\frac{4\pi}{3}} - \sqrt{\frac{4\pi}{3} - \frac{1}{2}\|du\|^2_{L^2(B_r)}}$, and $\beta = \sqrt{\frac{4\pi}{3}} + \sqrt{\frac{4\pi}{3} - \frac{1}{2}\|du\|^2_{L^2(B_r)}}$.

The purpose will now be to show that, in the second case, we have $\|df_r\|_{L^2(B_r)} < \alpha$.

Step 4 THE FUNCTION $r \longmapsto \|df_r\|^2_{L^2(B_r)}$ IS CONTINUOUS
We define the function $g : [0, 1] \longrightarrow \mathbb{R}$ by

$$\begin{aligned} g(0) &= 0 \\ g(r) &= \inf_{H^1(B_r, \mathbb{R})} F_r(\theta) = \|df_r\|^2_{L^2(B_r)}, \text{ if } r > 0 \end{aligned} \tag{5.43}$$

and show that g is continuous.

In order to achieve this, we will show that g is equal to the infimum of a uniformly Lipschitz family of functions.

For any $r \in [0, 1]$, we define $u_r \in \mathcal{C}^\infty(\overline{B^2}, Gr_2(\mathbb{R}^n))$ by

$$u_r(z) = u(rz), \forall z \in \overline{B^2}.$$

We will denote $S_r \in \mathcal{C}^\infty(\overline{B^2}, O(n))$, the map which to each z associates the orthogonal symmetry of $\mathbb{R}^n$ around $u_r(z)$.

Lemma 5.1.7 *There exists a constant $C_0 > 0$, depending only on u, such that for any $r_0 \in [0, 1]$ and any basis $b^{r_0} = (e_1^{r_0}, e_2^{r_0})$, a smooth section of $u_{r_0}^* \mathcal{F}$ over $\overline{B^2}$, there exists a family $b_r^{r_0} = (e_{r,1}^{r_0}, e_{r,2}^{r_0})$ of sections of $u_r^* \mathcal{F}$, parametrized by $r \in [0, 1]$, such that*

$$b_{r_0}^{r_0} = b^{r_0},$$

and

$$g_{b^{r_0}}(r) := \int_{B^2} |\langle de_{r,1}^{r_0}, e_{r,2}^{r_0}\rangle|^2 dxdy$$

is a Lipschitz function such that

$$|g_{b^{r_0}}(r) - g_{b^{r_0}}(r')| \leq C_0(1 + ||db^{r_0}||_{L^2})|r - r'|. \tag{5.44}$$

Proof We will construct $b_r^{r_0}$ by parallel transport along the deformation of u_{r_0} into u_r defined by $t \longmapsto u_{(1-t)r_0 + tr}$. For $r_0, r \in [0,1]$, let

$$R_r^{r_0} : B \longrightarrow SO(n)$$

be the unique solution of the system of equations

$$R_{r_0}^{r_0}(z) = \mathbb{1}, \ \forall z \in B, \tag{5.45}$$

$$\frac{\partial R_r^{r_0}}{\partial r}(z) = -\frac{1}{2} S_r^{-1}(z) \cdot \frac{\partial S_r}{\partial r}(z) \cdot R_r^{r_0}(z), \ \forall z \in B. \tag{5.46}$$

We can easily check, using (5.46), that $R_r^{r_0}$ takes values in $O(n)$, and that $(R_r^{r_0})^{-1} S_r R_r^{r_0}$ does not depend on r. Hence, taking into account (5.45), we deduce that $S_r R_r^{r_0} = R_r^{r_0} S_{r_0}$. This implies that

$$b_r^{r_0}(z) := (R_r^{r_0}(z)(e_1^{r_0}), R_r^{r_0}(z)(e_2^{r_0}))$$

is an orthonormal basis of $u_r(z)$.

We may then show that for every $r, r' \in [0,1]$,

$$|g_{b^{r_0}}(r) - g_{b^{r_0}}(r')|$$
$$\leq \left(2||db^{r_0}||_{L^2} + ||dR_r^{r_0}||_{L^2} + ||dR_{r_0}^{r_0}||_{L^2}\right) ||{}^tR_r^{r_0} dR_r^{r_0} - {}^tR_{r'}^{r_0} dR_{r'}^{r_0}||_{L^2},$$

which implies (5.44), using the fact that $S_r(z)$, and hence also $R_r^{r_0}(z)$, are C^∞ functions in all the variables $z \in \overline{B^2}$ and $r, r_0 \in [0, 1]$. □

A corollary of the preceding lemma is that

$$g(r) = \inf_{r_0, b^{r_0}} g_{b^{r_0}}(r).$$

Notice that, thanks to the existence of a section $\widetilde{b}$ of $u^*\mathcal{F}$ such that $M := ||d\widetilde{b}||_{L^2} < +\infty$, we know that $g(r) \leq M$, and hence that

$$g(r) = \inf_{\substack{r_0, b^{r_0} \\ ||db^{r_0}||_{L^2} \leq M}} g_{b^{r_0}}(r).$$

We leave to the reader the task of showing, using this identity and (5.44), that $|g(r) - g(r')| \leq C_0(1 + M)|r - r'|$.

Step 5 ESTIMATES FOR THE ENERGY OF f_r AND b_r
It follows from Step 3 that if

$$\|du\|_{L^2(B)}^2 \leq \frac{8\pi}{3} \tag{5.47}$$

then, $\forall r \in [0,1]$, $\|df_r\|_{L^2(B_r)} \in [0,\alpha] \cup [\beta,+\infty)$, where $\alpha < \beta$. Moreover, by Step 4, $r \longmapsto \|df_r\|_{L^2(B_r)}$ is a continuous function, which vanishes at 0. This implies for all $r \in [0,1]$

$$\|df_r\|_{L^2(B)} < \sqrt{\frac{4\pi}{3}} - \sqrt{\frac{4\pi}{3} - \frac{1}{2}\|du\|_{L^2(B_r)}^2}. \tag{5.48}$$

Going back to the identity (5.39), for $r = 1$, we now obtain

$$\|db\|_{L^2(B)}^2 \leq a(\|du\|_{L^2(B)}^2), \tag{5.49}$$

where $a(t) = \frac{16\pi}{3}\left(1 - \sqrt{1 - \frac{3}{8\pi}\|du\|_{L^2(B)}^2}\right)$.

Step 6 CONCLUSION
Suppose $u \in H^1(B, Gr_2(\mathbb{R}^n))$ is such that $\|du\|_{L^2(B)}^2 \leq \frac{8\pi}{3}$. Then (see [145]) we can construct a smooth approximating sequence $u_\epsilon \in H^1(B, G_{r_2}(\mathbb{R}^n)) \cap C^\infty$, such that

$$\|du_\epsilon\|_{L^2(B)}^2 < \frac{8\pi}{3}, \quad \forall \epsilon > 0 \tag{5.50}$$

and

$$u_\epsilon \to u \in H^1(B), \quad \text{when} \quad \epsilon \to 0. \tag{5.51}$$

By Step 4, it is possible to construct, for each $\epsilon > 0$, a frame b_ϵ, a section of $u_\epsilon^* \mathcal{F}$, whose energy is bounded according to (5.49). Using the weak compactness in $H^1(B)$, and the compact injection of $H^1(B)$ in $L^2(B)$, it is then simple to show that there exists a subsequence ϵ' of ϵ, and a map $(e_1, e_2) \in H^1(B, \mathbb{R}^n \times \mathbb{R}^n)$, such that

$$(e_{\epsilon',1}, e_{\epsilon',2}) \rightharpoonup (e_1, e_2) \text{ weakly in } H^1(B) \tag{5.52}$$

$$(e_{\epsilon',1}, e_{\epsilon',2}) \to (e_1, e_2) \text{ in } L^2(B) \text{ and a.e.} \tag{5.53}$$

$$\|d(e_1, e_2)\|_{L^2(B)}^2 \leq \liminf_{\epsilon' \to 0} \|d(e_{\epsilon',1}, e_{\epsilon',2})\|_{L^2(B)}^2. \tag{5.54}$$

5.1 Local results

In particular, the a.e. convergence implies that (e_1, e_2) is a section of $u^*\mathcal{F}$, and (5.54) implies that its energy is bounded.

Step 7 PROOF OF (5.24)
Using (5.52) an (5.53), it is easy to show that

$$\langle e_{\epsilon,1}, de_{\epsilon,2}\rangle \to \langle e_1, de_2\rangle \quad \text{in } \mathcal{D}'(B) \tag{5.55}$$

and hence, by (5.28), (e_1, e_2) satisfies

$$d(\star\langle e_1, de_2\rangle) = 0. \tag{5.56}$$

This concludes the proof of lemma 5.1.4. $\square$

Proof of theorem 5.1.1 Let $\gamma \leq \frac{8\pi}{3}$, and choose any $X \in \overline{\mathcal{C}_\gamma}$. Then, there exists a sequence of conformal immersions $(X_k)_{k\in\mathbb{N}}$, in $C^\infty(B, \mathbb{R}^n)$, such that

$$X_k \rightharpoonup X \quad \text{weakly in } H^1(B, H) \tag{5.57}$$

and

$$\int_B |d(u \circ X_k)|^2 dxdy \leq \frac{8\pi}{3}. \tag{5.58}$$

From lemma 5.1.4 and (5.58), it follows that it is possible to construct, for each k, a section $b_k := (e_{1,k}, e_{2,k})$ of $(u \circ X_k)^*\mathcal{F}$, whose energy is bounded by a $\left(\frac{8\pi}{3}\right)$. Moreover, the b_k are solutions of (5.56).
Since each immersion X_k is conformal, $\exists f_k, \theta_k \in C^\infty(\overline{B}, \mathbb{R})$ such that

$$dX_k = e^{f_k}[(\cos\theta_k e_{k,1} + \sin\theta_k e_{k,2})dx \\ + (-\sin\theta_k e_{k,1} + \cos\theta_k e_{k,2})dy]. \tag{5.59}$$

In particular, projecting the equation $d^2 X_k = 0$ along $e_{k,1}$ and $e_{k,2}$, we see that

$$\begin{cases} \dfrac{\partial \theta_k}{\partial x} + \dfrac{\partial f_k}{\partial y} = \omega^1_{k,2}\left(\dfrac{\partial}{\partial x}\right) \\ \dfrac{\partial \theta_k}{\partial y} - \dfrac{\partial f_k}{\partial x} = \omega^1_{k,2}\left(\dfrac{\partial}{\partial y}\right), \end{cases} \tag{5.60}$$

where $\omega_{k,2}^1 = \langle de_{k,2}, e_{k,1} \rangle$.

Step 1 ESTIMATES FOR f_k AND θ_k

Equation (5.60) has several consequences. On the one hand, because of (5.58), we have that

$$\Delta \theta_k = 0. \tag{5.61}$$

On the other hand,

$$-\Delta f_k = \{e_{k,1}.e_{k,2}\} = \sum_{j=1}^n \{e_{k,1}^j, e_{k,2}^j\}. \tag{5.62}$$

Decompose $f_k = f_{k,0} + f_{k,1} + \cdots + f_{k,n}$, where

$$\begin{cases} -\Delta f_{k,0} = 0 & \text{in } B \\ f_{k,0} = f_k & \text{on } \partial B \end{cases} \tag{5.63}$$

and $f_{k,j} = \widehat{(e_{k,1}^j)\,(e_{k,2}^j)}$. By theorem 3.1.2, we have that

$$\left\| \sum_{j=1}^n f_{k,j} \right\|_{L^\infty(B)} + \left\| \sum_{j=1}^n df_{k,j} \right\|_{L^2(B)} \le C \|de_{k,1}\|_{L^2(B)} \|de_{k,2}\|_{L^2(B)}. \tag{5.64}$$

ESTIMATE FOR $f_{k,0}$

To estimate $f_{k,0}$, we start by using the fact that $f_{k,0}$ is smooth harmonic in B and by applying the Poisson formula over every ball $B_r \subset B$, where $r \in (0,1]$. This gives

$$\forall z \in B_r, \quad rf_{k,0}(z) = \frac{1}{2\pi} \int_{\partial B_r} f_{k,0}(v) \frac{r^2 - |z|^2}{|v - z|^2} ds(v). \tag{5.65}$$

Integrating this formula for $1 - \delta < r < 1$, we obtain

$$\forall z \in B_r, \quad f_{k,0}(z) = \frac{1}{\pi(2-\delta)\delta} \int_{B \setminus B_{1-\delta}} f_{k,0}(v) \frac{|v|^2 - |z|^2}{|v-z|^2} dv^1 dv^2. \tag{5.66}$$

The second fact we will use is that if $f_{k,0}^+ = \sup(f_{k,0}, 0)$, then

$$\int_B f_{k,0}^+ dx dy \text{ is uniformly bounded in } k. \tag{5.67}$$

To prove (5.67), we remark that

$$\int_B e^{2f_k} dx dy = \frac{1}{2}\|dX_k\|^2_{L^2(B)} \tag{5.68}$$

is bounded, and that, by (5.64), $\lambda := \inf_{k \in \mathbb{N}} \sum_{j=1}^{\infty} f_{k,j} > -\infty$. It follows that

$$\int_B f_{k,0}^+ dx dy \leq \frac{1}{2e^{1+2\lambda}} \int_B e^{2(f_{k,0}+\lambda)} dx dy$$

$$\leq \frac{1}{2e^{1+2\lambda}} \int_B e^{2f_k} dx dy$$

$$= \frac{1}{4e^{1+2\lambda}} \|dX_k\|^2_{L^2(B)} \tag{5.69}$$

which, since X_k converges weakly to X in $H^1(B)$, implies (5.67).

We now derive the consequences of (5.66) and (5.67). First of all, for each compact $K \subset B$, there exists a constant $B(K)$ such that

$$\forall z \in K, \quad f_{k,0}(z) \leq B(K). \tag{5.70}$$

In fact, to see this it suffices to choose $\delta > 0$ such that $K \subset B_{1-2\delta}$, and to use (5.66). Then,

$$\forall z \in K, \quad f_{k,0}(z) \leq \frac{1}{\pi(2-\delta)\delta} \int_B f_{k,0}^+ dx dy \tag{5.71}$$

which yields (5.70).

Next, there are the following two alternatives.

(i) (Collapse) There exists a subsequence k' of k such that

$$\lim_{k' \to +\infty} \int_B f_{k',0}^- dx dy = -\infty, \tag{5.72}$$

where $f_{k',0}^- = \inf(f_{k',0}, 0)$. Then,

$$f_{k',0}(0) = \frac{1}{\pi} \int_B f_{k',0}(v) dv^1 dv^2 \to -\infty. \tag{5.73}$$

But since by (5.67) we also have that

$$f_{k',0}(0) = \frac{1}{\pi(2-\delta)\delta} \int_{B \setminus B_{1-\delta}} f_{k',0}(v) dv^1 dv^2. \tag{5.74}$$

We deduce that for all $\delta \in (0,1)$, $\int_{B \setminus B_{1-\delta}} f^-_{k',0}(v) dv^1 dv^2$ tends to $-\infty$, and hence, by (5.66) and (5.67), $f_{k',0}$ converges uniformly on every compact subset of $B_{1-\delta}$ to $-\infty$, and hence, δ being arbitrary, also on every compact $K \subset B$. Since $X_{k'} \rightharpoonup X$ in H^1, it follows that

$$\|dX\|^2_{L^2(K)} \leq \liminf_{k' \to \infty} \|dX_{k'}\|^2_{L^2(K)} = \liminf_{k' \to \infty} \int_K e^{2f_{k'}} dx dy = 0. \tag{5.75}$$

And this being true for every compact $K \subset B$, we deduce that $\|dX\|_{L^2(B)} = 0$, which means that X is a constant map.

(ii) $\int_B f^-_{k,0} dx dy$ is bounded and, by (5.66), this implies that for any compact $K \subset B$, $\exists A(K)$,

$$\forall z \in K, \quad A(K) \leq f_{0,k}(z). \tag{5.76}$$

From now on, we will exclude the case where X is the constant map. This implies that uniform lower bound over every compact (5.76) is valid. We remark that, since $f_{0,k}$ is harmonic, estimates (5.70) and (5.76) also imply that $\|df_{0,k}\|_{L^2(K)}$ will be a bounded function of k, for any compact $K \subset B$. Putting together this estimate and (5.64), we conclude that

$$\forall \text{ compact } K \subset B, \quad \|df_k\|_{L^2(K)} + \|f_k\|_{L^\infty(K)} \text{ is bounded.} \tag{5.77}$$

Now, since b_k is bounded in $H^1(B)$, $\omega^1_{k,2}$ is bounded in L^2, and hence, by (5.60) and (5.77), we have

$$\forall \text{ compact } K \subset B, \quad \|d\theta_k\|_{L^2(K)} \text{ is bounded.} \tag{5.78}$$

Step 2 PASSING TO THE LIMIT

Chose a compact $K \subset B$. Using the previous estimates and standard theorems on Sobolev spaces, we may extract a subsequence k' of k, such that

$$(b_{k'}, \theta_{k'}, f_{k'}) \rightharpoonup (b, \theta, f) \text{ weakly in } H^1(K) \tag{5.79}$$

$$(b_{k'}, \theta_{k'}, f_{k'}) \to (b, \theta, f) \text{ in } L^2(K) \tag{5.80}$$

$$(b_{k'}, \theta_{k'}, f_{k'}) \to (b, \theta, f) \text{ a.e. in } B. \tag{5.81}$$

Together with (5.77), this yields that we can pass to the limit in equation (5.59), which gives

$$dX = e^f [(\cos \theta e_1 + \sin \theta e_2)dx + (-\sin \theta e_1 + \cos \theta e_2)dy] \tag{5.82}$$

and because of the a.e. convergence (5.81), (e_1, e_2) will be a section of $(u \circ X)^*\mathcal{F}$.

This implies that X is a conformal immersion satisfying

$$e^{\inf_K f} \leq |dX| \leq e^{\sup_K f} \text{ on } K. \tag{5.83}$$

The pull-back by X of the Euclidean metric on $\mathbb{R}^n$ is $e^{2f}[(dx)^2 + (dy)^2]$. This metric is continuous since f is a solution of

$$-\Delta f = \{e_1 \cdot e_2\}, \tag{5.84}$$

and hence is continuous by Wente's lemma. This concludes the proof of theorem 5.1.1. □

5.2 Global results

We will now deduce, using the results of the previous section, more global versions of theorem 5.1.1 and lemma 5.1.4, without the hypothesis that the Gauss map's energy is bounded by a critical constant M (as, for instance, $M = \frac{8\pi}{3}$). However, we should be careful since, as is shown by example 5.2.2 below, if this energy is bigger than 8π (which corresponds more or less to the case where u "has the possibility to cover the sphere S^2 once"), then new phenomena may show up, and we can no longer control the energy of the moving frame. We start by a direct consequence of lemma 5.1.4.

Theorem 5.2.1 *Let Ω be a simply connected open subset of $\mathbb{C}$. Let $u \in H^1(\Omega, Gr_2(\mathbb{R}^n))$; then there is a moving frame b, a section of $u^*\mathcal{F}$, belonging to $H^1_{loc}(\Omega, \mathbb{R}^n \times \mathbb{R}^n)$.*

Proof

Step 1 COVERING OF Ω

We cover Ω by a finite family of simply connected open sets $U_1, ..., U_p$, such that for each $j = 1, ..., p$,

$$\int_{U_j} |du|^2 dx dy \leq \frac{8\pi}{3}. \tag{5.85}$$

To construct this covering we may, for instance, consider the sequence of compact $(K_m)_{m \in \mathbb{N}}$, such that $K_m \subset \Omega$, $\forall m \in \mathbb{N}$, and

$$K_l \text{ is contained in the interior of } K_m, \text{ if } l < m, \tag{5.86}$$

and $\lim_{l \to +\infty} K_l = \Omega$. There exists m such that

$$\int_{K_m} |du|^2 dx dy > \int_{\Omega} |du|^2 dx dy - \frac{8\pi}{3}. \tag{5.87}$$

Then we choose as the first open set, $V_0 = \Omega \setminus K_m$. Next, we cover the compact set K_{m+1} by open balls $V_1, ..., V_q$ of Ω such that $\int_{V_j} |du|^2 dx dy < \frac{8\pi}{3}$. In case certain open subsets are not diffeomorphic to balls, we cover them by smaller simply connected open sets. In this way we obtain a family $U_1, ..., U_p$ satisfying (5.85). Denote by u_j the restriction of u to U_j.

Step 2 WORK IN EACH U_j

Each open set U_j is conformally diffeomorphic to the ball B. Thus, we can apply lemma 5.1.4 to each of these open sets (the L^2 norms of the derivatives being invariant under conformal transformations), thanks to condition (5.85). We obtain that, for $j = 1, ..., p$, there exists a section $b_j := (e_{j,1}, e_{j,2})$ of $u_j^* \mathcal{F}$ with bounded energy. Define $U_{ij} = U_i \cap U_j$. Over each non-empty U_{ij}, there exists $\theta_j^i \in H^1(U_{ij}, \mathbb{R})$ such that

$$\text{in } U_{ij}, \quad b_j = b_i R_j^i, \quad \text{where } R_j^i = \begin{pmatrix} \cos \theta_j^i & -\sin \theta_j^i \\ \sin \theta_j^i & \cos \theta_j^i \end{pmatrix}. \tag{5.88}$$

This equation implies that

$$\langle de_{j,2}, e_{j,1} \rangle = \langle de_{i,2}, e_{i,1} \rangle - d\theta_j^i. \tag{5.89}$$

And hence, since $d(\star \langle de_{j,2}, e_{j,1} \rangle) = d(\star \langle de_{i,2}, e_{i,1} \rangle) = 0$, θ_j^i is a solution of

$$\Delta \theta_j^i = 0 \quad \text{in } U_{ij}. \tag{5.90}$$

Notice that $\theta_j^i + \theta_i^j = 0$ in U_{ij}, and that in every $U_{ijk} = U_i \cap U_j \cap U_k$, the θ_j^i satisfy the cocycle condition

$$\theta_j^i + \theta_k^j + \theta_i^k = 0. \tag{5.91}$$

Step 3 GLUING
We use an elementary result of sheaf theory: the Čech cohomology group of the sheaf of real harmonic functions over Ω is trivial, since Ω is a simply connected (see [73]). In our case this implies, by (5.91), that over each U_j there exists a real harmonic function h_j such that

$$\theta_j^i = h_i - h_j \quad \text{in } U_{ij}. \tag{5.92}$$

Moreover, since each $\theta_j^i \in H^1(U_{ij}, \mathbb{R})$, we deduce that for any compact subset K of Ω,

$$h_{j|K} \in H^1(U_j \cap K, \mathbb{R}). \tag{5.93}$$

Using (5.92), we may now rewrite (5.88) as

$$\text{in } U_{ij}, \quad b_j \begin{pmatrix} \cos h_j & -\sin h_j \\ \sin h_j & \cos h_j \end{pmatrix} = b_i \begin{pmatrix} \cos h_i & -\sin h_i \\ \sin h_i & \cos h_i \end{pmatrix} \tag{5.94}$$

and thus, over Ω we can define the section b of $u^*\mathcal{F}$ by

$$b := b_i \begin{pmatrix} \cos h_i & -\sin h_i \\ \sin h_i & \cos h_i \end{pmatrix} \quad \text{in } U_i, \tag{5.95}$$

and b has a finite energy over every compact K in Ω. □

Example 5.2.2 THE STEREOGRAPHIC PROJECTION
We will see in this example that the constant γ in inequality (5.21) of lemma 5.1.4 cannot be bigger than 8π, and that the function $\underline{a}$ such that

$$\inf_{\substack{b \text{ section of } u^*\mathcal{F} \\ \|du\|^2_{L^2(B)} = t}} \|db\|^2_{L^2(B)} = \underline{a}(t) \tag{5.96}$$

cannot be smaller than $\underline{a}(t) = -8\pi \log\left(1 - \frac{t}{8\pi}\right)$.

Define on $B_r \subset \mathbb{C}$, $u(z) = \dfrac{1}{1+|z|^2} \begin{pmatrix} 2x \\ 2y \\ 1-|z|^2 \end{pmatrix}$. This map takes its values in S^2, and is conformal. The inverse image of the standard metric on S^2 by u is

$$\frac{4}{(1+|z|^2)^2}[(dx)^2 + (dy)^2], \tag{5.97}$$

and

$$\|du\|^2_{L^2(B_r)} = 16\pi \int_0^R \frac{rdr}{(1+r^2)^2} = \frac{8\pi R^2}{1+R^2}. \tag{5.98}$$

Let $b = (e_1, e_2)$ be a minimizing section of $u^*\mathcal{F}$. We necessarily have, denoting by ω the volume form on S^2, that coincides with the curvature form of the bundle $\mathcal{F}$,

$$d\langle de_2, e_1\rangle = u^*\omega = \frac{4}{(1+|z|^2)^2}dx \wedge dy. \tag{5.99}$$

We integrate this identity over the ball B_r, for $0 < r < R$,

$$\begin{aligned}\frac{4\pi r^2}{1+r^2} &= \int_{B_r} u^*\omega \\ &= \int_{\partial B_r} \langle de_2, e_1\rangle \\ &\leq \sqrt{\int_{\partial B_r} ds} \sqrt{\int_{\partial B_r} |\langle de_2, e_1\rangle|^2 ds}. \end{aligned} \tag{5.100}$$

Recall that $df = \star\langle de_2, e_1\rangle$. Therefore, we deduce from inequality (5.100) that

$$\int_{\partial B_r} |df|^2 ds \geq \frac{8\pi r^3}{(1+r^2)^2}. \tag{5.101}$$

Integrating this inequality over $[0, R]$, we obtain

$$\begin{aligned}\int_{B_r} |df|^2 dxdy &\geq \int_0^R \frac{8\pi r^3 dr}{(1+r^2)^2} \\ &= 4\pi \log(1+R^2) - \frac{4\pi R^2}{1+R^2}.\end{aligned} \tag{5.102}$$

Hence, for any section b of $u^*\mathcal{F}$,

$$\|db\|^2_{L^2(B_r)} = 2\|df\|^2_{L^2(B_r)} + \|du\|^2_{L^2(B_r)} \geq 8\pi \log(1+R^2). \quad (5.103)$$

Eliminating R between (5.98) and (5.103), we see that

$$\|db\|^2_{L^2(B_r)^2} \geq -8\pi \log\left(1 - \frac{1}{8\pi}\|du\|^2_{L^2(B_r)}\right), \quad (5.104)$$

so that the function a cannot be smaller than $\underline{a}(t) = -8\pi \log\left(1 - \frac{t}{8\pi}\right)$. On the other hand, this also proves that the constant γ cannot be bigger than 8π, since here $\lim_{t \to 8\pi} \underline{a}(t) = +\infty$.

Notice that inequality (5.103) is optimal since for

$$\begin{cases} e_1 = \dfrac{1+|z|^2}{2} du \begin{pmatrix} 1 \\ 0 \end{pmatrix} = \begin{pmatrix} 1 \\ 0 \\ 0 \end{pmatrix} - \dfrac{2x}{1+|z|^2} \begin{pmatrix} x \\ y \\ 1 \end{pmatrix} \\[2em] e_2 = \dfrac{1+|z|^2}{2} du \begin{pmatrix} 0 \\ 1 \end{pmatrix} = \begin{pmatrix} 0 \\ 1 \\ 0 \end{pmatrix} - \dfrac{2y}{1+|z|^2} \begin{pmatrix} x \\ y \\ 1 \end{pmatrix} \end{cases} \quad (5.105)$$

we obtain equality in (5.103).

In fact, this example seems to be the most "efficient" to make $\|db\|^2_{L^2(B_r)}$ big, while keeping $\|du\|^2_{L^2(B_r)}$ constant. This leads us to the following problem.

Conjecture 5.2.3 *Prove that for any $u \in H^1(B, Gr_2(\mathbb{R}^n))$ such that $\|du\|^2_{L^2(B_r)} < 8\pi$, there exists a frame b in $H^1(B, \mathbb{R}^n \times \mathbb{R}^n)$, a section of $u^*\mathcal{F}$, such that*

$$\|db\|^2_{L^2(B_r)} \leq \underline{a}(\|du\|^2_{L^2(B_r)}), \quad (5.106)$$

where $\underline{a}(t) = -8\pi \log\left(1 - \frac{t}{8\pi}\right)$.

Let us come back to conformal immersions:

Theorem 5.2.4 *Suppose that S is a connected Riemannian surface, and $X \in H^1(S, \mathbb{R}^n)$ is such that there exists a sequence $(X_k)_{k \in \mathbb{N}}$ in $C^\infty \cap H^1(S, \mathbb{R}^n)$ satisfying the following hypotheses.*

(i) For each k, X_k is a conformal immersion.

(ii) For any $x \in S$, there is a ball $B(x, \rho_x)$ in S, such that $\forall k \in \mathbb{N}$,

$$\int_{B(x,\rho_x)} |A_k|^2 dx dy < \frac{8\pi}{3}. \tag{5.107}$$

(iii) $X_k \rightharpoonup X$, weakly in $H^1(S, \mathbb{R}^n)$.

Then X is either a constant map or a bilipschitz conformal immersion.

Proof Applying theorem 5.1.1 over each ball $B(x, \rho_x)$, we obtain right away that either $X_{|B(x,\rho_x)}$ is a constant map, or $X_{|B(x,\rho_x)}$ is a bilipschitz conformal immersion. Hence, we deduce that the set

$$\{x \in S \mid dX = 0\} \tag{5.108}$$

is both open and closed. Since S is connected, this yields our result. □

Remark 5.2.5 *Again, we can ask what happens when we do not control $\|A_k\|_{L^2}$ locally, as in the preceding result, but just assume that $\|A_k\|_{L^2}^2 \leq C$. Things are not clear. For instance, it could happen that the immersion X thus obtained has branch points, but this is not certain.*

5.3 Willmore surfaces

An important motivation for the preceding results is the study of immersed surfaces in $\mathbb{R}^3$, which are critical points of the functional

$$W(S) = \int_S H^2 d\mathcal{H}^2, \tag{5.109}$$

where $H = \frac{k_1 + k_2}{2}$ is the mean curvature and $d\mathcal{H}^2$ the 2-dimensional Hausdorff measure. This functional was proposed by Tom Willmore in 1960, and is also called the Willmore functional [182]. Its critical points are called Willmore surfaces, and satisfy the equation

$$\Delta H + 2H(H^2 - K) = 0, \tag{5.110}$$

where $K = k_1 k_2$ is the Gauss curvature. Notice that in dimension 3, we have

5.3 Willmore surfaces

$$H^2 = \frac{|A|^2}{4} + \frac{K}{2}, \tag{5.111}$$

where A is the second fundamental form. Thus, if the surface S has no boundary and genus $g \in \mathbb{N}$, we have, by the Gauss–Bonnet theorem,

$$W(S) = \frac{1}{4} \int_S |A|^2 d\mathcal{H}^2 + 2\pi(1-g), \tag{5.112}$$

and consequently, the problem is, from the variational point of view, equivalent to the study of the critical points of $\int_S |A|^2 d\mathcal{H}^2$. A natural question is to consider, for each $g \in \mathbb{N}$, the set $\mathcal{E}_g$ of smooth compact surfaces of genus g, without boundary and smoothly embedded in $\mathbb{R}^3$, and, inside each class $\mathcal{E}_g$, to look for the surfaces which minimize W. For $g = 0$, it is easy to show that all the round spheres in $\mathbb{R}^3$ minimize W in $\mathcal{E}_0$, using the identity

$$W(S) = \int_S \left(\frac{k_1 - k_2}{2}\right)^2 d\mathcal{H}^2 + \int_S K d\mathcal{H}^2. \tag{5.113}$$

For $g = 1$, Tom Willmore proved that the torus of revolution obtained by making a unit circle turn around a straight line contained in the plane of the circle, and located a distance $\sqrt{2}$ from the center of the circle, is a Willmore surface. This torus was given the name of its author, who conjectured that it minimizes W in $\mathcal{E}_1$.

In spite of some partial answers (see, for instance, [107] and [115]), this conjecture has still not been proved. On the other hand, a deep work of Leon Simon [154] shows the existence of a smooth surface in $\mathcal{E}_1$ minimizing W. However, we still do not know if this torus coincides with that of Willmore.

To conclude, we mention that this problem had been considered by Thomsen and Shadow in 1923 [168], Blaschke in 1929 [16] and Konrad Voss in 1950.

For a more detailed presentation of Willmore surfaces, see the last chapter of [183]. A study of Willmore surfaces in the framework of integrable systems and loop groups, like the theory for harmonic maps presented in the second chapter of this book, was made in [86].

5.4 Epilogue: Coulomb frames and conformal coordinates

As we saw at the beginning of this chapter, the existence of a Coulomb orthonormal frame field is closely connected to the existence of conformal coordinates on the same surface.

The following result sheds some light onto this connection.

Lemma 5.4.1 *Let Ω be an open subset of a smooth Riemannian surface. Then, for any Lipschitz orthonormal frame field (e_1, e_2) on Ω, and any function $f \in W^{1,\infty}(\Omega, \mathbb{R})$, we have, writing $\omega_2^1 = \langle \nabla e_2, e_1 \rangle$,*

$$[e^f e_1, e^f e_2] = e^{2f}[(\star\omega_2^1 - df)(e_2)e_1 - (\star\omega_2^1 - df)(e_1)e_2]. \quad (5.114)$$

Consequently, if Ω is simply connected, we have the equivalence (i) $\iff$ (ii), where

(i) *There exists a conformal diffeomorphism u, in $W_{loc}^{2,\infty}$, from an open subset ω in $\mathbb{R}^2$ to Ω, such that*

$$\begin{cases} \dfrac{\partial u}{\partial x} = \left|\dfrac{\partial u}{\partial x}\right| e_1(u) \\ \dfrac{\partial u}{\partial y} = \left|\dfrac{\partial u}{\partial y}\right| e_2(u). \end{cases} \quad (5.115)$$

(ii) *(e_1, e_2) is an orthonormal frame field belonging to $W_{loc}^{1,\infty}$ (i.e. locally Lipschitz) in Ω, such that*

$$d(\star\langle \nabla e_2, e_1 \rangle) = 0. \quad (5.116)$$

Proof
Step 1 We have

$$\begin{aligned}{}[e^f e_1, e^f e_2] &= e^{2f}([e_1, e_2] + df(e_1)e_2 - df(e_2)e_1) \\ &= e^{2f}[\nabla_{e_1} e_2 - \nabla_{e_2} e_1 + df(e_1)e_2 - df(e_2)e_1] \\ &= e^{2f}[(\omega_2^1(e_1) - df(e_2))e_1 + (\omega_2^1(e_2) + df(e_1))e_2]. \end{aligned} \quad (5.117)$$

We recall that if (α^1, α^2) is the basis of $T_m^*\Omega$, which is the dual of (e_1, e_2), we have

$$\star\alpha^1 = \alpha^2, \quad \star\alpha^2 = -\alpha^1, \quad (5.118)$$

5.4 Epilogue: Coulomb frames and conformal coordinates   245

which implies that

$$\star \omega_2^1 = \star(\omega_2^1(e_1)\alpha^1 + \omega_2^1(e_2)\alpha^2)$$
$$= \omega_2^1(e_1)\alpha^2 - \omega_2^1(e_2)\alpha^1. \tag{5.119}$$

It follows that

$$[e^f e_1, e^f e_2] = e^{2f}[(\star\omega_2^1(e_2) - df(e_2))e_1 + (-\star\omega_2^1(e_1) - df(e_1))e_2] \tag{5.120}$$

which implies (5.114).

Step 2 To show that (i) $\implies$ (ii), it suffices to apply (5.114) with $f = \log\left|\frac{\partial u}{\partial x}\right|$. This yields

$$\star\omega_2^1 = df \tag{5.121}$$

which implies (ii).

Step 3 Conversely, if (ii) is true, then there exists $f : \Omega \longrightarrow \mathbb{R}$ such that

$$df = \star\omega_2^1. \tag{5.122}$$

But since (e_1, e_2) is a pair of linearly independent locally Lipschitz vector fields, it is necessary that the metric g, for which (e_1, e_2) is an orthonormal basis, whose expression in local coordinates is

$$g_{kl} = (g^{ij})^{-1}, \quad \text{where } g^{ij} = e_1^i e_1^j + e_2^i e_2^j, \tag{5.123}$$

is itself locally Lipschitz and non-degenerate, and hence that the coefficients of the Levi-Civita connection ∇, associated to g, are locally bounded. Therefore,

$$\omega_2^1 = \langle \nabla e_2, e_1 \rangle \tag{5.124}$$

is also locally bounded, and (5.122) implies that f is a locally Lipschitz function on Ω. Thus, we consider the vector fields on Ω

$$X_1 = e^f e_1 \quad \text{and} \quad X_2 = e^f e_2. \tag{5.125}$$

By (5.114) and (5.122), X_1 and X_2 commute, i.e.

$$[X_1, X_2] = 0 \tag{5.126}$$

and moreover, these vector fields are locally Lipschitz. Hence, we can integrate them, and using the fact that Ω is simply connected, we deduce that there exists an unique map u from an open subset ω of $\mathbb{R}^2$ to Ω, such that

$$u(0,0) = m_0 \quad \text{(chosen point in } \Omega\text{)}$$

$$\frac{\partial u}{\partial x^1}(x) = X_1(u(x)) \tag{5.127}$$

$$\frac{\partial u}{\partial x^2}(x) = X_2(u(x)),$$

and hence u is a conformal diffeomorphism of class $W^{2,\infty}_{loc}$. □

Remark 5.4.2 *In the previous lemma, if we also suppose that there exists a constant $\delta > 0$ such that $\det(e_1, e_2) > \delta$ in Ω, then the result remains true if we replace all the locally Lipschitz quantities by globally Lipschitz ones.*

This result suggests that it should be possible to construct conformal coordinates over a Riemannian surface using "Coulomb" orthonormal moving frames. This idea is used in the following theorem, which is a sort of converse of the results in section 5.1.

Theorem 5.4.3 *Let (Ω, g) be an open subset of a Riemannian surface, whose non-degenerate metric written in some local coordinates satisfies*

  (i) *the coefficients of g belong to $H^1(\Omega) \cap L^\infty(\Omega)$;*
  (ii) *if K denotes the Gauss curvature of (Ω, g), then*

$$K\sqrt{\det g_{ij}} \in \mathcal{H}^1_{loc}(\Omega) \tag{5.128}$$

  (the local Hardy space).

Then, if Ω is simply connected, there exists a bilipschitz diffeomorphism from an open subset ω of $\mathbb{R}^2$ to (Ω, g), which is conformal.

Remark 5.4.4 *There is no difficulty in giving a sense to $K\sqrt{\det g_{ij}}$ if g is in $H^1 \cap L^\infty$. In fact, in this case, we can construct (by orthonormalizing) an orthonormal frame field (e_1, e_2) over Ω, which also belongs to $H^1 \cap L^\infty$, and we can define*

5.4 Epilogue: Coulomb frames and conformal coordinates

$$\omega_2^1 = g(\nabla e_2, e_1) = g_{ij}[e_1, e_2]^i dx^j, \tag{5.129}$$

which is clearly in L^2. Then, the equation

$$d\omega_2^1 = K\sqrt{\det g_{ij}}\, dx^1 \wedge dx^2 \tag{5.130}$$

enables us to calculate $K\sqrt{\det g_{ij}}$, and gives us that this quantity is in $H^{-1}(\Omega, \mathbb{R})$.

For a definition of the space $\mathcal{H}^1_{loc}(\Omega)$, see [157]. Notice that we can replace the hypothesis (5.128) by the hypothesis that $K\sqrt{\det g_{ij}}$ is the restriction to Ω of a $\mathcal{H}^1(\mathbb{R}^2)$ function.

Remark 5.4.5 *The following proof has some similarities with the one given by S.S. Chern [36]. Nevertheless, it does not involve integral equations in a complex variable, but elliptic estimates (although must admit that often, hidden behind an elliptic estimate, there is an estimate on a singular integral).*

Proof of theorem 5.4.3
Step 1 CONSTRUCTION OF A COULOMB FRAME
Starting from the simplest frame field, $(\frac{\partial}{\partial x^1}, \frac{\partial}{\partial x^2})$, we construct, using a standard algebraic procedure, a positively oriented orthonormal frame field for the metric g, which we denote by (e_1, e_2). Since $H^1 \cap L^\infty$ is an algebra and g is supposed to be non-degenerate, (e_1, e_2) also belongs to $H^1 \cap L^\infty$. For $\theta \in H^1(\Omega, \mathbb{R})$, we let

$$(e_{\theta,1}, e_{\theta,2}) = (e_1, e_2) \begin{pmatrix} \cos\theta & -\sin\theta \\ \sin\theta & \cos\theta \end{pmatrix}, \tag{5.131}$$

and write $\omega_{\theta,2}^1 = g(\nabla e_{\theta,2}, e_{\theta,2}) = g(\nabla e_2, e_1) - d\theta$. We choose, as in lemma 4.1.3, or lemma 5.1.4, a θ which minimizes the energy

$$\int_\Omega |\omega_{\theta,2}^1|^2 \sqrt{\det g_{ij}}\, dx^1 dx^2. \tag{5.132}$$

It follows that $\omega_{\theta,2}^1$ is of class L^2 in Ω, and is a solution of

$$\begin{cases} d(\star \omega_{\theta,2}^1) = 0 & \text{in } \Omega \\ \omega_{\theta,2}^1\left(\frac{\partial}{\partial n}\right) = 0 & \text{on } \partial\Omega, \end{cases} \tag{5.133}$$

where the Hodge $\star$ operator is that corresponding to the metric g. Henceforth, we denote by (e_1, e_2) the frame thus obtained.

Step 2 CONSTRUCTION OF A COMMUTING PAIR OF VECTORS
By (5.133), there exists a function $f \in H^1(\Omega, \mathbb{R})$, satisfying

$$\begin{cases} f = 0 & \text{on } \partial\Omega \\ df = \star\omega_2^1 & \text{in } \Omega. \end{cases} \qquad (5.134)$$

It follows from this equation that $-\sqrt{\det g_{ij}}\,\Delta_g f dx^1 \wedge dx^2 = d\omega_2^1$, and hence that

$$-\Delta_g f = K \quad \text{in } \Omega. \qquad (5.135)$$

Do not forget that the Laplacian used here is that for the metric g. This equation yields

$$-\sum_{i,j} \frac{\partial}{\partial x^i}\left(g^{ij}\sqrt{\det g^{kl}}\,\frac{\partial f}{\partial x^j}\right) = K\sqrt{\det g_{ij}} \quad \text{in } \Omega \qquad (5.136)$$

The r.h.s. of this equation belongs to local Hardy space. This implies that f is locally bounded and continuous. To show this we use a generalization of theorem 3.1.2 to the case of elliptic operators with non-constant coefficients which are bounded in L^∞; see [33] (see also [14]). Let

$$X_1 := e^f e_1 \quad \text{and} \quad X_2 := e^f e_2. \qquad (5.137)$$

We are tempted to integrate these two vector fields. However, there is a difficulty that shows up: although it is simple to check that X_1 and X_2 belong to $H^1 \cap L^\infty$, it is in general not true that these fields are Lipschitz, which poses a problem.

Step 3 INTEGRATION
Consider the basis of $T^*\Omega$, dual of (e_1, e_2), which we denote by (α^1, α^2). Again, it is easy to show that the 1-forms α^1 and α^2 are in $H^1 \cap L^\infty$. We let

$$A^1 = e^{-f}\alpha^1 \quad \text{and} \quad A^2 = e^{-f}\alpha^2. \qquad (5.138)$$

Likewise, A^1 and A^2 are in $H^1 \cap L^\infty$. Now we remark that since

5.4 Epilogue: Coulomb frames and conformal coordinates

$$df \wedge \alpha^1 = (\star df) \wedge (\star \alpha^1) = -\omega_2^1 \wedge \alpha^2, \quad (5.139)$$

we have

$$dA^1 = e^{-f}(d\alpha^1 - df \wedge \alpha^1) = e^{-f}(d\alpha^1 + \omega_2^1 \wedge \alpha^2) = 0. \quad (5.140)$$

[Recall that by Cartan's formula,

$$\begin{aligned} d\alpha^1(e_1, e_2) &= e_1.\alpha^1(e_2) - e_2.\alpha^1(e_1) - \alpha^1([e_1, e_2]) \\ &= -\alpha^1(\nabla_{e_1} e_2 - \nabla_{e_2} e_1) \\ &= -\omega_2^1(e_1), \quad (5.141) \end{aligned}$$

and hence, $d\alpha^1 + \omega_2^1 \wedge \alpha^2 = 0$].

Likewise, $dA^2 = 0$. This implies that there exists a map $F : \Omega \longrightarrow \mathbb{R}^2$, of class $H^2 \cap W^{1,\infty}$, such that

$$dF = \begin{pmatrix} A^1 \\ A^2 \end{pmatrix}. \quad (5.142)$$

It is now easy to conclude that F is a conformal Lipschitz diffeomorphism, whose inverse is the parametrization we were looking for. □

Remark 5.4.6 *It would be interesting to take advantage of the geometrical properties of Coulomb frames in other situations, for instance on Riemannian manifolds of dimension larger than 2. As a curiosity, we will conclude with a study of "special" Coulomb frames, over a symplectic surface (the structure group involved being $SL(2, \mathbb{R})$).*

SPECIAL COULOMB FRAMES OVER A SYMPLECTIC SURFACE

We start by giving some definitions. A symplectic surface is a surface equipped with a symplectic form which — in this special case — is the same as a volume form. Let $(\mathcal{M}, \omega)$ be such a symplectic surface: $\mathcal{M}$ is a surface and ω is a volume form on $\mathcal{M}$. If Ω is an open subset of $\mathcal{M}$, we define a *special frame* on Ω to be the assignment, to each point z in Ω, of a basis $(e_1(z), e_2(z))$ of $T_z\mathcal{M}$ such that

$$\omega(e_1, e_2) = 1.$$

In spite of the small amount of geometric structure we assume, it is possible to define a functional on the set of special frames on (Ω, ω), which generalizes what we saw for Riemannian surfaces. For any special frame $e := (e_1, e_2)$, we define

$$F(e) := \int_\Omega \left(\omega(e_1, [e_1, e_2])^2 + \omega([e_1, e_2], e_2)^2 \right) \omega.$$

We leave to the reader the task of checking that if $(\mathcal{M}, g)$ is a Riemannian surface, $\omega = \mathrm{dvol}_g$ and e is an orthonormal frame field, then $F(e) = \int_\Omega |\langle de_1, e_2\rangle|^2 \mathrm{dvol}_g$.

It will be convenient to replace a frame by its dual coframe, i.e. at each point z, the basis of $T_z^*\mathcal{M}$ dual to $(e_1(z), e_2(z))$: if we denote this basis by $\alpha = (\alpha^1, \alpha^2)$, we then have

$$\alpha^a(e_b) = \delta^a_b, \text{ for } a, b = 1, 2.$$

The condition of e being special is equivalent to

$$\alpha^1 \wedge \alpha^2 = \omega.$$

We define $\mathcal{S}(\Omega) := \{\alpha = (\alpha^1, \alpha^2) \text{ coframe field on } \Omega, \text{ such that } \alpha^1 \wedge \alpha^2 = \omega\}$. We show, using the Cartan formula given above, that if we define the quantities λ^1 and λ^2 such that

$$\begin{cases} d\alpha^1 &= \lambda^1 \omega \\ d\alpha^2 &= \lambda^2 \omega, \end{cases}$$

then

$$\begin{aligned} \lambda^1 &= -\omega([e_1, e_2], e_2) \\ \lambda^2 &= -\omega(e_1, [e_1, e_2]) \end{aligned}$$

and thus,

$$F(e) = F^\star(\alpha) := \int_\Omega ((\lambda^1)^2 + (\lambda^2)^2) \omega.$$

We will say that $\alpha \in \mathcal{S}(\Omega)$ is a critical point of $F^\star$, if $F^\star(\alpha)$ is stationary w.r.t. infinitesimal compactly supported variations in Ω.

Lemma 5.4.7 *Suppose $\alpha \in \mathcal{S}(\Omega)$ is a smooth critical point of $F^\star$. Then, there exists a function $f : \Omega \longrightarrow \mathbb{R}$, and a chart $(u, v) : \Omega \longrightarrow \mathbb{R}^2$, such that*

$$(e^f \alpha^1, e^f \alpha^2) = d(u, v). \tag{5.143}$$

5.4 Epilogue: Coulomb frames and conformal coordinates

Moreover, we may consider the metric g defined over Ω by

$$g = \alpha^1 \otimes \alpha^1 + \alpha^2 \otimes \alpha^2, \tag{5.144}$$

and this metric has a constant Riemannian curvature.

Proof
Step 1 EXISTENCE OF (u,v)
It consists of showing that if $\alpha \in \mathcal{S}(\Omega)$ is a critical point of $F^\star$, then there exists f such that $d(e^f \alpha^1) = d(e^f \alpha^2) = 0$. To show this, we simply use the fact that$= F^\star(\alpha)$ is stationary for a transformation of the type

$$\begin{aligned}(\alpha^1, \alpha^2) &\longmapsto (\alpha^1 \cos(\epsilon\phi) + \alpha^2 \sin(\epsilon\phi), -\alpha^1 \sin(\epsilon\phi) + \alpha^2 \cos(\epsilon\phi))\\ &= (\alpha^1 + \epsilon\phi\alpha^2, \alpha^2 - \epsilon\phi\alpha^1) + o(\epsilon).\\ &= (\alpha^1_\epsilon, \alpha^2_\epsilon),\end{aligned}$$

where $\phi \in \mathcal{C}^\infty_c(\Omega, \mathbb{R})$ (action of $SO(2)$). We see that

$$d\alpha^1_\epsilon = d\alpha^1 + \epsilon(d\phi \wedge \alpha^2 + \phi\lambda^2 \alpha^1 \wedge \alpha^2) + o(\epsilon),$$

$$d\alpha^2_\epsilon = d\alpha^2 + \epsilon(\alpha^1 \wedge d\phi - \phi\lambda^1 \alpha^1 \wedge \alpha^2) + o(\epsilon),$$

and hence that the first variation of $F^\star(\alpha)$ is

$$\begin{aligned}\delta F^\star(\alpha)(\phi\alpha^2, -\phi\alpha^1) &= 2\int_\Omega \lambda^1 d\phi \wedge \alpha^2 + \lambda^2 \alpha^1 \wedge d\phi\\ &= 2\int_\Omega d(\phi(\lambda^1 \alpha^2 - \lambda^2 \alpha^1)) - \phi d(\lambda^1 \alpha^2 - \lambda^2 \alpha^1).\end{aligned} \tag{5.145}$$

We deduce from (5.145) that

$$d(\lambda^1 \alpha^2 - \lambda^2 \alpha^1) = 0, \tag{5.146}$$

and hence, that there exists $f : \Omega \longrightarrow \mathbb{R}$ such that

$$\lambda^1 \alpha^2 - \lambda^2 \alpha^1 = df. \tag{5.147}$$

It is now easy to see that this yields that $d(e^f \alpha^1) = d(e^f \alpha^2) = 0$ and also the existence of (u, v) satisfying (5.143). We consider the metric g defined by (5.144). The moving frame $e = (e_1, e_2)$ (dual basis to α) will then be an orthonormal moving frame for this scalar product. The associated connection form is $\omega^1_2 = -\omega^2_1$ such that $de_a = \omega^b_a e_b$. This relation is also equivalent to

$$d\alpha^1 + \omega_2^1 \wedge \alpha^2 = d\alpha^2 + \omega_1^2 \wedge \alpha^1 = 0.$$

Then, we can easily deduce that

$$\omega_2^1 = -\lambda^1 \alpha^1 - \lambda^2 \alpha^2. \qquad (5.148)$$

Step 2 THE COMPLETE EULER–LAGRANGE SYSTEM
We can apply two other types of infinitesimal variations $\alpha \longmapsto \alpha_\epsilon$ where

$$\alpha_\epsilon = (\alpha^1 + \epsilon\phi\alpha^2, \alpha^2 + \epsilon\phi\alpha^1) + o(\epsilon), \qquad (5.149)$$

and

$$\alpha_\epsilon = (\alpha^1 + \epsilon\phi\alpha^1, \alpha^2 - \epsilon\phi\alpha^2) + o(\epsilon). \qquad (5.150)$$

For the first type (5.149), we have

$$\delta F^\star(\alpha)(\phi\alpha^2, \phi\alpha^1) = 2\int_\Omega d\phi \wedge (\lambda^1\alpha^2 + \lambda^2\alpha^1) + 2\phi\lambda^1\lambda^2\alpha^1 \wedge \alpha^2.$$

For the second type of variation (5.150), we obtain

$$\delta F^\star(\alpha)(\phi\alpha^1, -\phi\alpha^2) = 2\int_\Omega d\phi \wedge (\lambda^1\alpha^1 - \lambda^2\alpha^2) + \phi((\lambda^1)^2 - (\lambda^2)^2)\alpha^1 \wedge \alpha^2.$$

From these two relations we deduce two Euler equations which should be added to equation (5.146) obtained before,

$$d(\lambda^1\alpha^2 + \lambda^2\alpha^1) = 2\lambda^1\lambda^2\alpha^1 \wedge \alpha^2, \qquad (5.151)$$

$$d(\lambda^1\alpha^1 - \lambda^2\alpha^2) = ((\lambda^1)^2 - (\lambda^2)^2)\alpha^1 \wedge \alpha^2. \qquad (5.152)$$

Step 3 INTERPRETATION OF THE EULER EQUATIONS
Let K be the curvature of the Riemannian surface (Ω, g). This function is defined on Ω by

$$-d(\lambda^1\alpha^1 + \lambda^2\alpha^2) = d\omega_2^1 = K\alpha^1 \wedge \alpha^2. \qquad (5.153)$$

From (5.146), (5.151), (5.152) and (5.153), we deduce

5.4 Epilogue: Coulomb frames and conformal coordinates

$$\begin{cases} d(\lambda^1 \alpha^1) &= \frac{1}{2}((\lambda^1)^2 - (\lambda^2)^2 - K)\alpha^1 \wedge \alpha^2 \\ d(\lambda^2 \alpha^1) &= \lambda^1 \lambda^2 \alpha^1 \wedge \alpha^2 \\ d(\lambda^1 \alpha^2) &= \lambda^1 \lambda^2 \alpha^1 \wedge \alpha^2 \\ d(\lambda^2 \alpha^2) &= \frac{1}{2}(-(\lambda^1)^2 + (\lambda^2)^2 - K)\alpha^1 \wedge \alpha^2. \end{cases} \quad (5.154)$$

Developing the relations $d(\lambda^a \alpha^b) = d\lambda^a \wedge \alpha^b + \lambda^a d\alpha^b$, and using (5.154), we obtain

$$\begin{cases} d\lambda^1 &= \frac{1}{2}(|\lambda|^2 + K)\alpha^2 \\ d\lambda^2 &= -\frac{1}{2}(|\lambda|^2 + K)\alpha^1, \end{cases} \quad (5.155)$$

where $|\lambda|^2 = (\lambda^1)^2 + (\lambda^2)^2$.

We remark that $\alpha^1 = e^{-f} du$ and $\alpha^2 = e^{-f} dv$, by (5.143). Therefore, (5.155) imply that if we consider λ^1 and λ^2 as functions of (u, v),

$$\frac{\partial \lambda^1}{\partial u} = \frac{\partial \lambda^2}{\partial v} = 0,$$

or equivalently, $\lambda^1 = \lambda^1(v)$ and $\lambda^2 = \lambda^2(u)$.

Still because of (5.155), we have

$$\frac{d\lambda^1}{dv}(v) = -\frac{d\lambda^2}{du}(u) = \frac{1}{2}(|\lambda|^2 + K)e^{-f},$$

which implies that this quantity is equal to a constant γ. Hence, (5.155) yields $d\lambda^1 = \gamma dv$ and $d\lambda^2 = -\gamma du$, and integrating (denoting by u_0 and v_0 certain constants)

$$\begin{cases} \lambda^1 &= \gamma(v - v_0) \\ \lambda^2 &= -\gamma(u - u_0). \end{cases}$$

Substituting this into (5.147), we obtain

$$df = e^{-f}\gamma[(u - u_0)du + (v - v_0)dv].$$

We deduce from this equation that

$$e^f = \frac{\gamma}{2}((u - u_0)^2 + (v - v_0)^2) + \delta,$$

where δ is an integration constant. Knowing that $\frac{1}{2}(|\lambda|^2 + K)e^{-f} = \gamma$, we conclude that

$$K = 2\gamma\delta.$$

$\square$

References

[1] U. Abresch, *Constant mean curvature tori in terms of elliptic functions*, J. reine angew. Math. 374 (1987), 189–192.
[2] L. Almeida, *The regularity problem for generalized harmonic maps into homogeneous spaces*, Calc. Var. 3 (1995), 193–242.
[3] M.F. Atiyah, *Geometry of Yang–Mills fields*, Lezione Fermiane, Scuola Normale Superiore, Pisa 1979.
[4] P. Baird, *Harmonic maps with symmetry, harmonic morphisms and deformation of metrics*, Research Notes in Mathematics 87, Pitman 1983.
[5] P. Baird, J. Eells, *A conservation law for harmonic maps*, Symp. Utrecht (1980), Lec. Notes Maths. 894, Springer 1981, 1–25.
[6] S. Baraket, *Estimations of the best constant involving the L^∞ norm in Wente's inequality*, Ann. Fac. Sciences Toulouse, vol. V, n. 3 (1996), 373–385.
[7] J. Bergh, J. Löfström, *Interpolation spaces, an introduction*, Springer Verlag, Berlin, 1976.
[8] F. Bethuel, *The approximation problem for Sobolev mappings between manifolds*, Acta Mathematica 167 (1991), 167–201.
[9] F. Bethuel, *Un résultat de régularité pour des solutions de l'équation des surfaces à courbure moyenne prescrite*, C. R. Acad. Paris 314 (1992), 1003–1007.
[10] F. Bethuel, *On the singular set of stationary harmonic maps*, Manuscripta Math. 78 (1993), 417–443.
[11] F. Bethuel, *Weak limits of Palais–Smale sequences for a class of critical functionals*, Calc. Var. 1 (1993), 267–310.
[12] F. Bethuel, H. Brezis, F. Hélein, *Ginzburg–Landau vortices*, Birkhäuser, Boston 1994.
[13] F. Bethuel, F. Demengel, *Extensions for Sobolev mappings between manifolds*, Calc. Var. 3 (1995), 475–491.
[14] F. Bethuel, J.-M. Ghidaglia, *Improved regularity for solutions of elliptic equations involving Jacobian and applications*, J. Maths. Pures et Appliquées 72 (1993), 441–475.
[15] F. Bethuel, X. Zheng, *Density of smooth functions between two manifolds in Sobolev spaces*, J. Funct. Anal. 80 (1988), 60–75.

[16] W. Blaschke, *Vorlesungen über Differentialgeometrie III*, Springer, Berlin 1929.
[17] A.I. Bobenko, *Surfaces in terms of 2 by 2 matrices. Old and new integrable systems*, in [62].
[18] O. Bonnet, *Note sur une propriété de maximum relative à la sphère*, Nouvelles Annales de mathématiques, t. XII (1853), p. 433.
[19] H. Brezis, *Analyse fonctionnelle*, Masson, Paris 1983.
[20] H. Brezis, *Laser beams and limiting cases of Sobolev inequalities*, Nonlinear Partial Differential Equations and their Applications, Collège de France Seminar, Vol. II, H. Brezis, J.L. Lions ed., Pitman 1982.
[21] H. Brezis, J.-M. Coron, *Multiple solutions of H-systems and Rellich's conjecture*, Comm. Pure Appl. Math. 37 (1984), 149–187.
[22] H. Brezis, J.-M. Coron, E.H. Lieb, *Harmonic maps with defects*, Comm. Math. Phys. 107 (1986), 649–705.
[23] H. Brezis, S. Wainger, *A note on limiting cases of Sobolev embeddings and convolutions inequalities*, Comm. Partial Diff. Equations 5 (1980), 773–789.
[24] F. Burstall, D. Ferus, F. Pedit, U. Pinkall, *Harmonic tori in symmetric spaces and commuting Hamiltonian systems on loop algebra*, Annals of Math. 138 (1993), 173–212.
[25] F. Burstall, F. Pedit, *Harmonic maps via Adler–Kostant–Symes theory*, in [62].
[26] F. Burstall, F. Pedit, *Dressing orbits of harmonic maps*, Duke Math. J. 80 (1995), 353–382.
[27] F.E. Burstall, J.H. Rawnsley, *Twistor theory for Riemannian symmetric spaces with applications to harmonic maps of Riemann surfaces*, Lecture Notes in Math. 1424, Springer, Berlin 1990.
[28] G. Carbou, *Regularity for a nonlinear variational problem*, Manuscripta Math. 78 (1993), 37–56.
[29] K.C. Chang, W.Y. Ding, R. Ye, *Finite-time blow-up of the heat flow of harmonic maps from surfaces*, J. Diff. Geometry 36 (1992), 507–515.
[30] S.-Y.A. Chang, L. Wang, P.C. Yang, *Regularity of harmonic maps*, Comm. Pure Appl. Math. 52, No. 9 (1999), 1099–1111.
[31] S.-Y.A. Chang, L. Wang, P.C. Yang, *A regularity theory for bi-harmonic maps*, Comm. Pure Appl. Math. 52, No. 9 (1999), 1113–1137.
[32] S. Chanillo, *Sobolev inequalities involving divergence free maps*, Comm. Partial Diff. Equations 16 (1991), 1969–1994.
[33] S. Chanillo, Y. Li, *Continuity of solutions of uniformly elliptic equations in R^2*, Manuscripta Math. 77 (1992), 415–433.
[34] L.L. Chau, G.M. Lin, W.Y. Shi, *Kac–Moody algebra in the self-dual Yang–Mills equation*, Phys. Rev. D(3) 25 (1982), 1086–1094.
[35] Y.M. Chen, *The weak solutions to the evolution problems of harmonic maps*, Math. Zeitschrift 201 (1989), 69–74.
[36] S.S. Chern, *An elementary proof of the existence of isothermal parameters on a surface*, Proc. Amer. Math. Soc. (1955), 771–782.
[37] S.S. Chern, *Moving frames*, Astérisque, hors série, Société Mathématique de France, 1985, p. 67–77.
[38] P. Choné, *A regularity result for critical points of conformally*

invariant functional, Potential Anal. 4 (1995), 269–296.

[39] R. Coifman, P.-L. Lions, Y. Meyer, S. Semmes, *Compensated compactness and Hardy spaces*, J. Math. Pures Appl. 72 (1993), 247–286.

[40] R. Coifman, R. Rochberg, *The molecular characterization of certain Hardy spaces*, Astérisque 77, Société Mathématique de France, 1980.

[41] L. Crane, *Action of the loop group on the self-dual Yang–Mills equation*, Comm. Math. Phys. 110 (1987), 301–335.

[42] G. Darboux, *Leçons sur la théorie générale des surfaces*, livre VII, chapitre X, Gauthier-Villars 1884 or Jacques Gabay 1993.

[43] G. Dell'Antonio, D. Zwanziger, *Every gauge orbit passes inside the Gribov horizon*, Probabilistic methods in quantum field theory and quantum gravity, 1990, NATO Adv. Sci. Inst. Ser. B: Phys., 224.

[44] J.-M. Delort, *Existence de nappes de tourbillons sur R^2*, C. R. Acad. Sci. Paris 312 (1991), 85–88.

[45] L. Dolan, *Kac–Moody algebra is hidden symmetry of chiral models*, Phys. Rev. Lett. 49 (1981), 1371–1374.

[46] J. Dorfmeister, F. Pedit, H. Wu, *Weierstrass type representation of harmonic maps into symmetric spaces*, Comm. Anal. and Geom. 6 (1998), 633–668.

[47] B. Doubrovine, S. Novikov, A. Fomenko, *Géométrie contemporaine*, tomes 1 et 2, éditions Mir, Moscou, 1982; or *Modern Geometry*, Springer Verlag.

[48] J. Douglas, *Solution to the problem of Plateau*, Trans. Amer. Math. Soc. 33 (1931), 263–321.

[49] C.J. Earle, J. Eells, *A fibre bundle description of Teichmüller theory*, J. Diff. Geom. 3 (1969), 19–43.

[50] J. Eells, L. Lemaire, *A report on harmonic maps*, Bull. London Math. Soc. 10 (1978), 1–68.

[51] J. Eells, L. Lemaire, *Another report on harmonic maps*, Bull. London Math. Soc. 20 (1988), 385–524.

[52] J. Eells, J. Wood, *A conservation law for harmonic maps*, Springer Lecture Notes in Math. 894 (1981), 1–25.

[53] L.C. Evans, *Weak convergence methods for nonlinear partial differential equations*, CBMS regional conference n. 74, AMS 1988.

[54] L.C. Evans, *Partial regularity for stationary harmonic maps into spheres*, Arch. Rat. Mech. Anal. 116 (1991), 101–163.

[55] L.C. Evans, S. Müller, *Hardy spaces and the two-dimensional Euler equations with nonnegative vorticity*, J. Amer. Math. Soc. 7 (1994), 199–219.

[56] L.D. Faddeev, L.A. Takhtajan, *Hamiltonian methods in the theory of solitons*, Springer-Verlag 1987.

[57] C. Fefferman, *Characterization of bounded mean oscillations*, Bull. Amer. Math. Soc. 77 (1971), 585–587.

[58] C. Fefferman, E. Stein, *H^p spaces of several variables*, Acta Math. 129 (1972), 137–193.

[59] A. Floer, *Morse theory for Lagrangian intersection*, J. Diff. Geom. 28 (1988), 513–547, *The unregularized gradient flow of the symplectic action*, Comm. Pure Appl. Math. 41 (1988), 775–813; *A relative Morse index for the symplectic action*, Comm. Pure Appl. Math. 41 (1988), 393–407; *Symplectic fixed points and holomorphic spheres*,

Comm. Math. Phys. 120 (1989), 575–611, *Witten's complex and infinite dimensional Morse theory*, J. Diff. Geom. 30 (1989), 207–221.
[60] A. Freire, *Uniqueness for the harmonic map flow in two dimensions*, Calc. Var. 3 (1995), 95–105.
[61] A. Freire, S. Müller, M. Struwe, *Weak convergence of wave maps from (1+2)-dimensional space to Riemannian manifolds*, Inventiones Math. 130 (1997), 589–617.
[62] A. Fordy, J.C. Wood, *Harmonic maps and integrable systems*, Aspects of Mathematics, E23, Vieweg, 1994. Texts available in www.amsta.leeds.ac.uk/Pure/staff/wood/Fordy/Wood/contents.html
[63] Y. Ge, *An elliptic variational approach to immersed surfaces of prescribed Gauss curvature*, Calc. Var. 7, No.2 (1998), 173–190, *Immersed surfaces of prescribed Gauss curvature into Lorentzian spaces*, preprint to appear in Proc. Amer. Math. Soc.
[64] Y. Ge, *Estimations of the best constant involving the L^2 norm in Wente's inequality and compact H-surfaces in Euclidean spaces*, Control, Optimization and Calculus of Variations, vol. 3 (1998), 263–300.
[65] Y. Ge, *A remark on generalized harmonic maps into the sphere*, Nonlinear analysis, T.M.A., 36, 495–506, 1999.
[66] Y. Ge, F. Hélein, *A remark on compact H-surfaces into $\mathbb{R}^3$*, to appear in Math. Zeitschrift.
[67] P. Gérard, *Résultats récents sur les fluides incompressibles bidimensionnels*, Séminaire Bourbaki 44ème année (1991-92) n. 757.
[68] M. Gerstenhaber, H.E. Rauch, *On extremal quasiconformal mappings I, II*, Proc. Nat. Acad. Sci. USA 40 (1954), 808–812 and 991–994.
[69] M. Giaquinta, *Multiple integrals in the calculus of variations and nonlinear elliptic systems*, Princeton University Press, 1983.
[70] M. Giaquinta, E. Giusti, *On the regularity of minima of variational integrals*, Acta Math. 148 (1982), 31–46; *The singular set of the minima of certain quadratic functionals*, Ann. Scuola Norm. Sup. Pisa Cl. Sci. 9 (1984), 45–55.
[71] M. Giaquinta, G. Modica, J. Souček, *Cartesian currents in the calculus of variations I and II*, Ergebnisse des Mathematik und ihrer Grenzgebiete, 3. Folge, Vol. 37 and 38, Springer 1998.
[72] J. Ginibre, G. Vélo, *The Cauchy problem for the $O(N)$, $\mathbb{C}P(N-1)$ and $G\mathbb{C}(N,P)$ models*, Ann. Physics 142, 2 (1982), 393–415.
[73] P. Griffiths, J. Harris, *Principles of algebraic geometry*, Wiley Interscience, New York, 1978.
[74] M. Gromov, *Partial differential relations*, Springer, 1986.
[75] C.H. Gu, *On the Cauchy problem for harmonic maps defined on two-dimensional Minkowski space*, Comm. Pure Appl. Math. 33, 6 (1980), 727–737.
[76] M. Guest, *Harmonic maps, loop groups and integrable systems*, Cambridge University Press, Cambridge 1997.
[77] M. Günther, *Isometric embeddings of Riemannian manifolds*, Proc. of the International Congress of Mathematicians, Kyoto, Japan 1990, The Mathematical Society of Japan, 1991.
[78] M. Grüter, *Regularity of weak H-surfaces*, J. reine angew. Math. 329 (1981), 1–15.
[79] P. Hajlasz, P. Strzelecki, *Subelliptic p-harmonic maps into spheres*

[80] R. Hamilton, *Harmonic maps of manifolds with boundary*, Lecture Notes in Math. 471, Springer-Verlag, 1975 (chapter IV, section 5).

[81] R. Hardt, *Singularities of harmonic maps*, Bull. Amer. Math. Soc., vol. 34 n.1, January 1997, 15–34.

[82] R. Hardt, F.H. Lin, *Mappings minimizing the L^p norm of the gradient*, Comm. Pure Appl. Math. 40 (1987), 555–558.

[83] F. Hélein, *Régularité des applications faiblement harmoniques entre une surface et une sphère*, C. R. Acad. Sci. Paris 311 (1990), 519–524.

[84] F. Hélein, *Regularity of weakly harmonic maps from a surface into a manifold with symmetries*, Manuscripta Math. 70 (1991), 293–218.

[85] F. Hélein, *Régularité des applications harmoniques entre une surface et une variété riemannienne*, C. R. Acad. Sci. Paris 312 (1991), 591–596.

[86] F. Hélein, *Willmore surfaces and loop groups*, J. Diff. Geom., 50 (1998), 331–385.

[87] F. Hélein, *Symétries dans les problèmes variationnels et applications harmoniques*, Istituti editoriale e poligrafici internazionali, Pisa-Roma, 1998.

[88] F. Hélein, *Problèmes variationnels invariants par transformation conforme en dimension 2*, preprint arXiv: math.DG/0101237.

[89] F. Hélein, *Constant mean curvature surfaces, harmonic maps and integrable systems*, Lectures in Mathematics, ETH Zuerich, Birkhäuser, 2001.

[90] S. Helgason, *Differential geometry, Lie groups and symmetric spaces*, Academic Press, New York, 1978.

[91] D. Hilbert, *Die Grundlagen der Physik*, Nachr. Ges. Wiss. Göttingen (1915) 395–407, (1917) 53–76, (1924) 1–32.

[92] S. Hildebrandt, H. Kaul, K.O. Widman, *An existence theorem for harmonic mappings of Riemannian manifolds*, Acta Math. 138 (1977), 1–16.

[93] S. Hildebrandt, K.O. Widman, *Some regularity results for quasi-linear elliptic systems of second order*, Math. Zeitschrift 142 (1975), 67–86.

[94] N. Hitchin, *Harmonic maps from a 2-torus to the 3-sphere*, J. Diff. Geom. 31 (1990), 627–710.

[95] R. A. Hunt, *On $L(p,q)$ spaces*, L'enseignement mathématique XII (1966), 249–276.

[96] F. John, L. Nirenberg, *On functions of bounded mean oscillations*, Comm. Pure Appl. Math. 14 (1961), 415–426.

[97] J. Jost, *Two-dimensional geometric variational problems*, Wiley (1991).

[98] J. Jost, *Riemannian geometry and geometric analysis*, Universitext, Springer 1995.

[99] J. Keller, J. Rubinstein, P. Sternberg, *Reaction-diffusion processes and evolution to harmonic maps*, SIAM J. Appl. Math. 49 n. 6 (1989), 1722–1733.

[100] S. Klainerman, I. Rodnianski, *On the global regularity of wave maps in the critical Sobolev norm*, arXiv: math.AP/0012034.

[101] S. Kobayashi, K. Nomizu, *Foundations of differential geometry*, volumes 1 and 2, Wiley Interscience, New York 1963.

[102] N. Kuiper, *On C^1-isometric embeddings I*, Proc. Konikl. Neder. Akad.

References

[103] O. Ladyzhenskaya, N. Ural'tseva, *Linear and quasilinear elliptic equations*, Academic Press, New York and London, 1968.

[104] H.B. Lawson, *Complete minimal surfaces in S^3*, Ann. Math. 92 (1970), 335–374.

[105] P.D. Lax, R.S. Phillips, *Scattering theory*, Academic Press.

[106] L. Lemaire, *Applications harmoniques de surfaces riemanniennes*, J. Diff. Geom. 13 (1978), 51–78.

[107] P. Li, S.T. Yau, *A new conformal invariant and its applications to the Willmore conjecture and first eigenvalues of compact surfaces*, Invent. Math. 69 (1982), 269–291.

[108] F.H. Lin, *Une remarque sur l'applications $x/|x|$*, C. R. Acad. Sci. Paris 305 (1987), 529–531.

[109] F.H. Lin, *Gradient estimate and blow-up analysis for stationary harmonic maps*, C. R. Acad. Sci. Paris, t. 323, Série I (1996), 1005–1008 and Ann. of Math. 149 (1999), 785–829.

[110] F.H. Lin, T. Rivière, *Energy quantization for harmonic maps*, to appear in Duke Math. J.

[111] J.-L. Lions, J. Peetre, *Sur une classe d'espaces d'interpolation*, Inst. Hautes Etudes Sci. Publ. Math. 19 (1964), 5–68.

[112] I. McIntosh, *Global solutions of the elliptic 2D periodic Toda lattice*, Nonlinearity 7 (1994), 85–108.

[113] M. Melko, I. Sterling, *Integrable systems, harmonic maps and the classical theory of surfaces*, in [62].

[114] Y. Meyer, *Ondelettes*, Hermann 1988, tome 1.

[115] S. Montiel, A. Ros, *Minimal immersions of surfaces by the first eigenvalues and conformal area*, Inventiones Math. 83 (1986), 153–166.

[116] C.B. Morrey Jr., *The problem of Plateau on a Riemannian manifold*, Ann. Math. 49 (1948), 807–951.

[117] C.B. Morrey Jr., *Multiple integrals in the calculus of variations*, Springer-Verlag, Berlin, 1966.

[118] J. Mossino, *Inégalités isopérimétriques et applications en physique*, Hermann, Paris 1984.

[119] L. Mou, P. Yang, *Regularity for n-harmonic maps*, J. Geom. Anal. 6 (1996), 91–112.

[120] S. Müller, *Higher integrability of determinants and weak convergence in L^1*, J. Reine Angew. Math. 412 (1990), 20–34.

[121] S. Müller, V. Šverák, *On surfaces of finite total curvature*, J. Diff. Geom. 42 n.2 (1995), 229–258.

[122] F. Murat, *Compacité par compensation : condition nécessaire et suffisante de continuité faible sous une hypothèse de rang constant*, Annali Scuola Normale Superiore, Pisa, Serie VIII n. 1 (1981), 69–102.

[123] J. Nash, *The imbedding problem for Riemannian manifolds*, Ann. Math. 63 (1956), 20–63.

[124] P. Olver, *Applications of Lie groups to differential equations*, second edition, Graduate text in Math. 107, Springer-Verlag, New York, 1993.

[125] R. Penrose, *The twistor program*, Rep. Mathematical Phys. 12 (1977), 65–76.

[126] R. Penrose, W. Rindler, *Spinors and space-time, volumes 1 and 2*, Cambridge Monographs in Mathematical Physics, Cambridge University Press 1984.

[127] A.I. Pluzhnikov, *Some properties of harmonic mappings in the case of spheres and Lie groups*, Sov. Math. Dokl. 27 (1983), 246–248.

[128] K. Pohlmeyer, *Integrable Hamiltonian systems and interactions through quadratic constraints*, Comm. Math. Phys. 46 (1976), 207–221.

[129] S. Pohozaev, *Eigenfunctions of the equation $\Delta u = \lambda f(u)$*, Soviet Math. Dokl. 6 (1965), 1408–1411 (translation of Dokl. Akad. Nauk SSSR 165 (1965), 33–36).

[130] A.N. Pressley G. Segal, *Loop groups*, Oxford Math. Monographs, Clarendon Press, Oxford 1986.

[131] P. Price, *A monotonicity formula for Yang–Mills fields*, Manuscripta Math. 43 (1983), 131–166.

[132] J. Qing, *Boundary regularity of weakly harmonic maps from surfaces*, J. Funct. Anal. 114 (1993), 458–466.

[133] J. Qing, *On singularities of the heat flow for harmonic maps from surfaces into spheres*, Comm. Anal. Geom. 3 (1994), 297–317.

[134] J.H. Rawnsley, *Noether's theorem for harmonic maps*, Diff. Geom. Methods in Math. Phys., Reidel 1984, 197–202.

[135] E. Reifenberg, *Solutions to the Plateau problem for m-dimensional surfaces of varying topological type*, Acta Math. 104 (1966), 1–92.

[136] T. Rivière, *Everywhere discontinuous harmonic maps into spheres*, Acta Math. 175 (1995), 197–226.

[137] T. Rivière, *Flot des applications harmoniques en dimension deux*, in Thèse de Doctorat, Université de Paris VI, 1993.

[138] T. Rivière, *Dense subsets of $H^{1/2}(S^2, S^1)$*, Ann. Global Anal. and Geom. 18 (2000), 517–528.

[139] T. Rivière, *Interpolation space and energy quantization for Yang–Mills fields*, to appear in Comm. Anal. Geom.

[140] H. Rund, *The Hamilton-Jacobi theory in the calculus of variations, its role in mathematics and physics*, Krieger, Huntington, 1973.

[141] J. Sacks, K. Uhlenbeck, *The existence of minimal immersions of 2-spheres*, Ann. of Math. 113 (1981), 1–24.

[142] D. Salamon, *Morse theory, the Conley index and Floer homology*, Bull. London Math. Soc. 22 (1990), 113–140.

[143] J.H. Sampson, result communicated to J. Eells.

[144] R. Schoen, *Analytic aspects of the harmonic map problem*, Seminar on Nonlinear Partial Differential Equations (S.S. Chern editor), MSRI Publications, vol. 2, Springer-Verlag, New York, 1984.

[145] R. Schoen, K. Uhlenbeck, *Approximation theorems for Sobolev mappings*, preprint (1986).

[146] R. Schoen, K. Uhlenbeck, *A regularity theory for harmonic maps*, J. Diff. Geom. 17 (1982), 307–335, and 18 (1983), 329.

[147] R. Schoen, K. Uhlenbeck, *Boundary regularity and the Dirichlet problem for harmonic maps*, J. Diff. Geom. 18 (1983), 253–268.

[148] R. Schoen, K. Uhlenbeck, *Regularity of minimizing harmonic maps into the sphere*, Invent. Math. 78 (1984), 89–100.

[149] H. Sealey, *Some conditions ensuring the vanishing of harmonic*

differential forms with applications to harmonic maps and Yang–Mills theory, Math. Proc. Camb. Phil. Soc. 91 (1982), 441–452.
- [150] G. Segal, G. Wilson, *Loop groups and equations of KdV type*, Inst. Hautes Etudes Sci. Publ. Math., 61 (1985), p. 5–65.
- [151] S. Semmes, *Chord-arc surfaces with small constants I, II*, Adv. Math. 85 (1991), 198–233, and Adv. Math. 88 (1991), 403–412, *Hypersurfaces in R^n whose normal has small BMO norm*, Proc. Amer. Math. Soc. 112 (1991), 403–412.
- [152] S. Semmes, *A primer on Hardy spaces and some remarks on a theorem of Evans and Müller*, Comm. Partial Diff. Equations 19 (1994), 277–319.
- [153] J. Shatah, *Weak solutions and developments of singularities of the $SU(2)$ σ-model*, Comm. Pure Appl. Math. 41 (1988), 459–469.
- [154] L. Simon, *Existence of Willmore surfaces*, Canberra Miniconference on Geometry and partial differential equations, Proc. Cent. Math. Anal., Australian National University 10 (1986), 187–216; *Existence of surfaces minimizing the Willmore functional*, Comm. Anal. Geom. Vol.1, n° 2 (1993), 281–326.
- [155] G. Stampacchia, *Equations elliptiques du second ordre à coefficients discontinus*, Presses Univ. de Montréal, 1966.
- [156] E. Stein, *Singular integrals and differentiability properties of functions*, Princeton University Press, Princeton 1970.
- [157] E. Stein, *Harmonic analysis*, Princeton University Press, Princeton 1993.
- [158] E. Stein, G. Weiss, *Introduction to Fourier analysis on Euclidean spaces*, Princeton University Press, Princeton 1970.
- [159] W. Strauss, *On weak solutions of semi-linear hyperbolic equations*, An. Acad. Brasil Cienc. 42 (1970), 643–651.
- [160] M. Struwe, *On the evolution of harmonic mappings of Riemannian surfaces*, Comment. Math. Helv. 60 (1985), 558–581.
- [161] P. Strzelecki, *Regularity of p-harmonic maps from p-dimensional ball into a sphere*, Manuscr. Math. 82 (1994), 407–415.
- [162] H. Takeuchi, *Some conformal properties of p-harmonic maps and a regularity for sphere valued p-harmonic maps*, J. Math. Soc. Japan 26 (1994), 217–334.
- [163] T. Tao, *Global regularity of wave maps, I*, arXiv:math.AP/0010068, to appear in I.M.R.N.
- [164] T. Tao, *Global regularity of wave maps, II*, arXiv:math.AP/0011173, submitted to Comm. Math. Phys.
- [165] L. Tartar, *Remarks on oscillations and Stokes equations*, in "Macroscopic modelling of turbulent flow" (Nice 1984), Lecture Note in Physics 230, Springer 1985, 24–31.
- [166] L. Tartar, *The compensated compactness method applied to systems of conservation laws*, in "Systems of nonlinear partial differential equations", J.M. Ball ed., NATO ASI Series C111, Reidel, New York 1983, 263–285.
- [167] L. Tartar, *Imbedding theorems in Lorentz spaces and applications*, preprint 1995.
- [168] G. Thomsen, *Über konforme Geometrie I: Grundlagen der konformen Flächentheorie*, Abh. Math. Sem. Hamburg (1923), 31–56.
- [169] G. Tian, *Gauge theory and calibrated geometry I*, Ann. Math. 151

(2000), 193–268.
- [170] P. Topping, *The optimal constant in Wente's L^∞ estimate*, Comment. Math. Helv. 72, No. 2 (1997), 316–328.
- [171] T. Toro, *Surfaces with generalised second fundamental form in L^2 are Lipschitz manifolds*, J. Diff. Geom. 39 (1994), 65–101.
- [172] T. Toro, *Geometric conditions and existence of bi-Lipschitz parametrisations*, Duke Math. J. vol. 77 n. 1 (1995), 193–227.
- [173] T. Toro, C. Wang, *Compactness properties of weakly p-harmonic maps into homogeneous spaces*, Indiana Univ. Math. J. 44 (1995), 87–113.
- [174] K. Uhlenbeck, *Harmonic maps into Lie groups (classical solutions of the chiral model)*, J. Diff. Geom. 30 (1989), 1–50.
- [175] R.S. Ward, *On self-dual gauge fields*, Phys. Lett. 61A (1977), 81–82.
- [176] R.S. Ward, R.O. Wells, *Twistor geometry and field theory*, Cambridge University Press, 1989.
- [177] H. Wente, *An existence theorem for surfaces of constant mean curvature*, J. Math. Anal. Appl. 26 (1969), 318–344.
- [178] H. Wente, *Large solutions to the volume constrained Plateau problem*, Arch. Rat. Mech. Anal. 75 (1980), 59–77.
- [179] H. Wente, *Counter-example to a conjecture of H. Hopf*, Pacific J. Math. 121 (1986), 193–243.
- [180] B. White, *Homotopy classes in Sobolev spaces and energy minimizing maps*, Bull. Amer. Math. Soc. 13 (1985), 166–168.
- [181] B. White, *Infima of energy functionals in homotopy classes of mappings*, J. Diff. Geom. 23 (1986), 127–142.
- [182] T.J. Willmore, *Note on embedded surfaces*, Ann. Ştiint. Univ. "Al. I. Cuza", Iaşi. Ia. Mat. (1965), 493–496.
- [183] T.J. Willmore, *Riemannian geometry*, Oxford Science Publications, Clarendon Press, Oxford, 1993.
- [184] J.C. Wood, *Nonexistence of solutions to certain Dirichlet problems for harmonic maps*, preprint, Leeds University, 1981.
- [185] V.E. Zhakarov, A.B. Shabat, *Integration of nonlinear equations of mathematical physics by the method of inverse scattering II*, Funktsional Anal. i Prilozhem 13 (1978), 13–22.
- [186] W. Ziemer, *Weakly differentiable functions*, Springer-Verlag 1989.

Index

$\mathcal{A}(S^2, U(n))$, 73
$\mathcal{A}(S^2, u(n))$, 78

Bäcklund transformation, 50
BMO, xxiv, 130, 155, 162, 163, 201
B_r, 73

Cartan immersion, 95
Christoffel symbols, 6
$\mathcal{C}^k(\Omega)$, xxiv
$\mathcal{C}^{k,\alpha}(\Omega)$, xxiv, 46
complexified Jacobi field, 71
conformal coordinates, 5, 244
conformal immersion, 224
conformal Killing field, 28
conformal transformation, 3, 7, 115
connection, 6, 62, 84, 212
conservation law, 11, 14
constant Gauss curvature surfaces, 52
constant mean curvature surfaces, 51, 63, 120
Coulomb frame, 165, 167, 173, 188, 227, 244
covariant derivative, 6, 62
curvature, 66

Dirichlet functional, 7
Dirichlet integral, 3
Dirichlet problem, 35
$\mathcal{D}'(\Omega)$, xxiv
dressing action, 72

ϵ-regularity, 151, 193
extended harmonic lifting, 88, 89
extended harmonic map, 68
extended Jacobi fields, 79

fiber bundle, 61, 227

Gauge group, 174

Gauss map, 225
Grassmannian, 225

$H^1(\Omega)$, xxiii
$\mathcal{H}^1$, xxiv, 128, 162, 179, 201, 223
Hardy space, xxiv, 128, 162, 179, 201, 223
harmonic maps, 5, 8
Hausdorff measure, 150
$H^k(\Omega)$, xxiii
Hodge decomposition, 140, 195
Hodge star operator, 86, 196
Hopf differential, 18, 30, 31, 134

infinitesimal symmetries, 70
isoperimetric inequality, 124, 241
Iwasawa decomposition, 91

Jacobi field, 70

Kac–Moody algebra, 71
Killing vector field, 15, 24

Laplacian, 2
Lax pair, 61
$L\mathfrak{G}$, 89
$L\mathfrak{G}^{\mathbb{C}}$, 73
$L\mathfrak{g}^{\mathbb{C}}$, 74
$L^+\mathfrak{G}^{\mathbb{C}}$, 73, 75
$L^+\mathfrak{g}^{\mathbb{C}}$, 74
$L\mathfrak{G}_\tau$, 90, 92
$L\mathfrak{g}^{\mathbb{C}}_\tau$, 92
Lie algebra, 55, 68
lifting, 83
locally minimizing maps, 42
loop group, 67, 89
Lorentz space, xxiv, 135, 167, 179
$L^p(\Omega)$, xxiii
$\mathcal{L}^{q,\lambda}(\Omega)$, xxiv, 155, 183
$L^{(p,q)}(\Omega)$, xxiv, 135, 167, 179

$LSO(n+1)_\tau$, 96

metric, 2
monotonicity formula, 26, 153, 154, 194
Morrey–Campanato space, xxiv, 155, 183

Nash–Moser theorem, 8
Noether current, 14
Noether harmonic maps, 42
Noether's theorem, 11, 41, 50, 102, 177

$\Omega\mathfrak{G}$, 67
$\Omega\mathfrak{g}$, 68
$\Omega\mathfrak{G}^{\mathbb{C}}$, 67
$\Omega\mathfrak{g}^{\mathbb{C}}$, 68
$\Omega_{hol}\mathfrak{g}^{\mathbb{C}}$, 68
orthonormal frame bundle, 173
orthonormal moving frame, 165

quaternions, 57

second fundamental form, 9, 223, 224

section, 61
Sobolev space, xxiii, 31
$SO(n)^{\mathbb{C}}$, 67
spectral parameter, 81
stationary maps, 43, 150, 193
stereographic projection, 44, 239
stress–energy tensor, 18, 20, 154
symmetries on $\mathcal{N}$, 12

twisted loop group, 91

unit sphere, 10

$v_{a,r}$, 155
vector bundle, 211

wave equation, 103
weakly harmonic maps, 37, 101, 166
Weierstrass type representation, 91
Wente's inequality, 115, 145
$W^{k,p}(\Omega)$, xxiii

Yang–Mills, 71